Coulomb Excitations and Decays in Graphene-Related Systems

T0239585

Coulomb Excitations and Decays in Graphene-Related Systems

Chiun-Yan Lin, Jhao-Ying Wu, Chih-Wei Chiu, and Ming-Fa Lin

CRC Press
Taylor & Francis Group
Boca Raton London New York

CRC Press is an imprint of the
Taylor & Francis Group, an **informa** business

CRC Press
Taylor & Francis Group
6000 Broken Sound Parkway NW, Suite 300
Boca Raton, FL 33487-2742

First issued in paperback 2021

ISBN 13: 978-0-367-77963-4 (pbk)
ISBN 13: 978-0-367-21861-4 (hbk)

**Visit the Taylor & Francis Web site at
http://www.taylorandfrancis.com**

**and the CRC Press Web site at
http://www.crcpress.com**

Contents

Preface

This book presents a new theoretical framework of electronic properties and electron–electron (e–e) interactions, and it fully explores the Coulomb excitations and decays in graphene-related systems. It has covered a plenty of critical factors related to different lattice symmetries, layer numbers, dimensions, stacking configurations, orbital hybridizations, intralayer and interlayer hopping integrals, spin–orbital couplings, temperatures, electron/hole (e–h) dopings, electric field, and magnetic quantization. Apparently, there exist a rich and unique electronic excitation phenomena due to the distinct energy bands and wave functions in various condensed-matter systems, as clearly illustrated in the diverse (momentum, frequency)-phase diagrams. The complex Coulomb excitations, the intraband and interband e–h excitations, and the undamped and damped plasmon modes provide effective deexcitation scatterings and thus dominate the Coulomb decay rates. Graphene, silicene, and germanene differ from one another in Coulomb excitations and decays, as a result of the distinct strengths of spin–orbital couplings and hopping integrals. The calculated results, accompanied with the concise physical pictures, obviously show the very important roles of the inelastic Coulomb scattering. Of course, part of the theoretical predictions could explain up-to-date experimental measurements, and further examinations are required for most of them.

The generalized tight-binding model, which is built from the multiorbital tight-binding functions on the distinct sublattices, is developed to solve the electronic properties under the composite effects arising from orbital hybridizations, the spin–orbital couplings, the one-site energies of electric field, and the magnetic quantization. Furthermore, the random-phase approximation (RPA) is thoroughly modified to agree with the layer-dependent Coulomb potentials obtained from the Dyson equation. As a result, these two methods can match with each other for the various electric and magnetic fields. Concerning few-layer graphene systems, the interlayer and intralayer hopping integrals and Coulomb interactions, based on the layer-dependent A and B sublattices, are taken into account simultaneously by the direction combination of the generalized tight-binding model and the modified RPA. Specifically, the dimensionless energy loss function (the intrinsic screened response function) is first defined for layered materials. A similar framework could be generalized to the other emergent 2D materials under the detailed calculations/investigations, such as the layered tinene, phosphorene, antimonene, and bismuthene. Further studies would provide significant differences among these systems and be very in thoroughly understanding the close/complicated relations of essential physical properties. On the other hand, the theoretical models should be derived to solve the Coulomb excitations in 1D and 0D systems without good spatial translation symmetry. For example, 1D graphene nanoribbons and 0D graphene quantum dots have open boundary conditions, so that they, respectively, possess many energy subbands and discrete energy levels. Maybe, the dielectric function tensor, being characterized by the subband/level index, is an effective way to understand the excitation properties.

The developed theoretical models in this book provide new research categories, such as (I) the dynamic charge oscillations, (II) the static impurity screenings, and (III) the diverse Coulomb decay behaviors in layered graphene systems. (I) The bare and screened response functions are thoroughly clarified using the e–e inelastic scatterings dynamically. Specifically, the dimensionless energy loss function is very useful in further understanding of the time-dependent carrier propagation in 2D planar structures. It is sensitive to the magnitude and direction of transferred momenta and various frequencies. From the theoretical point of view, one must perform the 3D Fourier transform for it to explore the time- and position-induced collective oscillation phenomenon, accompanied with Landau dampings. (II) Apparently, the layer-dependent effective Coulomb potentials could be gotten from the Dyson equation, and similar calculations are done for the induced charge densities. The free carrier density and the Fermi surface, or the energy gap, are expected to play critical roles on the unusual screening behaviors. Whether the long-range effective Coulomb interactions decay quicker than the inverse of distance is the standard criteria for charge screening ability. Moreover, the 2D Friedel oscillations in doped graphene systems are the studying focuses. (III) Under the layer-projection scheme, the screened exchange self-energy, which is combined with the effective e–e interactions, can deal with the Coulomb decay rates in any layered graphene systems/emergent 2D materials. More complicated scattering mechanisms, being closely related to the distinct single-particle and collective excitations, will be presented in the accurate calculations and diverse deexcitation phenomena.

Acknowledgment

This work was supported in part by the National Science Council of Taiwan, the Republic of China, under Grant Nos. NSC 98-2112-M-000-013-MY4, NSC 99-2112-M-165-001-MY3 and NSC 107-2112-M-017-001.

Authors

Chiun-Yan Lin earned a PhD in physics in 2014 at the National Cheng Kung University (NCKU), Taiwan. Since 2014, he has been a postdoctoral researcher in the Department of Physics at NCKU. His main scientific interests include the field of condensed matter physics, modeling, and simulation of nanomaterials. Most of his research focuses on the electronic and optical properties of two-dimensional nanomaterials.

Jhao-Ying Wu earned a PhD in physics in 2009 at NCKU (Tainan, Taiwan). After that, he was a postdoctoral fellow until 2016. He became a professor at the National Kaohsiung University of Science and Technology. His main scientific interests focus on theoretical condensed matter physics, including the electronic and optical properties of low-dimensional systems, Coulomb excitations, and quantum transport.

Chih-Wei Chiu is an associate professor in the Department of Physics, National Kaohsiung Normal University, Taiwan. He earned a PhD in 2005 at NCKU, Taiwan. His research work mainly deals with the physical properties of graphene-related nanosystems using numerical simulations.

Ming-Fa Lin is a distinguished professor in the Department of Physics, NCKU, Taiwan. He earned a PhD in physics in 1993 at the National Tsing-Hua University, Taiwan. His scientific interests focus on essential properties of carbon-related materials and low-dimensional systems.

1

Introduction

Graphene-related systems have attracted a lot of theoretical and experimental researches, mainly owing to the unique hexagonal symmetry, unusual layered structures/nanoscaled thickness, and various stacking configurations [1–12]. Such systems are suitable for exploring the basic science and high-potential applications [1–9,13–17]. Electronic excitations/deexcitations arising from the electron–electron (e–e) Coulomb interactions are one of the essential physical properties, being closely related to the geometric and band structures. They are determined by the intrinsic many-particle properties and play a critical role in all condensed-matter systems with different dimensions. A new theoretical framework is developed for the layered systems; a modified random-phase approximation (RPA) is conducted on 2D materials. In this book, a systematic and thorough investigation clearly indicates the diverse Coulomb excitation/decay phenomena, in which it covers many sp^2-bonding carbon-created materials, e.g., monolayer graphene, double-layer graphene, AA-, AB-, ABC-, & AAB-stacked graphenes, sliding bilayer graphene, simple graphite, Bernal graphite, rhombohedral graphite, metallic/semiconducting carbon nanotubes, and monolayer silicene/germanene. The composite effects due to lattice symmetry, layer number, dimensionality, stacking configuration, temperature, doping, electric field, and magnetic field will be discussed in detail. Specially, the magnetoelectronic excitations, which are associated with the magnetic quantization, require a combination of the generalized tight-binding mode and the modified RPA. The theoretical predictions are fully compared with the experimental measurements from the electron energy loss spectroscopy (EELS), and part of them are consistent with the latter. The measured EELS spectra have confirmed the diverse collective excitations (plasmon modes) of the free carriers and the π & $\pi + \sigma$ valence electrons at different frequency ranges (the detailed discussions in Section 2.5).

The dielectric function (ϵ) and the dimensionless energy loss ($\text{Im}[-1/\epsilon]$) function, which, respectively, represent the bare and screened response ability of charged carriers under the Coulomb field perturbation, are critical in fully understanding the intrinsic excitation and decay properties of condensed-matter materials. In general, the imaginary part of ϵ and the prominent peaks in $\text{Im}[-1/\epsilon]$, respectively, correspond to the single-particle and collective excitations (electron–hole (e–h) excitations and plasmon modes). ϵ is strictly defined for bulk graphites [18], monolayer graphene [18], and cylindrical carbon nanotubes [19], since such systems possess a good translational symmetry, respectively, in the 3D, 2D, and 1D spaces. That is to say, the bare Coulomb e–e interactions mainly determined by the transferred momentum (**q**) exhibit well-behaved/dimension-determined forms. On the other hand, the longitudinal dielectric function is expressed as a tensor form for the layered graphenes. The layer-projection method, which is closely related to the Coulomb scattering of the initial and final electronic states, will be introduced to deal with electric polarizations. As a result, ϵ is a layer-dependent tensor function with double indices ($\epsilon_{ll'}$; l the layer index), in which the band-structure effects on the excitation spectra and Coulomb matrix elements are fully taken into account in the theoretical model. For example, an N-layer graphene possesses N pairs of valence and conduction bands due to the carbon $2p_z$ orbitals, and the sublattice-decomposed wave functions are used to accurately evaluate the layer-dependent $P_{ll'}$ or $\epsilon_{ll'}$. Under the Born approximation, the effective energy loss function is characterized by the inelastic scattering probability of the incident electron beam. It has been successful in understanding the diverse electronic excitation spectra in layered systems, e.g., AA- [20–23], AB- [24–26], and ABC-stacked graphenes [27]. The magnetoplasmon is discussed in this book by the development of the modified RPA and the generalized Peierls tight-binding model [28–31]. It is suitable for studying the inter-Landau-level (inter-LL), single-particle excitations and magnetoplasmon modes [32–34]. Moreover, electronic excitation spectra are quite efficient decay channels by the inelastic Coulomb scatterings, being the strong effects on the energy widths of the quasiparticle states (the excited electrons or holes). The Coulomb decay rates are evaluated from the self-energy method by detailed derivations. They will show certain important differences among monolayer graphene, silicene, and germanene with the electron/hole doping. As to the experimental side, EELS and inelastic X-ray scatterings (IXS) are two very powerful techniques in providing Coulomb excitation phenomena. Their recent developments [35,36] and measurements on the screened response functions of graphene-related materials [37–40] are explored in detail. Also, the quasiparticle energy spectra under the measurements of the angle-resolved photoemission spectroscopy (ARPES) are useful in determining the Coulomb decay rates [41–43].

Monolayer graphene, as displayed in Figure 1.1a, has a planar honeycomb lattice composed of two equivalent sublattices, so that it exhibits the linear and isotropic Dirac-cone band structure near the Fermi level (E_F) [2]. A pristine system is only a zero-gap semiconductor, since density of states (DOS) is vanishing at E_F. The low-lying DOS is a linear V-shape form centered at $E_F = 0$ [2]. It leads to the specific interband e–h excitations and the absence of intraband ones [2]. Temperature could induce the thermal excitations between the gapless valence and conduction states; that is, it generates conduction electrons and valence holes. The free carrier density per area is identified to reveal a simple T^2-dependence, and their magnitude is estimated to $\sim 10^{11}$ e/cm^2 at room temperature. The intraband single-particle excitations are induced/enhanced by temperature, while a great reduce is observed in the interband ones. Most importantly, temperature can create intraband plasmon modes at sufficiently high T, in which they belong to 2D acoustic collective excitations defined by a specific $\sqrt{q}$-dependence at long wavelength limit. By means of alkali–adatom absorptions or the applications of gate voltages [44–47], there exists a very high free carrier density ($\sim 10^{13}$ e/cm^2) in an extrinsic graphene system. The intraband & interband excitations and the intraband plasmons are expected to be very much pronounced in the bare and screened excitation spectra, respectively. This has been clearly confirmed from EELS measurements [40]. Monolayer band structure is also reliable when the interlayer distance is very large (e.g., more than double that of graphite) [48]. The effects, which purely arise from the interlayer Coulomb interactions, could be investigated for a double-layer graphene system. The acoustic and optical plasmon modes are clearly revealed in the EELS spectra, and they, respectively, correspond to the collective oscillations of two-layer charge carriers in phase and out of phase [49,50]. The experimental measurements are required to verify these two modes.

The layered graphenes present the various stacking configurations, such as AAA, ABA, ABC, and AAB ones. The symmetries of geometric structures have strong effects on band structures and thus greatly diversify Coulomb excitations/decays. The AA stacking means that carbon atoms in different layers possess the same (x, y) projections, as shown in Figure 1.1b. From the first-principles evaluations using Vienna Ab initio simulation package (VASP) [51], the ground state energy of the AA-stacked graphene is predicted to be the highest among all the stacking configurations. This clearly indicates that it is relatively difficult to synthesize the AA stacking in laboratory experiments [52]. According to both VASP and tight-binding model calculations [30,53], there exist N pairs of Dirac-cone structures in the N-layer AA stacking; furthermore, such linear energy bands are intersecting at the K/K′ valley even in the presence of vertical and nonvertical interlayer hopping integrals. For example, two pairs of valence Dirac cones in bilayer system are identified from the measurements of ARPES on the quasiparticle energy spectrum [54,55]. The valence and conduction bands strongly overlap one another so that a lot of free electrons and holes are created by the interlayer atomic interactions. This semimetallic property is in sharp contrast with the semiconducting behavior of monolayer graphene. Such carriers further induce collective excitations in which the number of plasmon modes is identical to that of layer [20]. Also, single-particle excitation channels are complicated/enriched by more energy bands. From the theoretical point of view, the interlayer hopping integrals and the interlayer Coulomb interactions need to be simultaneously included in the analytic formulas; that is, the band-structure effects are thoroughly covered in the electronic excitations/deexcitations. The main features of plasmon modes and e–h dampings, the sensitive dependences on the transferred momentum and energy [$(\mathbf{q}, \omega)$], are worthy of a systematic investigation. In addition, the experimental measurements are absent till now.

The AB-stacked graphenes frequently appear in the experimental observations using various methods, e.g., the successful syntheses by the mechanical exfoliation [1, 56], the chemical vapor deposition (CVD) [57, 58], and the electrostatic manipulation of scanning tunneling microscopy (STM) [59, 60]. This shows that their ground state energies are much lower than those of the AA stackings [56]. The AB stacking is the natural periodical sequence of Bernal graphite [61, 62]. The neighboring layers are attracted together by the weak but significant van der Waals interactions, so that the few-layer AB stacking could be obtained from the natural graphite under the mechanical action [1, 56]. They only possess half of carbon atoms in the same (x, y) projections (Figure 1.1c), or they are characterized by a relative shift of the C–C bond length (b) along the armchair direction about the initial AA stacking. There are more complex interlayer hopping integrals and extra site energies due to the different chemical environments experienced by the A^l and B^l sublattices (l layer index), compared with the AA stacking. For AB bilayer stacking, the Dirac-cone energy bands become two pairs of parabolic valence and conduction bands, in which the latter are also initiated from the K/K′ valley (the corners of the hexagonal first Brillouin zone). Furthermore, the trilayer system presents an extra Dirac cone with a slight distortion. Such band-structure characteristics of few-layer AB stackings have been confirmed by the ARPES measurements [63–66]. In short, an even-N (an odd-N) AB stacking exhibits N pairs of parabolic bands (accompanied with a weakly separated Dirac-cone structure). In general, a small overlap of valence and conduction bands is revealed in the first pair of few-layer systems [64]. Apparently, the 2D free electron/hole density purely arising from the interlayer interactions is very low, leading to the absence of collective excitations in pristine AB-stacked systems [67, 68]. However, extrinsic few-layer AB stackings are expected to present the unusual Coulomb excitation behaviors (Chapter 5).

The ABC stacking, corresponding to Figure 1.1d), is predicted to have the lower ground state energy than that of the ABA stacking [53]. The ABC-stacked few-layer graphenes are easily synthesized in the experimental growth [56–58]. However, its 3D counterpart, rhombohedral graphite, is presented with a low-concentration arrangement in natural graphite; that is, most of the 3D graphite belongs to the ABA (Bernal) stacking [61, 62]. The unique geometric symmetry induces rich hopping integrals and unusual electronic structures. The hopping integrals, which arise from the neighboring and next-neighboring layers, cover the vertical and nonvertical interlayer atomic interactions. The band structures exhibit three pairs of partially flat, sombrero-shaped, and linear energy bands; the distinct energy dispersions centered at K point have been verified from ARPES experiments [64–66]. Specifically, the first pair just overlaps/touches at E_F, leading to a sharp DOS. Such electronic wave functions are ascribed to the special surface states due to the main contributions of carbon atoms in the outmost two layers [2, 30, 69]. Although the ABC stacking is a semimetal with a very low free carrier density, it is predicted to exhibit the low-frequency plasmon mode closely related to the localized states [27]. Furthermore, the novel momentum dependence of pristine plasmons is never observed in other stackings. On the other hand, the doping carriers in extrinsic systems will strongly compete with the original surface states. It is expected to create dramatic changes in the characteristics of e–h dampings and plasmon modes during the variation of E_F.

The AAB stacking is the direct combination of AA and AB stackings, as displayed in Figure 1.1e. Such systems have been successfully synthesized and

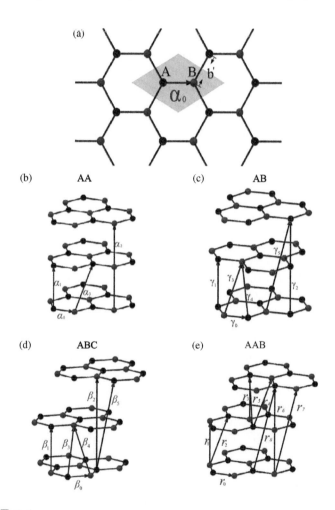

FIGURE 1.1
(a) Honeycomb lattice structure of monolayer graphene. The geometric structure of (b) AA-, (c) AB-, (d) ABC-, and (e) AAB-stacked graphenes. Sublattices A and B are shown by black and red colors, respectively. b' represents the C-C bond length. α's, β's, and γ's are the atomic hopping integrals indicated by blue arrows.

experimentally observed by distinct methods, e.g., the mechanical exfoliation directly by a scalper or scotch tape [70], the CVD growth on SiC substrate [71] and Ru(0001) surface [72], the liquid-phase exfoliation of natural graphite in N-methyl-2-pyrrolidone [73], and the STM [60, 74–79]. In particular, the AAB stacking could be obtained by the rotating or horizontal shifting of the top graphite layer along the armchair direction. That is to say, the stacking configuration is continuously changed under the electrostatic modulation of STM. Furthermore, the corresponding DOS is also measured, indicating a narrow energy gap in trilayer ABA stacking [72, 80]. According to the first-principles method, the ground state energies per unit cell of six carbon atoms in trilayer graphenes are evaluated for the stacking configurations. They are estimated as follows: −55.832866, −55.857749, −55.862386, and −55.864039 eV for AAA, AAB, ABA, and ABC stackings, respectively [53]. The theoretical calculations predict that the AAB stacking is more stable than the AAA one, or the former presents a more promising future in experimental syntheses. The lower-symmetry AAB stacking possesses the most complicated interlayer hopping integrals, in which this special property is clearly identified from a consistent/detailed comparison between the tight-binding model and VASP calculations in the low-lying energy bands [81, 82]. For example, the AAB-stacked trilayer graphene exhibits three pairs of energy bands with the oscillatory, sombrero-shaped, and parabolic dispersions, in which the first ones determine a bandgap of <10 meV. Of course, the semiconducting pristine system only creates inter-π-band e–h excitations, but not the low-frequency plasmons. The doping effects on the Coulomb excitations are the first theoretical study, and the greatly diversified phenomena are described in Chapter 7.

How to create the dramatic transitions of essential properties is one of the mainstream topics in pristine graphenes. The continuous stacking configurations, which possess high and low geometric symmetries, could enrich and diversify the physical phenomena. The sliding bilayer graphene presents the transformation between highly symmetric stackings, being an ideal system for fully exploring the electronic topological transitions. There are some experimental successful syntheses, such as the stacking boundaries including the relative shifts between neighboring graphene layers by the CVD method [83], the sliding of graphene flakes on graphene substrate/micrometer-size graphite flakes initiated by the STM tip [74, 75], and atomic force microscope (AFM) tip [76] and the electrostatic-manipulation STM performed on a highly oriented pyrolytic graphite (HOPG) surface [60, 77–79]. Specifically, the last method has generated a continuous and large-scaled movement of the top graphene layer, so that the sliding bilayer graphene is expected to be achieved by this method. The theoretical model calculations are focused on the electronic [84–90], magnetic [91], optical [92, 93], transport [94–96], and phonon [97–100] properties. For example, two low-lying isotropic Dirac cones are dramatically transformed into two pairs of parabolic bands during the variation of AA→AB [85, 91]. Furthermore, the sliding bilayer graphene exhibits three kinds of LLs – the well-behaved, perturbed, and undefined LLs – and three magneto-optical selection rules of $\Delta n = 0$, 1 & 2 (n: the quantum number for each LL) [28, 91]. During the continuous variation of stacking configuration, energy bands present the serious distortions and free carrier density show the drastic changes [28, 91]. These are predicted to induce the novel Coulomb excitation phenomena in pristine and extrinsic systems.

The external electric and magnetic fields are one of the critical factors in the creation of diverse electronic excitations. A uniform perpendicular electric field ($E_z \hat{z}$) could be achieved by applying a gate voltage on a layered graphene system. There are a lot of such experimental setups up to now, being verified to have strong effects on transport properties and thus potential applications in nanoelectronic devices [101–104]. E_z creates distinct Coulomb potential energies on the layer-dependent sublattices. Apparently, energy dispersions and bandgap are drastically changed by E_z, in which the semimetal–semiconductor transitions might occur as E_z varies [105, 106]. The effects of E_z on band structures are also diversified by the stacking configurations; that is, the E_z-enriched energy bands are sensitive to AAA, ABA, ABC, and AAB stackings (the detailed discussions on Chapter 9). From the current model calculations of the modified RPA, a pristine N-layer AAA stacking has one acoustic plasmon mode and ($N-1$) optical ones [21, 22], in which the former and the latter, respectively, originate from the intraband and interband electronic excitations. An electric field obviously results in the charge transfer among different graphene layers; therefore, it could modulate the number, frequency, and intensity of collective excitation modes, e.g., the emergence of new plasmon modes and the decline of threshold plasmon frequency. As for the AB bilayer stacking, the low-frequency plasmon, which corresponds to the intraband single-particle excitations, is generated by a sufficiently high E_z [26]. This greatly contrasts with a pristine system with any plasmon mode [23]. Such results directly illustrate that the E_z-induced oscillatory parabolic bands, with the high DOSs, could create new plasmon modes, or the main features of band structures could determine the single-particle and collective excitations. Similar effects are revealed in the trilayer ABA and ABC stackings [27, 67]. This work will cover the complete results in the E_z-enriched excitation spectra of the ABA, ABC, and AAB stackings.

A uniform magnetic field ($B_z \hat{z}$) could flock together the neighboring electronic (k_x, k_y) states and thus generate the highly degenerate LLs and the special wave functions in the oscillatory forms. Apparently, the magnetic quantization is directly reflected in magnetoelectronic excitations, belonging to a new topic of the Coulomb interactions. The LL characteristics, which are thoroughly explored by the generalized tight-binding model [28–31], cover the normal or abnormal spatial probability distributions with the regular/irregular zero-point numbers, the definition of quantum number from the domination sublattice, the dependences of energy spectrum on n and B_z, the noncrossing/crossing/anticrossing behaviors, and the specific magneto-optical selection rules [28, 31]. They strongly depend on the number of layer and stacking configuration, i.e., they are greatly diversified by the geometric factors.

The charged particles under $B_z \hat{z}$ experience a transverse magnetic force. The cyclotron motions present rather strong competitions with the longitudinal plasma oscillations due to the e–e interactions. This critical mechanism is responsible for the unusual excitation phenomena. Up to date, there are only few theoretical predictions on the magnetoplasmon modes of monolayer and AA/AB bilayer graphenes [32–34]. The direct combination of the generalized tight-binding model and the modified RPA is a developed theoretical framework. In addition to temperature, a magnetic field in monolayer graphene could drive a lot of discrete magnetoplasmons even under the low-energy range, mainly owing to the inter-LL excitations. Their oscillation frequencies exhibit the critical transferred momenta and the nonmonotonous momentum dependence. Moreover, the quantized magnetoplasmons and the 2D acoustic plasmon mode might coexist in the AA bilayer stacking, but not in the AB one. This difference lies in the stacking-dependent free carrier densities.

Graphite is one of the most extensively investigated materials, experimentally and theoretically, for much more than 100 years. This layered system is suitable/ideal for studying the diverse 3D and 2D physical phenomena; furthermore, it possesses a lot of up-to-date applications. Graphite crystals consist of a series of stacked graphene plane. Generally, there exist three kinds of stacking configurations in the layered graphites and compounds, namely AA, AB, and ABC stackings. Simple hexagonal, Bernal, and rhombohedral graphites exhibit unusual essential properties as a result of the honeycomb lattice and stacking sequence. Among them, the AB-stacked graphite is predicted to be the most stable system, according to the first-principles calculations on the ground state energy [51]. Natural graphite is composed of the dominating AB stacking and the partial ABC one [61,62]. The AA-stacked graphite, which possesses the highest-symmetry crystal structure, does not survive in nature. The periodical AA stacking is frequently observed in the Li-intercalation graphite compounds [107]. Simple hexagonal graphite has been successfully synthesized using the dc plasma in hydrogen–methane mixtures [108]. For the AA-, AB-, and ABC-stacked graphites, the first/third system owns the highest/lowest 3D free carrier density [31]. Their low-energy band structures, respectively, present the (k_x, k_y)-plane Dirac cones, the monolayer/bilayer characteristics at the H/K point, and the spiral Dirac-cone structure [31]. The ARPES measurements on Bernal graphite have verified the linear and parabolic energy dispersions near the H and K valleys [109–112]. The theoretical studies of Coulomb excitations are focused on the free-carrier-induced, π, and $\pi + \sigma$ plasmons, in which they are, respectively, revealed in the distinct frequency ranges of $\omega_p \sim 0.1$ eV, 5 eV; 20 eV. The current investigation is the first theoretical study on whether the low-frequency plasmons exist in rhombohedral graphite [113–115]. The low-frequency plasmons belong to the optical mode because of the 3D bare Coulomb potential. Apparently, dimensionality and stacking configuration/interlayer hopping integrals play critical roles in determining its characteristics, the existence, and momentum & temperature dependence. On the experimental side, the reflection energy loss spectroscopy

(REELS; details in Chapter 2), with very high energy resolution, has been utilized to thoroughly examine the T-dependent low-frequency optical plasmon in Bernal graphite. Furthermore, both REELS and transmission energy loss spectroscopy (TEELS) could accurately identify the middle-frequency π plasmon modes.

Carbon nanotubes, with very strong σ bondings, are first discovered by Iijima using an arc-discharge evaporation method in 1991 [116,117]. Each nanotube is a hollow cylinder that could be regarded as a rolled-up graphitic sheet in a cylindrical form. Its geometric structure is characterized by a primitive lattice vector ($\mathbf{R}$) of monolayer graphene (details in Chapter 12). Both radius (r) and chiral angle (θ), which correspond to $\mathbf{R}$, would dominate the essential physical properties. From the theoretical calculations of VASP and tight-binding model [118–120], there are three kinds of carbon nanotubes, namely, metallic, narrow-gap, and middle-gap ones, being determined by (r, θ). Only armchair nanotubes belong to 1D metals, since DOSs due to the linear bands are finite at E_F. The (r, θ)-dependent energy gaps are thoroughly verified from the scanning tunneling spectroscopy (STS) measurements on the low-energy DOSs [121]. Any cylindrical nanotubes present the semiconductor–metal transitions during the variation of a uniform axial magnetic field. The double degeneracy of electronic states is destroyed under this field. Furthermore, the periodical oscillations of essential properties under a flux quantum $\phi_0 = hc/e$ are the so-called Aharonov–Bohm effect, as clearly observed in optical [122] and transport properties [123]. As to a lot of 1D energy subbands of a cylindrical carbon nanotube, they are defined by the angular momenta (J's) and axial wave vectors (k_y's). The e–e Coulomb interactions would satisfy the conservation of the transferred angular momentum and axial momentum (L, q) [19,124–126]. Specifically, both single-particle and collective excitations possess the L-decoupled modes, i.e., there exist many intra-π-band and inter-π-band transitions. The previous theoretical studies show that the free carriers in a metallic armchair nanotube create a 1D $L = 0$ acoustic plasmon mode with a specific momentum dependence. All the carbon nanotubes exhibit several inter-π-band plasmons of $\omega_p \sim 1 - 4$ eV and one π plasmon mode of $\omega_p > 5$ eV, being consistent with experimental measurements [39].

The monoelement IV-group condensed-matter systems have attracted a lot of experimental [127–129] and theoretical studies [34,130–133], especially for those combined with 2D & 3D structures. The emergent 2D materials are excellent candidates in exploring the unique physical phenomena, such as the Dirac-cone band structure or the multiconstant-energy loops [134–136], the magnetically quantized LLs [137,138], the ultrahigh carrier mobility [139,140], the novel quantum Hall effects [130, 141], and the optical selection rules [137, 142]. These systems are expected to present high potentials in the near-future technological applications [143,144]. Since the successful exfoliation of few-layer graphenes in 2004 [1], silicene, germanene, and tinene are, respectively, grown on distinct substrate surfaces, e.g., Si on Ag(111), Ir(111) & ZrB$_2$ surfaces [145–147], Ge on Pt(111), Au(111) & Al(111) surfaces [128,148–150],

and Sn on Bi_2Te_3 surface. The latter three systems possess the buckled honeycomb structures, in which a strong competition between the sp^2 and sp^3
bondings accounts for optimal geometries. However, graphene is a hexagonal
plane. Their spin–orbit couplings (SOCs) are significant and much stronger
than those of pure carbon systems. These two characteristics will dominate the
essential physical properties. From the VASP [136] and tight-binding model
calculations [131], the low-lying electronic structures of monolayer silicene
and germanene appear at the K/K' valley; furthermore, they are mainly
determined by the outermost $3p_z/4p_z$ orbitals even in the mixing of two kinds
of chemical bondings. The non-negligible SOCs create the separated Dirac-
cone structures, with narrow energy gaps, e.g., $E_g \sim 7.9$ meV for silicene and
$E_g \sim 93$ meV for germanene for the model predictions [131, 136]. Compared
with monolayer graphene, it is relatively easy to reveal the anisotropic energy
dispersions as a result of smaller intralayer hopping integrals. The application of a uniform perpendicular electric field further induces splitting of spin-
related energy bands [149–152]. Also, this E_z-field causes the energy gap to
change from finite to zero values. On the other hand, monolayer tine exhibits
very strong multiorbital hybridizations and SOCs [133], so that the outside
four orbitals $(5s, 5p_x, 5p_y, 5p_z)$ need to be considered in the low-energy band
structures. The calculated electronic energy spectra could be verified from
ARPES measurements [54, 55, 63–66, 109–112], as done for few-layer graphenes
and graphites.

For monolayer silicene and germanene, there are some theoretical studies
on Coulomb excitations/decay rates [34, 132, 152–154]. These two systems are
different from monolayer graphene in certain many-particle properties, mainly
owing to the existence of SOCs and buckled structures. Energy gap, electric
field, magnetic field, and doping would induce diversities of excitation phenomena, in which the diverse momentum-frequency phase diagrams cover the
various single-particle excitation boundaries and plasmon modes, e.g., four
kinds of E_F-, SOC-, and E_z-dependent plasmon modes in germanene [34].
Such excitations might become effective deexcitation channels when the occupied electrons/holes are excited into the unoccupied states under the perturbation of an incident electron beam or an electromagnetic field. That is to
say, the excited electrons or holes in conduction/valence bands could further
decay by the inelastic Coulomb scatterings. The decay rates of excited states
have been explored by the screening exchange using the Matsubara Green's
functions. The dynamic Coulomb responses from the valence and conduction electrons are taken into account simultaneously [34, 132, 152–154]. The
decay processes and their dependence on the wave vector, valence/conduction
states, and Fermi energy/doping density will be investigated in detail. A comparison with monolayer graphene is also made. The current work indicates
that the intraband & interband single-particle excitations, and the distinct
plasmon modes are responsible for the deexcitation behaviors. The rich and
unique Coulomb decay rates appear as a consequence of oscillatory energy
dependence, strong anisotropy on wave vectors, nonequivalent valence and

conduction states/Dirac points, and the similarity with 2D electron gas for low-energy conduction electrons and holes. The predicted Coulomb decay rates could be directly examined from the high-resolution ARPES measurements by the energy widths of quasiparticle states at low temperature [54, 55, 110–112].

This book is focused on the recent progresses of graphene-related systems in Coulomb excitations/deexcitations under e–e interactions. The whole content is organized as follows. Chapter 2 covers the developed theoretical framework and the detailed experimental techniques. The bare and screened response functions are derived in the analytic forms, especially for those in few-layer graphenes using the modified RPA. The geometric symmetry, layer number, stacking configuration, dimension, and electric and magnetic fields are taken into account simultaneously. How to accurately measure the energy loss spectra is thoroughly discussed for the high-resolution REELS and TEELS. The previous experimental measurements on carbon-sp^2 condensed-matter systems provide useful information about the geometry-enriched electronic excitation behaviors. The Coulomb decay rates in monolayer graphene, silicene, and germanene are calculated by using the self-energy method of Matsubara's Green functions. The inelastic Coulomb scatterings from the various valence and conduction states play critical roles.

The focuses in Chapter 3 are the effects of the critical factors, such as temperature and doping, on creating the intraband single-particle excitations and low-frequency 2D acoustic plasmons. The low-frequency analytic formula of the polarization function could be obtained in the vicinity of K/K'. Also, the pure Coulomb coupling effects are explored for a double-layer system with a sufficiently large interlayer distance. The composite effects, which are due to the interlayer hopping integrals and the interlayer Coulomb interactions in few-layer graphene systems, will create the diverse excitation phenomena in the spectra of (frequency,momentum)-phase diagrams. Their systematic studies of the AAA-, ABA-, ABC-, and AAB- graphenes are, respectively, conducted in Chapters 4, 5, 6, and 7. In addition, Chapter 8, as fully investigated for the sliding bilayer graphene, clearly illustrates the continuous transformation of the geometric symmetry and thus the dramatic changes in electronic properties and Coulomb excitations. The layer-dependent Coulomb potentials, being induced by a uniform perpendicular electric field, could greatly diversify band structures and excitation behaviors, as clearly indicated in Chapter 9. The magnetic quantization is elaborated in Chapter 10; we define a vector-potential-dependent Peierls phase/period in the hopping integrals/the real crystal [31]. There exist new excitation channels, the inter-LL single-particle and collective excitations, which display unusual momentum dependences. The generalized tight-binding model and the modified RPA are combined to fully comprehend the AA and AB bilayer stackings. Also, the strong competitions among the longitudinal Coulomb interactions, transverse magnetic forces, and stacking configurations are discussed in detail.

The dimension- and geometry-enriched electronic excitations are, respectively, explored in Chapters 11 and 12 for 3D graphites and 1D carbon nanotubes. The simple hexagonal, Bernal, and rhombohedral graphites are in sharp contrast with one another for the low-lying band structures, covering energy dispersions, isotropic/anisotropic behaviors, and free electron/hole densities. How to directly reflect in the low-frequency electronic excitations is worthy of a systematic investigation. As for each carbon nanotube, the cylindrical symmetry creates a lot of angular-momentum-decoupled excitation modes, being absent in other condensed-matter systems. A close relationship between 1D band structures and excitation spectra is proposed to fully understand the unusual Coulomb excitations. The important differences among 1D, 2D, and 3D graphene-related systems require a detailed discussion. Chapters 13 and 14, respectively, correspond to electronic excitations and Coulomb decay rates in the emergent monolayer silicene/germanene, in which the significant SOCs and buckled structures are the critical factors in determining the diversified Coulomb excitations and decay rates, or distinguishing from monolayer graphene. Finally, Chapter 15 includes concluding remarks and future perspectives.

2

Theories for Electronic Excitations in Layered Graphenes, 3D Graphites, and 1D Carbon Nanotubes: Experimental Equipments

We develop the theoretical modes for the dielectric responses of graphene-related systems with various dimensionalities. The electron–electron (e–e) Coulomb interactions due to the π electrons induce the dynamic and static charge screenings within the middle frequency ($\omega \leq 10$ eV). Within the linear response, the random-phase approximation under the external electric and magnetic fields is modified to satisfy the layered structures/geometric symmetries. The energy loss function, which is directly related to the experimental measurements, is very useful in exploring the unusual electronic excitations in each system. It can be further defined by an analytical formula. In general, there are two kinds of experimental equipments for examining the theoretical predictions. The significant characteristics of distinct dimensional systems in the measured energy loss spectra are discussed thoroughly. Moreover, the inelastic Coulomb decay rates are also investigated in detail.

2.1 Dielectric Functions of Layered Graphenes

When monolayer graphene is present in an external Coulomb potential, the π electrons due to the $2p_z$ orbitals will effectively screen this perturbation. The charge redistribution directly reflects the dynamic/static carrier screening and thus creates to the induced potential. Within the linear response, the dimensionless dielectric function is defined as the ratio between bare potential and effective potential

$$\epsilon(\mathbf{q}, \omega) = \lim_{V^{ex} \to 0} \frac{V^{ex}(\mathbf{q}, \omega)}{V^{eff}(\mathbf{q}, \omega)}. \tag{2.1}$$

It can also be characterized by charge densities and longitudinal electric fields, namely, ρ^{ex}/ρ^{tot} and D_l/E_l. By the momentum-dependent Poisson equations and the self-consistent-field approach, the induced Coulomb potential is the product of the bare Coulomb potential and the induced charged density, in which the latter is proportional to the effective Coulomb potential, and the coefficient is bare response function (P) under the linear response. As a result, the dielectric function is given by

$$\epsilon\,(q,\phi\,,\omega\,) = \epsilon_0 \,-\, V_q P(q,\phi\,,\omega\,), \qquad (2.2)$$

where

$$P(q,\phi\,,\omega\,) = \sum_{h,h'=c,v} \langle \mathbf{k}; h|e^{-i\mathbf{q}\cdot\mathbf{r}}|\mathbf{k}+\mathbf{q}; h' \rangle \times \frac{f(E^{h'}(\mathbf{k}+\mathbf{q})) - f(E^h(\mathbf{k}))}{E^{h'}(\mathbf{k}+\mathbf{q}) - E^h(\mathbf{k}) - (\omega+i\Gamma)}. \qquad (2.3)$$

In general, a 2D system has an electronic state expressed by $(k_x, k_y; h)$. $h = v/c$ corresponds to valence/conduction state. $\epsilon_0 (=2.4)$ is the background dielectric constant due to the high-energy σ-electron excitations. $V_q = 2\pi\, e^2/q$ is the 2D bare Coulomb potential of the 2D electron gas. The band-structure effect on the bare Coulomb interactions are included in the bare response function by the square of the inner product between the initial and final states in the momentum transfer (the first term inside the integration on the first Brillouin zone of Eq. (2.3)). f is the Fermi–Dirac distribution function, and Eq. (2.2) is suitable under any temperatures in intrinsic and extrinsic systems. Γ is the broadening phenomenological parameter arising from various deexcitation channels, depending on the frequency range of electronic excitation. The transferred momentum and frequency are conserved during the e–e Coulomb interactions; that is, they are necessary in describing the electronic excitations, the single- and many-particle excitations. In inelastic experimental measurements, the energy loss spectra are associated with $\mathbf{q}$ and ω. The direction of the former is ϕ between $\mathbf{q}$ and KM, and $0° \leq \phi \leq 30°$ is sufficient because of the hexagonal symmetry. The dielectric function in Eqs. (2.2) and (2.3) is similar for monolayer silicene and germanene with buckled honeycomb lattices, while electronic states are modified by the significant spin–orbital interactions (details in Chapter 13).

The dielectric function could be used to explore the effective Coulomb interactions between two charges and thus understand the screening length. First, we calculate the static dielectric functions that depend on the Fermi level of monolayer graphene. For an intrinsic (extrinsic) system, the Fermi level is located at the Dirac point (the conduction/valence cone), so monolayer graphene is a zero-gap semiconductor (a metal with the free electron/hole density roughly proportional to the square of the Fermi momentum). As a result, the dielectric function is finite/divergent under the long wavelength limit ($q \to 0$) for an intrinsic/extrinsic graphene. Second, the momentum-dependent effective Coulomb potential is transformed into the real-space e–e interaction by the standard 2D Fourier transform. The long-range behavior, the effective interaction inversely proportional to the distance, is deduced to remain in an intrinsic graphene. However, for an extrinsic graphene, the e–e interaction close to the charged impurity will decline quickly and exhibit an effective screening length closely related to the free carrier density. Third, the similar Fourier transform is done for the real-space induced charge density with the well-known Friedel oscillations associated with the dimensionality and Fermi momentum.

The dielectric responses of N-layer graphenes become more complicated, compared with the monolayer system. There exist perturbed Coulomb potentials and induced charges arising from all the layers; that is, the external and induced Coulomb potentials due to each layer need to be taken into consideration simultaneously. The intralayer & interlayer hopping integrals and the intralayer & interlayer Coulomb interactions are covered in the modified random-phase approximation (RPA). The full band structure can provide the exact and reliable electronic excitations, in which the rich and unique $(\mathbf{q}, \omega)$-phase diagrams are very sensitive to the stacking configuration and the number of layers.

The incident electron beam is assumed to be uniform on each layer, so the π electrons on the distinct layers experience similar bare Coulomb potentials. Such carriers exhibit the dielectric screening closely related to two different layers/the same layer. Specifically, the excited electron and hole in each excitation pair, which is due to the Coulomb perturbation, frequently occur on distinct layers. The Feynman diagram of the Coulomb excitations between two layers is depicted in Figure 2.1. By the Dyson equation, the effective Coulomb potential for two electrons on the l-th and l'-th layers is expressed as

$$\epsilon_0 V_{ll'}^{eff}(\mathbf{q}, \omega) = V_{ll'}(\mathbf{q}) + \sum_{mm'} V_{lm}(\mathbf{q}) P_{mm'}^{(1)}(\mathbf{q}, \omega) V_{m'l'}^{eff}(\mathbf{q}, \omega). \qquad (2.4)$$

Equation (2.4) clearly reveals the bilayer-dependent effective Coulomb potential as the $N \times N$ matrix form, in which the Coulomb potential, induced charge density, and response function are described by any two layers. The first term is also useful in understanding the Coulomb decay rates in layered systems. Specifically, the bilayer-created response function is

$$P_{mm'}(\mathbf{q}, \omega) = 2 \sum_{k} \sum_{nn'} \sum_{h,h'=c,v} \left(\sum_{i} u_{nmi}^{h}(\mathbf{k}) u_{n'm'i}^{h'*}(\mathbf{k}+\mathbf{q}) \right)$$
$$\times \left(\sum_{i'} u_{nmi'}^{h*}(\mathbf{k}) u_{n'm'i'}^{h'}(\mathbf{k}+\mathbf{q}) \right) \times \frac{f(E_n^h(\mathbf{k})) - f(E_{n'}^{h'}(\mathbf{k}+\mathbf{q}))}{E_n^h(\mathbf{k}) - E_{n'}^{h'}(\mathbf{k}+\mathbf{q}) + \hbar\omega + i\Gamma}. \qquad (2.5)$$

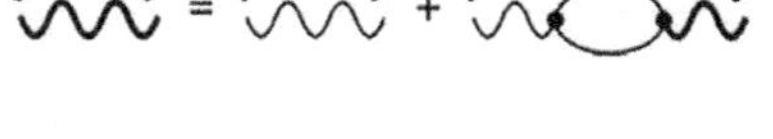

FIGURE 2.1
The Feynman diagram of the Coulomb excitations between two layers under the modified RPA.

Any electronic states, which agree with the conservations of the transferred momentum and frequency, can be decomposed into the layer-dependent contributions by analyzing their wave functions, and so does the response function. We have included all the significant intralayer and interlayer hopping integrals in Eq. (2.5). This decomposition concept is critically important to match with the layer-dependent Coulomb potential in Eq. (2.4), so that the modified RPA could be generalized to the layered systems even in the presence of external electric and magnetic fields. Moreover, the layer-dependent dielectric function becomes a tensor form:

$$\epsilon_{ll'}(\mathbf{q},\omega) = \epsilon_0 \delta_{ll'} - \sum_m V_{lm}(\mathbf{q}) P_{m,l'}(\mathbf{q},\omega). \tag{2.6}$$

The zero points of the dielectric function tensor are available in understanding the plasmon modes, while the spectral intensities of the collective and single-particle excitations are absent. The definition of energy loss function is necessary in the further calculated formulas. The effective Coulomb potential directly links with the bare one through the following relationship:

$$\sum_{l''} \epsilon_{ll''}(\mathbf{q},\omega) V_{l''l'}^{eff}(\mathbf{q},\omega) = V_{ll'}(\mathbf{q}). \tag{2.7}$$

The inelastic scattering rate, which the probing electrons transfer the specific momentum and frequency $(\mathbf{q},\omega)$ to the N-layer 2D materials, is delicately evaluated from the Born approximation [23, 25, 27, 261]. It is used to define the dimensionless energy loss function:

$$\mathbf{Im}[-1/\epsilon] \equiv \sum_l \mathbf{Im}\left[-V_{ll}^{eff}(\mathbf{q},\omega)\right]\Big/\left(\sum_{lm} V_{lm}(q)/N\right). \tag{2.8}$$

The denominator is the average of all the external Coulomb potentials on the different layers. Equation (2.8) can be applied to any emergent 2D systems, such as the layered graphene, silicene, germanene, tinene, phosphorene, antimonene, and bismuthene (the group-IV and group-V 2D materials). It is the screened response function responsible for the experimental inelastic energy loss spectra. The dimensionless loss function is useful in exploring the various plasmon modes in the specific systems, and the imaginary of the bare response function describes the single-particle electron–hole excitations. All the equations developed in this section are suitable for a layered condensed-matter system under a uniform perpendicular electric field. It only needs to modify the band-structure changes due to the layer-dependent Coulomb site energies.

2.2   AA-, AB-, and ABC-Stacked Graphites

The bulk graphites possess infinite graphene layers, so that their energy bands have an extra wave vector along the k_z-direction. Electronic states are described by (k_x, k_y, k_z) within the first Brillouin zones. Energy dispersions are dominated by the honeycomb lattice on the (x, y) plane, stacking configuration; intralayer and interlayer hopping integrals. All the graphites are semimetals because of the interlayer van der Waals interactions. However, the AA-stacked graphite (ABC-stacked one) exhibits the largest (smallest) overlap between the valence and conduction bands and thus the highest (lowest) free electron and hole densities, directly reflecting the geometric symmetry. Band structures and free carrier densities are quite different for three kinds of graphites, and electronic excitations are expected to behave so (details in Chapter 11) [113 115]. For example, the low-frequency plasmon due to the interlayer atomic interactions is very sensitive to the AA, AB, or ABC stacking.

The 3D transferred momentum (q_x, q_y, q_z) in graphites is conserved during the e–e Coulomb interactions, as observed in 3D electron gas. The analytic form of the dielectric function, which is similar for any graphites, is directly evaluated from the RPA

$$\epsilon(q_x, q_y, q_z, \omega) = \epsilon_0 - \sum_{h,h'=c,v} \int_{1stBZ} \frac{e^2 d^2\mathbf{k}_\parallel dk_z}{q^2\pi^2} |\langle \mathbf{k}_\parallel + \mathbf{q}_\parallel, k_z + q_z; h'|e^{i\mathbf{q}\cdot\mathbf{r}}$$

$$\times |\mathbf{k}_\parallel, k_z; h\rangle|^2 \frac{f(E^{h'}(\mathbf{k}_\parallel + \mathbf{q}_\parallel, k_z + q_z)) - f(E^h(\mathbf{k}_\parallel, k_z))}{E^{h'}(\mathbf{k}_\parallel + \mathbf{q}_\parallel, k_z + q_z) - E^h(\mathbf{k}_\parallel, k_z) - (\omega + i\Gamma)}.$$

$$(2.9)$$

The k_z-integration of the first Brillouin zone is distinct in three kinds of stacking configurations. There exist the low-frequency and π plasmons. The former could survive at small transfer momenta, while it might be difficult to observe it at large ones. Under this case, the anisotropic dependence on the $(q_x, q_y)-$ plane is negligible; that is, $\mathbf{q} = [\mathbf{q}_\parallel, \mathbf{q_z}]$. Since the low-lying energy bands are almost isotropic near the K/K' point, the 3D integration could be reduced to the 2D integration, i.e., $\int_{1stBZ} d^2\mathbf{k} dk_z \rightarrow \int_{1stBZ} 2\pi k_\parallel dk_\parallel dk_z$.

Both 3D graphite and 2D monolayer graphene have similar dielectric functions in Eqs. (2.2) and (2.2), while their electronic excitations quite differ from each other. The dimensionality and band structure are responsible for the significant differences. Under the long wavelength limit, the bare Coulomb potentials, respectively, approach to $1/q^2$ and $1/q$ for the former and the latter. The stronger Coulomb potential in graphite clearly indicates that it is relatively easy to observe the low-frequency plasmon modes due to free carriers and the screening charge distributions. Moreover, graphites possess the extra k_z-dependent energy dispersions, compared with that of graphene. The larger overlap of valence and conduction bands arising from the interlayer hopping integrals results in free electrons and holes simultaneously. These are useful in understanding why three kinds of graphites exhibit the rich and unique $(\mathbf{q}, \omega)$-phase diagrams (Chapter 11).

2.3 Carbon Nanotubes

Electronic states in a cylindrical carbon nanotube, with the nanoscaled radius (r), are characterized by the longitudinal wave vector (k_y) and the azimuthal angular momentum $(k_x = J/r, J = 1, 2, ..., N_{u/2}, N_u)$ the atom number in a primitive cell; discussed in Chapter 12). As a result of cylindrical symmetry, the transferred longitudinal momentum and transverse angular momentum are conserved during the e–e Coulomb interactions. Electronic excitations are well defined by (q, L), and so does the dielectric response. So within RPA, the dielectric function of a single-walled carbon nanotube, which includes all the intra- and inter-π-band excitations at any temperatures, is

$$\epsilon'(q, L, \omega + i\Gamma) = \epsilon_0 + 2 \sum_J \int_{1stBZ} \frac{dk_y}{(2\pi)^2} \frac{2\omega_{vc}(J, k_y; q, L)}{(\omega_{vc}(J, k_y; q, L))^2 - (\omega + i\Gamma)^2}$$

$$\times V(q, L) |\langle J + L, k_y + q; h' | e^{iqy} e^{iL\phi'} | J, k_y; h \rangle|^2, \quad (2.10)$$

where

$$|\langle J + L, k_y + q; h' | e^{iqy} e^{iL\phi'} | J, k_y; h \rangle|$$

$$= \frac{1}{4} \{1 + [q^2 + (L/r)^2]/36\}^{-6}$$

$$\times \left| 1 - \frac{H_{12}(J + L, k_y + q) H_{12}^*(J, k_y)}{|H_{12}(J + L, k_y + q) H_{12}^*(J, k_y)|} \right|^2. \quad (2.11)$$

$\omega_{vc}(J, k_y; q, L) = E^{h'}(J + L, k_y + q)\text{-}E^h(J, k_y)$ is the excitation energy between the final and initial states. $\epsilon_0 - 2.4$ is identical to that in mono-layer graphene. $V(q, L) = 4\pi e^2 I_L(qr) K_L(qr)$ is the bare Coulomb potential of an electron gas in a cylindrical tubule [48,124]. I_L (K_L) is the first (second) kind of modified Bessel function of the order L. The Coulomb interaction in a carbon nanotube is characterized by the second term in the integrand of Eq. (2.10) and the third term corresponds to the Coulomb matrix element. The wave functions of energy bands greatly modify the Coulomb interactions and produce noticeable effects on the physical properties of carbon nanotubes, e.g., electron energy loss spectroscopy (EELS) and impurity screenings [125]. Electronic excitations are expected to strongly depend on the metallic and semiconducting behavior, being sensitive to the radius and chiral angle of a cylindrical carbon nanotube.

A single-walled carbon nanotube is similar to monolayer graphene in geometric structures, but the former needs to satisfy the periodical boundary condition. As a result, the important differences are revealed in Coulomb excitations, as clearly indicated the dielectric functions in Eqs. (2.2) and (2.10). The 1D and 2D bare Coulomb potentials are divergent in logarithmic and linear forms. In general, the 1D and 2D parabolic (linear) bands DOSs exhibit the square-root divergent form and the shoulder structure (the shoulder structure and the V-shape form). Moreover, carbon nanotubes possess the L-decoupled single-particle excitations and plasmon modes, mainly owing to the cylindrical symmetry. It is thus expected to have the L-dependent diverse phase diagrams. e.g., the L-decoupled inter-π-band plasmons (Chapter 12).

For a multiwalled carbon nanotube, there are complicated dynamic/static charge screenings from the different layers. The layer-dependent Dyson equation and polarization function, as done for layered graphenes, are available in exploring excitation spectra due to multiwalled systems [126]. The Coulomb excitations are greatly diversified by the relative stacking, layer number and chiral angle of coaxial carbon nanotubes. For example, the (5,5)-(10,10) bilayer nanotube has three kinds of rotational symmetries, namely, C_5, D_{5h} and S_5, in which they create the diverse (q, L)-phase diagram with the distinct plasmon modes and electron–hole excitations (Landau dampings).

2.4   Electron Excitations under a Uniform Perpendicular Magnetic Field

Monolayer graphene consists of two sublattices of A and B atoms with a C–C bond length of b=1.42 Å. Only the hopping integral between two nearest-neighbor atoms, $\gamma_0 = 2.598$eV, is used in the tight-binding model. Based on the two tight-binding functions of the periodic $2p_z$ orbitals, the zero-field Hamiltonian is a 2×2 Hermitian matrix. Monolayer system exits in a perpendicular uniform magnetic field $\mathbf{B} = B_z\hat{z}$. The magnetic flux through a hexagon is $\Phi = (3\sqrt{3}b^2 B_z/2)/\phi_0$, where the flux quantum $\phi_0 = hc/e$ $4,14 \times 10^{-15}$ T·m^2. The vector potential, being chosen as $\mathbf{A} = B_z x\hat{y}$, creates a new periodicity along the armchair direction, as clearly indicated in Figure 2.2. The magnetic Hamiltonian matrix element could be obtained by multiplication

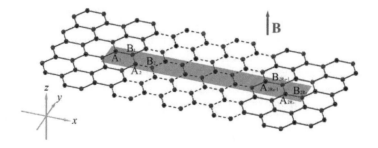

FIGURE 2.2
The magnetically enlarged unit cell under a perpendicular magnetic field $\mathbf{B} = B_z\hat{z}$.

of the zero-field Hamiltonian matrix element by a Peierls [31]. Such phase is assumed to have an integer period $R_B = 1/\Phi$, so that the enlarged rectangular unit cell contains $4R_B$ carbon atoms and the magnetic Hamiltonian becomes a $4R_B \times 4R_B$ Hermitian matrix. The first Brillouin zone is greatly reduced along $\hat{k}_x$. The nearest-neighbor hopping integral related to the extra position-dependent Peierls phase is changed into

$$|\langle\ B_k|H|A_j\rangle| = \gamma_0 \exp[i\mathbf{k} \cdot (R_{A_j} - R_{B_k})]$$

$$\times \exp\left(i2\pi\phi_0 \int_{R_{A_j}}^{R_{B_k}} \mathbf{A}d\mathbf{r} \right) \qquad (2.12)$$

$$= \gamma_0 t_j \delta_{j,k} + \gamma_0 s \delta_{j,k+1},$$

where $t_j = \exp\{i[-k_x b/2 - k_y(\sqrt{3}b)/2 + \pi\Phi(j - 1 + 1/6)]\} + \exp\{i[-k_x b/2 + k_y(\sqrt{3}b)/2 - \pi\Phi(j-1+1/6)]\}$ and $s = \exp[i(-k_x b)]$. The Hamiltonian dimension is 32,000 under the magnetic field strength of $B_z = 10$ T. The huge magnetic Hamiltonian could be efficiently solved using the band-like matrix by the rearrangement of the R_B tight-binding functions (details in Refs. [28,29]). Monolayer graphene exhibits the highly degenerate Landau levels (LLs) and the well-behaved wave functions with spatial symmetries, in which the significant quantum number, n/m, is determined by the number of zero points in the latter.

Magnetoelectronic excitations are characterized by the transferred momentum q and the excitation frequency ω, which determine the longitudinal dielectric function. They are independent of the direction of the momentum transfer, since the LLs possess the isotropic characteristics [32,33]. The magnetic dielectric function and bare response function have the same forms identical to Eqs. (2.2) and (2.3), respectively. In the presence of magnetic field, electronic states become fully quantized. The summation in Eq. (2.3) corresponds to all possible single-particle transitions between Landau states $|m\rangle$ and $|n\rangle$. The response function is now expressed as

$$P(\mathbf{q},\omega) = \frac{1}{3bR_B\pi} \sum_{n,m;k} |\langle n; \mathbf{k} + \mathbf{q}|e^{i\mathbf{q}\cdot\mathbf{r}}|m; \mathbf{k}\rangle|^2_{\mathbf{q}=q_y, \mathbf{k}=k_y}. \qquad (2.13)$$

Only q_y and k_y components are under the numerical calculations, and the evaluated results remain the same along the other direction of q_x and k_x. Since all the π-electronic states are covered in the magnetoelectronic excitations, the strength and frequency of the resonances in $\text{Im}(-1/\epsilon)$ could be correctly defined. Moreover, effects due to temperatures and high dopings are allowed.

The magnetoelectronic excitation spectra are expected to be greatly diversified by the stacking configurations in layered graphene systems, as discussed later in Chapter 10. The significant effects, which arise from the magnetic field, the interlayer/intralayer hopping integrals, and the interlayer/intralayer Coulomb interactions, could be taken into account simultaneously by using the layer-dependent RPA in Eqs. (2.4)–(2.7). The main reason is that the generalized/magnetic tight-binding model is consistent with the modified RPA under the concept of layer projection. That is to say, the LL energy spectra and wave functions in few-layer systems are first evaluated from the generalized tight-binding model. The layer-dependent response functions, which determine the single-particle magnetoexcitations, are investigated by the layer projections of LL wave functions. And then, the sensitive dependence of energy loss spectra on the magnetic field strength could be explored in detail. A similar method is suitable for other emergent layered materials, such as silicene and germanene.

2.5   EELS and Inelastic X-Ray Scatterings

EELS [35, 37–40, 155–213] and inelastic X-ray scattering (IXS) [36, 165, 214–226] are the two kinds of efficient methods in examining/verifying excitation spectra in condensed-matter systems. They have been successfully utilized to identify Coulomb excitations and phonon dispersion spectra of any dimensional (0D–3D) materials. There exist both inelastic transmission and reflection EELS, in which the latter is suitable for thoroughly exploring the low-energy excitations lower than 1eV. [180–182, 192, 195, 196, 199–205] A longtime development for EELS is done since the first measurement on bulk graphite in 1969 [37,38]. The resolutions of transferred energies and momenta are under investigation. IXS just starts to experience a rapid growth in recent decades, so this technique needs to be greatly enhanced by various manners. As a result, most of the electronic excitations in carbon/graphene-related systems are accurately measured by EELS. These two techniques have their own advantages, being reliable in the different environments, as discussed later.

The reflection energy loss spectroscopy (REELS) instrumentation extracts the bulk and surface energy loss functions of the backscattered electrons from the sample surface [35]. If the incident electron beams have a kinetic energy with a few hundred eV, the relatively simple technique can provide loss spectra with an energy resolution of a few meV, which is sufficient to resolve vibrational and electronic excitation modes [155]. REELS is widely used for investigating the physical and chemistry properties of material surfaces [156]. It typically operates with 25 meV energy resolution in the energy range between 15 and 70 eV [157], while controlling the momentum resolution down to 0.013 Å^{-1} (better than 1% of a typical Brillouin zone) [227]. However, the energy resolution can be possibly close to 1 meV under the condition where much weaker electron beams are adopted using a high-resolution monochromator at an ultrahigh vacuum base pressure ($\sim 2\times10^{-10}$ Torr) [156]. On the other hand, in transmission EELS instruments, the incident electron beams pass through the sample and can be adopted for use in transmission electron microscopy (TEM) to detect the material structure. This technique is commonly understood as EELS. The spectroscopy uses higher energy electron beams, typically 100–300 keV, as employed in the TEM. The energy loss is appreciable, typically varying from a few eV up to hundreds of eV, with an energy resolution <1 eV and momentum resolution 0.01 Å^{-1}. In particular, the improvements for electron monochromators and spectrometers make the achievement of an energy resolution <50 meV [158,228]. The dedicated transmission energy loss spectroscopy (TEELS) instruments have excellent momentum resolution, and at best 80 meV energy resolution; they are suitable for measuring the electronic excitations down to 0.5 eV under the consideration of the interference from the zero-loss line [159,160]. The spectral resolution should be sufficient for studying collective excitations in most metallic and doping semiconductor materials. The dispersion relations of the plasmons can be measured by an angle-resolved EELS, which is performed with low-energy electrons and using an analyzer to detect the scattered electrons. The analyzer is a magnetic-prism system, as shown in Figure 2.3, where the commercially available Gatan spectrometer is installed beneath the camera and the basic interface and ray paths are shown as well. The surface of the prism is curved to reduce the spherical and chromatic aberrations. Scattered electrons in the drift tube are deflected by the magnetic field into a variable entrance aperture (typically variable from 1 to 5 mm in diameter). All the electrons in any direction are focused on the dispersion plane of the spectrometer; electrons that lose more energy deflect further away from the zero-energy loss electrons according to the Lorenz force law. The magnetic prism projects energy-loss spectrum of the electrons onto a Charge-coupled device (CCD) camera, which is straightforward to capture the whole energy distribution simultaneously. It is possible to modulate the resolution of the transferred momentum by varying the half-angle of the incident beam in TEM and the scattering beam in the spectrometer. The momentum resolution is typically in the order of $\sim 0.01\text{Å}^{-1}$, based on the angle variation of a few milliradians.

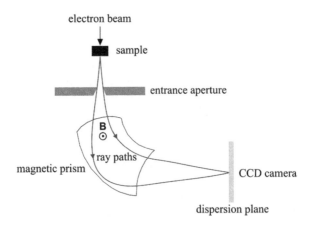

FIGURE 2.3
The magnetic-prism analyzer of an inelastic electron energy loss spectroscopy.

IXS can directly probe the microscopic dynamic behavior in nanoscale systems. It has been successfully utilized to detect a wide range of physical phenomena, such as phonon dispersion in solids, dynamics of disordered materials and biological systems, as well as electronic excitations in condensed matter systems [36]. The transferred energy and momentum are independent variables and cover the full spectrum of the dielectric response. The medium inelastic X-ray beam line is designed to provide high photon flux over the typical Brillouin zone sizes; the photon energy is distributed from 4.9 to 15 keV, with an energy resolution of ∼70 meV and momentum resolution of $\sim 0.02 - 0.03 \text{Å}^{-1}$. In particular, the instrument in the Swiss Light Source has an extremely good energy resolution of 30 meV [229]. The experimental resolution is possible to achieve with few meV with the development new synchrotron sources. Furthermore, IXS can be used to measure all kinds of electronic excitations because the electronic charges can interact with X-rays. Using hard X-ray synchrotron sources, the spectroscopy is a powerful technique to detect the interior properties of bulk materials, and it can also be applied to the systems under external electric and magnetic fields [230]. Depicted in Figure 2.4, the analyzer, built on the basis of Bragg optics, efficiently collects and analyzes the energies and momenta of the scattered photons in a small space, and provides detailed information on the intrinsic electronic properties of the system. To maximize the scattered photon intensity, a spherically bent analyzer (typically 10 cm in diameter) is used to capture the scattered radiation of the momentum-transfer photons in a small solid angle. The transferred energy is projected onto a CCD detector and the full energy-loss spectrum is scanned by varying the Bragg angle of the crystal. Operated in the Rowland circle geometry, the measured double differential scattering cross section describes the elementary excitations in the characteristic energy-loss regime via the dissipation–fluctuation theorem.

There are important differences between EELS and IXS techniques in measuring environments. The incident particle beam could be focused into ∼ 10 and 100 Å for the former and the latter, respectively [35, 155–157].

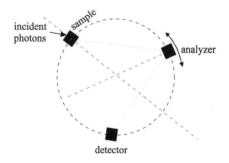

FIGURE 2.4
Schematic diagram of an inelastic light scattering spectroscopy.

Furthermore, EELS has more excellent resolutions in transferred energies and momenta, compared with IXS. Much more inelastic scattering events could be measured using EELS within a short time; that is, the EELS measurers for the complete excitation spectra would be done more quickly and accurately. EELS is suitable for low-dimensional systems and nanoscaled structures, since the electron beam simultaneously provides the information of material size and position. However, IXS from the continuous synchrotron radiation exhibits a strong intensity with tunable energy and momentum. The extreme surrounding environments, accompanied with the applications of magnetic/electric fields and different temperatures/pressures, could be overcome by the inelastic light scatterings. The external fields strongly affect the incident charges, and the sample chamber is too narrow so that EELS cannot work under such environments. For example, the IXS measurements are useful in examining/verifying the inter-LL excitations and magnetoplasmons in graphene-related systems.

The high-resolution EELS could serve as a powerful experimental technique to explore the Coulomb excitations in carbon-related systems and emergent 2D materials. The experimental measurements have successfully confirmed certain electronic excitations due to the π and σ carriers in the sp^2 bonding systems, such as graphites [37, 38, 167, 168, 231–233], graphite intercalation compounds [169–172], single- and multi-walled carbon nanotubes [174–179], single- and few-layer graphenes [40, 183–205], C_{60}-related fullerenes [206–208], carbon onions [208–212], and graphene nanoribbons [213]. In general, the interlayer interactions, dimensions, geometric symmetries, stacking configurations, and chemical dopings might induce many/some/few free conduction electrons/valence holes, leading to the low-frequency acoustic or optical plasmon modes with frequency about $\omega_p < 1$ eV. For example, under room temperature, the 3D Bernal graphite possesses low-frequency optical plasmons at $45-50$ and 128 meV under the long wavelength limit, respectively, corresponding to the electric polarizations parallel and perpendicular to the z-axis [113]. Specifically, the detailed temperature-dependent EELS show that the former are sensitive to temperature effects [113]. The high-density free electrons and holes, which are, respectively, created in the donor- and accepter- type graphite intercalation compounds (layered superlattice systems), possess the ~ 1-eV optical plasmons due to the coupling of collective excitations on infinite layers, in which they strongly depend on the transferred momenta [234]. On the other hand, for the layered graphene systems with adatom chemisorptions [235], the low-frequency plasmons become 2D acoustic modes; that is, they have the $\sqrt{q}$ dependence at small transferred moment [236].

The π plasmons, which are due to all the valence π electrons, are found to exist in carbon-related systems, except for diamond. Their frequencies are higher than 5 eV with a strong momentum dependence [37–39, 167, 169–174, 183, 184, 186, 192, 194, 195, 197]. For example, the π-plasmon frequencies in Bernal graphite increase from ~ 7 to 11 eV as the transferred momentum is in the range of $0 < q < 1.4$ Å^{-1} [37, 38, 167]. Similar π-plasmon modes are revealed in layered graphenes and carbon nanotubes [39, 169–174, 183, 184, 186, 192, 194, 195, 197]. Their frequencies are enhanced by the increase of graphene layers [186]. Moreover, there exist several discrete interband plasmon modes in cylindrical nanotube systems [177], in which their momentum dependences are very weak. They belong to optical modes and possess the frequencies of 0.85, 1.25, 2.0, 2.55, and 3.7 eV, indicating the inter-π-band excitation mechanisms closely related to the low-lying occupied valence and unoccupied conduction energy subbands. It should be noticed that such plasmons are in sharp contrast with the free-carrier-induced acoustic modes (discussed earlier). As to the $\pi + \sigma$ plasmons, they arise from the collective excitations of the $\pi + \sigma$ occupied electrons. Apparently, the superhigh carrier density will create pronounced fluctuations accompanied with large resonance frequencies. In general, the $\pi + \sigma$ plasmon frequencies are higher than 20 eV and strongly dependent on the transferred momenta [194]. However, they might appear at a lower frequency of $\sim$ 15 eV, e.g., those in few-layer graphenes. That all the π electrons and part of σ ones take part in such collective excitations could explain this result. In short, the aforementioned four kinds of plasmon modes are examined/identified by the high-resolution EELS, while the inelastic light scattering measurements are absent till now. The latter are suitable for the direct verification of π and $\pi + \sigma$ plasmons with higher frequencies.

2.6 Coulomb Decay Rates and Angle-Resolved Photoemission Spectroscopy (ARPES)

The many-body self-energy is derived to characterize the valence quasiparticle properties in layered graphene-related systems with valence and conduction bands. It is suitable under any temperature and doping. Furthermore, the relation between the quasiparticle energy widths and ARPES measurements is discussed in detail. In addition, three kinds of femtosecond pump-probe spectroscopies are useful in comprehending the lifetimes/energy widths of the specific electronic states are also discussed.

2.6.1 Coulomb Decay Rates in Layered Graphene-Related Systems

The incident electron beam and electromagnetic field, which act on the layered graphene-related systems, will be dynamically screened by conduction and valence electrons, or will have strong interactions between the external perturbations and the charge carriers. During the complicated screening processes, they create excited electrons (holes) above (below) the Fermi level. Such intermediate states could further decay by the inelastic e–e and electron–phonon scatterings. At low temperature, we only focus on the former mechanisms. The Coulomb decay rate $(1/\tau)$ is fully determined by the effective interaction potential (V^{eff}) between two charge carriers, in which the dynamic e–e interactions could be understood from the layer-dependent modified RPA. By using Matsubara Green's functions [237], $1/\tau$ of monolayer graphene is calculated from the quasiparticle self-energy, the screened exchange energy (the RPA self-energy as clearly shown in Figure 2.5)

$$\Sigma(\mathbf{k}, h, ik_n) = -\frac{1}{\beta} \sum_{\mathbf{q}, h', i\omega_m} V^{eff}(\mathbf{k}, h, h'; \mathbf{q}, i\omega_m) G^{(0)}(\mathbf{k} + \mathbf{q}, h', ik_n + i\omega_m),$$

(2.14)

where $\beta = (k_B T)^{-1}$, $ik_n = i(2n + 1)/\pi/\beta$ (complex fermion frequency), $i\omega_m = i2m\pi/\beta$ (complex boson frequency) and $G^{(0)}$ is the noninteracting Matsubara Green's function. $V^{eff}(\mathbf{k}, h, h'; \mathbf{q}, i\omega_m) = V(\mathbf{k}, \mathbf{q}, h, h')/\epsilon(\mathbf{q}, i\omega_m) = V_q |\langle h', \mathbf{k} + \mathbf{q}| e^{i\vec{q}\cdot\vec{r}} |h, \mathbf{k}\rangle|^2/[\epsilon(\mathbf{q}, i\omega_n)]$ is the screened Coulomb interactions with the band-structure effect, in which the intraband and interband deexcitation channels need to be taken into account simultaneously. V_q is the 2D bare Coulomb potential energy and $\epsilon(\mathbf{q}, i\omega_n)$ is the RPA dielectric function. This

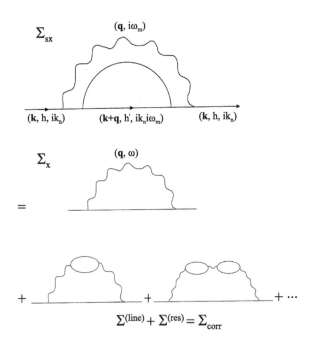

FIGURE 2.5
The electron self-energy in RPA.

equation is also suitable for monolayer silicene and germanene under the spin-degenerate states [238], although they possess significant spin–orbital couplings. It does not need to solve the spin-up- and spin-down-related Coulomb decay rates separately, since they make the same contributions. That is, it is sufficient in fully exploring the wave-vector-, conduction/valence-, and energy-dependent self-energies (Eq. 2.14).

Under the analytic continuation $ik_n \to E_h(\mathbf{k})$, the carrier self-energy could be divided the bare exchange energy, line part and residue part:

$$\Sigma_{sx}(\mathbf{k}, h, E^h(\mathbf{k})) = \Sigma_x(\mathbf{k}, h, E^h(\mathbf{k})) + \Sigma^{(line)}(\mathbf{k}, h, E^h(\mathbf{k})) + \Sigma^{(res)}(\mathbf{k}, h, E^h(\mathbf{k})), \tag{2.15}$$

in which

$$\Sigma_x(\mathbf{k}, h, E^h(\mathbf{k})) = -\sum_{\mathbf{q}, h'} V(\mathbf{k}, \mathbf{q}, h, h') f(E^{h'}(\mathbf{k} + \mathbf{q})), \tag{2.16}$$

$$\Sigma^{(line)}(\mathbf{k}, h, E^h(\mathbf{k})) = -\frac{1}{\beta} \sum_{\mathbf{q}, h', i\omega_m} [V^{eff}(\mathbf{k}, h, h'; \mathbf{q}, i\omega_m) - V(\mathbf{k}, h, h'\mathbf{q})]$$
$$\times G^{(0)}(\mathbf{k} + \mathbf{q}, h', E^h(\mathbf{k}) + i\omega_m), \tag{2.17}$$

and

$$\Sigma^{(res)}(\mathbf{k}, h, E^h(\mathbf{k})) = -\frac{1}{\beta} \sum_{\mathbf{q}, h', i\omega_m} [V^{eff}(\mathbf{k}, h, h'; \mathbf{q}, i\omega_m) - V(\mathbf{k}, h, h'\mathbf{q})]$$
$$\times [G^{(0)}(\mathbf{k} + \mathbf{q}, h', ik_n + i\omega_m)$$
$$- G^{(0)}(\mathbf{k} + \mathbf{q}, h', E^h(\mathbf{k}) + i\omega_m)]. \tag{2.18}$$

The summation of the line and residue parts is the so-called correlation self-energy. The imaginary part of the residue self-energy determines the Coulomb decay rate, being characterized as

$$Im\Sigma^{(res)}(\mathbf{k}, h, E^h(\mathbf{k})) = \frac{-1}{2\tau(\mathbf{k}, h)}$$
$$= \sum_{\mathbf{q}, h'} Im[-V^{eff}(\mathbf{k}, h, h'; \mathbf{q}, \omega_{de})]$$
$$\times \{n_B(-\omega_{de})[1 - n_F(E^{h'}(\mathbf{k} + \mathbf{q}))] \tag{2.19}$$
$$- [n_F(E^{h'}(\mathbf{k} + \mathbf{q}))]\}$$
$$= \frac{-1}{2\tau_e(\mathbf{k}, h)} + \frac{-1}{2\tau_h(\mathbf{k}, h)}.$$

$\omega_{de} = E^h(\mathbf{k}) - E^{h'}(\mathbf{k} + \mathbf{q})$ is the deexcitation/decay energy. n_B and n_F are, respectively, the Bose–Einstein and Fermi–Dirac distribution functions. Equation (2.17) clearly means that an initial state of $(\mathbf{k}; h)$ can be deexcited to all available $(\mathbf{k} + \mathbf{q}; h')$ states under the Pauli exclusion principle and the conservations of energy and momentum. The excited states above or below the Fermi level, respectively, exhibit the electron and hole decay rates (the first and second terms in Eq. (2.20)). By the detailed derivations, the zero-temperature Coulomb decay rates of the excited electrons and holes are

$$
\begin{aligned}
\frac{1}{\tau_e(\mathbf{k}, h)} + \frac{1}{\tau_h(\mathbf{k}, h)} = & -2 \sum_{\mathbf{q}, h'} Im[-V^{eff}(\mathbf{k}, h, h'; \mathbf{q}, \omega_{de})] \\
& \times [-\Theta(\omega_{de})\Theta(E^{h'}(\mathbf{k} + \mathbf{q}) - E_F)) \\
& + [\Theta(-\omega_{de})\Theta(E_F - E^{h'}(\mathbf{k} + \mathbf{q}))].
\end{aligned}
\tag{2.20}
$$

where E_F is the Fermi energy for a pristine system/an extrinsic system with carrier doping. Θ is the step function that limits the available deexcitation channels. The Coulomb decay rate is double the energy width of a quasiparticle state. Equations (2.19) and (2.20) could be generalized to a single-walled carbon nanotube with a cylindrical symmetry [239].

The layer-projection method could be developed to thoroughly investigate the Coulomb decay rates in few-layer graphene-related systems. Any electronic states are composed of tight-binding functions localized at the different layers, so that their inelastic Coulomb scatterings are closely related to the effective layer-dependent Coulomb potentials ($V_{ll'}^{eff}$'s). First, we need to evaluate $V_{ll'}^{eff}$ in Eq. (2.4) using the analytic and numerical forms simultaneously. And then, Eq. (2.20) is directly suitable for studying the decay rates. The various deexcitation channels are similar in the layer-dependent $V_{ll'}^{eff}$, but they might exhibit distinct weights. However, the calculations become heavy even under the tight-binding model. On the other hand, there are only few studies on bilayer graphenes up to now [240], in which the main decay mechanisms are not clear in the first-principles method because of the numerical resolution.

2.6.2   ARPES Measurements on Occupied Quasiparticle Energy Widths

ARPES is the most efficient & reliable equipment in studying the quasiparticle band dispersions and energy widths for the occupied electronic states within the first Brillouin zone. Their measurements could examine the band-structure calculations by the tight-binding model and the first-principles method, and the predicted Coulomb decay rates under the screened exchange self-energy. In general, the ARPES chamber is combined with the instruments of sample synthesis to measure in situ quasiparticle states. When a specific condensed-matter system is illuminated by the soft X-ray (Figure 2.6), the occupied valence states are excited to the unoccupied intermediate ones under electric-dipole perturbation. Photoelectrons are excited by incident photons and escape outside of the material surface into the vacuum, and then they are measured by an angle-resolved (energy,momentum) analyzer. The total momenta of photoelectrons are evaluated from the electron gas model, in

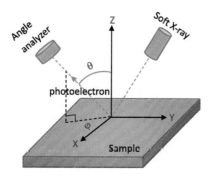

FIGURE 2.6
Schematic diagram of ARPES. Soft X-ray is used as the light source. θ and φ, respectively, represent the polar and azimuthal angles of detected photoelectrons.

which the parallel and perpendicular components depend on the polar and azimuthal angles, as shown in Figure 2.6 by θ and φ, respectively. The former is conserved through the photoemission process, while the conservation law is not reliable for the latter because of the destruction of translation symmetry along the direction normal to the surface. As a result, ARPES measurements mainly focused on two- and quasi-two-dimensional systems with negligible energy dispersions perpendicular to surface. However, the nonconservation issue might be solved using the important characteristics of the k_z-dependent band structure, as done for the 3D band structure of layered graphite by that at $k_z = 0$ [109–112]. Specifically, ARPES measurements could provide energy widths of valence states, directly reflecting the many-particle deexcitation scatterings arising from e–e and electron–phonon interactions. Improvements in energy and momentum resolutions have become a critical factor for studying the emergent low-dimensional materials. Up to date, the best resolutions for energy and angular distribution are, separately, $\sim$1 meV and 0.1° in the ultraviolet region.

The high-resolution ARPES is the only experimental instrument for directly measuring the wave-vector-dependent valence/occupied energy spectra. The experimental measurements have confirmed the geometry-enriched band structures in the graphene-related condensed-matter systems, as verified for various dimensions, layer numbers, stacking symmetries, and adatom/molecule chemisorptions. There exist (k_x, k_y, k_z)-dependent 3D band structures of Bernal graphite [109–112], 1D parabolic energy subbands in graphene nanoribbons [241, 242], the linearly isotropic Dirac-cone structure in monolayer graphene [43, 243], and few-layer AA-stacked graphene [54, 55], two pairs of parabolic dispersions in AB-stacked bilayer graphene [63–66], the coexistent linear and parabolic bands in symmetry-broken bilayer graphene [244], the linear and parabolic bands in trilayer graphene with ABA stacking [64–66], the linear, partially flat, and sombrero-shaped bands in ABC-stacked trilayer graphene [64–66], the metal–semiconductor transitions and the tunable low-lying energy bands after the molecule/adatom absorptions on graphene surface [245]. On the other side, the predicted Coulomb decay rates could be examined from the high-resolution ARPES measurements on the energy widths, as clearly revealed in potassium chemiadsorption on monolayer graphene. The ARPES energy spectra are done along KM and KΓ directions under various doping concentrations of monolayer electron-doped graphene, obviously indicating the linewidth variation with wave vector. They are further utilized to analyze the doping-dependent momentum distribution curves (MDCs). The Lorentzian peak forms are centered at quasiparticle energies; furthermore, they exhibit the full width at half-maximum intensity identified as $-2Im\Sigma^{(res)}$ (just the scattering rate). The single-particle excitation and plasmon modes, as well as the electron–phonon scatterings at finite temperatures, are proposed to comprehend the unusual energy dependences of the MDC linewidths. The ARPES measurements at low temperatures could provide the Coulomb-scattering-dominated MDCs to verify theoretical calculations.

In addition to the direct ARPES measurements on the energy widths of the valence quasiparticle states, the lifetimes (the inverse of the former) of the specific states, including the Fermi-momentum states and the excited valence and conduction band-edge states, could be examined by three kinds of pump-probe (also see Section 14.1). The femtosecond photoelectron spectroscopy is available for fully exploring the carrier relaxation near the Fermi level; that is, it is suitable for the semimetallic and metallic systems. For example, the measured lifetimes, which correspond to the Fermi-momentum states in the Bernal graphite and the metallic single-walled and multiwall carbon nanotubes, are, respectively, $\tau \sim 0.5$ and $\tau \sim 0.2$ ps at room temperature [41–43,246–249]. As for the femtosecond optical absorption/transmission/reflectivity and fluorescence spectroscopies, they are designed for the semiconducting systems, such as type-II narrow-gap and type-III moderate-gap carbon nanotubes. The latter nanotube systems are identified to exhibit the lifetimes of $\tau \sim 0.3 - 1.5$ and $\tau \sim 0.3 - 1.5$ ps, respectively, being associated with the first and second prominent absorption peaks [250–257]. Such decay rates are attributed to the intraband inelastic Coulomb scatterings. The experimental measurements on carbon nanotubes are consistent with the theoretical predictions [239]. The time- and temperature-dependent photoluminescence spectra have been made on very small $(6, 4)$ carbon nanotubes in the range of $48 - 182$ K, revealing the band-edge-state lifetimes due to the first part of energy bands about $100 - 20$ ps [258, 259]. Such femtosecond spectroscopies are the critical tools in studying the generation, relaxation, and recombination of nonequilibrium charge carriers, i.e., they can probe and verify the time-dependent carrier dynamics.

3

Monolayer Graphene

Plasmons in monolayer graphene have attracted fundamental interest stemming from electronic excitations of massless Dirac fermions. At zero temperatures, there are no intrinsic plasmons in a monolayer graphene due to the absence of free carriers. However, because of the zero gap, monolayer graphene is predicted to be the first undoped system that could display the low-frequency plasmons purely influenced by temperature. This is manifested by the fact that temperature could induce the thermal Fermi–Dirac distribution of electrons in the conduction band and holes in the valence band. The free carriers further induce the intraband single-particle excitations (SPEs) and the low-frequency plasmons. On the other hand, the low-frequency plasmons can also be triggered in an extrinsic monolayer graphene, where the doping-free carriers can create rich electronic excitations. The low-frequency graphene plasmons have attracted considerable interest from both theoretical and experimental research; they have been found to be strongly dependent on temperature and doping free carrier density by inelastic X-ray scattering (IXS) and electron energy loss spectroscopy (EELS). Whether such plasmons exist are mainly determined by temperature, the doping free carrier density, and transferred momentum. This is manifested by the plasmon dispersion in the excitation phase diagrams under the long wavelength limit.

3.1 Without Doping: Temperature and Doping Effects

The essential physical properties of a monolayer graphene are mainly determined by the $2p_z$ orbitals of carbon atoms. In the framework of the generalized tight-binding model, the Bloch wave function is characterized by the linear combination of the tight-binding functions φ_A and φ_B on two sublattices A and B in a unit cell:

$$\Psi(k_x, k_y) = C_A \varphi_A(k_x, k_y) + C_B \varphi_B(k_x, k_y), \tag{3.1}$$

where C_A and C_B are the normalization factors. $\varphi_{A(B)}$ is based on the atomic $2p_z$ orbital of sublattices A(B) spanned over all periodical carbon atoms in the crystal and expressed as

$$\varphi_A = \frac{1}{\sqrt{N_A}} \sum_{i=1}^{N_A} \exp(i\mathbf{k} \cdot \mathbf{R}_{A_i}) \chi(\mathbf{r} - \mathbf{R}_{A_i})$$
$$\varphi_B = \frac{1}{\sqrt{N_B}} \sum_{i=1}^{N_B} \exp(i\mathbf{k} \cdot \mathbf{R}_{B_i}) \chi(\mathbf{r} - \mathbf{R}_{B_i}), \tag{3.2}$$

where N_A (N_B) is the total number of A (B) atoms, and $\chi(\mathbf{r} - \mathbf{R}_A)$ ($\chi(\mathbf{r} - \mathbf{R}_B)$) represents the normalized orbital $2p_z$ wave function for an isolated atom at coordinate $\mathbf{R}_A$ ($\mathbf{R}_B$).

The 2×2 tight-binding Hamiltonian in the nearest-neighbor approximation takes the form:

$$\begin{bmatrix} 0 & \gamma_0 h(k_x, k_y) \\ \gamma_0 h^*(k_x, k_y) & 0 \end{bmatrix}, \tag{3.3}$$

where $h(k_x, k_y) = e^{ik_x b/\sqrt{3}} + 2e^{-ik_x b/2\sqrt{3}} \cos(\frac{k_y b}{2})$, $\gamma_0(=-2.569$ eV$)$ is the nearest-neighbor hopping integral, and the site energies of A and B sublattices and the overlap integral between A and B sublattices are set to zero [260]. The π-electronic energy dispersions are given by

$$E^{c,v}(k_x, k_y) = \pm\gamma_0\{1 + 4\cos(\frac{3bk_y}{2})\cos(\frac{\sqrt{3}bk_x}{2}) + 4\cos^2(\frac{\sqrt{3}bk_x}{2})\}^{\frac{1}{2}}, \tag{3.4}$$

and the corresponding Bloch functions are

$$\Psi^{c,v}_{\pm}(k_x, k_y) = \frac{1}{\sqrt{2}}[\varphi_A(k_x, k_y) \pm \frac{h^*(k_x, k_y)}{|h(k_x, k_y)|}\varphi_B(k_x, k_y)]. \tag{3.5}$$

The superscript c (v) represents the conduction (valence) states, corresponding to the bonding (antibonding) energies. The energy dispersions are depicted in Figure 3.1a. According to Eq. (3.4), one can derive a double-degenerate state at the K point, i.e., $E^{c,v}(K) = E^{c,v}(2\pi/3a, 2\pi/3\sqrt{3}a) = 0$, which means that the two linear subbands cross at the K point, the so-called Dirac point. At zero temperature, the occupied conduction band is symmetric to the unoccupied valence band about the Fermi level, which is located at the Dirac point, as shown in Figure 3.1a. The DOS is vanishing at the Dirac point and proportionally increased with conduction and valence energies (Figure 3.1c). With the isotropic Dirac cone dispersion at low energies, it indicates that the low-frequency excitations are identical with respect to different directions of the transferred momentum q. By substituting the energies and wave functions of the electronic states derived in Eqs. (3.4) and (3.5), the dielectric function in Eq. (3.3) can be expressed as

$$\epsilon(q, \omega) = \epsilon_0 - V_q P(q, \omega), \tag{3.6}$$

where the bare polarization function is

$$P(q, \omega) = 2 \sum_{h,h'=c,v} \int_{1stBZ} \frac{dk_x dk_y}{(2\pi)^2} |\langle k_x + q_x, k_y + q_y; h'|e^{iq_x x}e^{iq_y y}|k_x, k_y; h\rangle|^2$$

$$\times \frac{f(E^{h'}(k_x + q_x, k_y + q_y)) - f(E^h(k_x, k_y))}{E^{h'}(k_x + q_x, k_y + q_y) - E^h(k_x, k_y) - (\omega + i\Gamma)}. \tag{3.7}$$

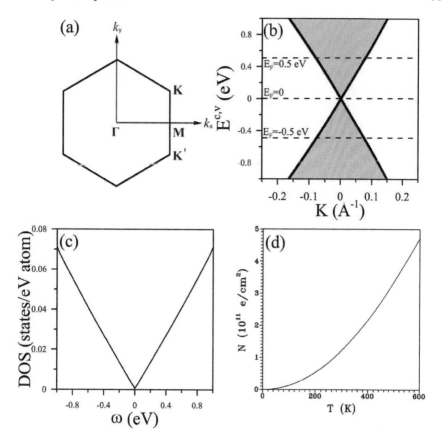

FIGURE 3.1
(a) The hexagonal first Brillouin zone of graphene. (b) The Dirac-cone band structure and (c) density of states of monolayer graphene. (d) Temperature-dependent total free carrier density of monolayer graphene.

The matrix element in the earlier equation is simply reduced as

$$\langle k_x + q_x, k_y + q_y; h' | e^{iq_x x} e^{iq_y y} | k_x, k_y; h \rangle = \frac{1}{N_A + N_B} \sum_{\mathbf{R}=\mathbf{R_A},\mathbf{R_B}}$$

$$\times \langle \chi(\mathbf{r} - \mathbf{R}) | e^{-i\mathbf{q}\cdot(\mathbf{r}-\mathbf{R})} | \chi(\mathbf{r} - \mathbf{R}) \rangle \frac{1}{2} \left[1 \pm \frac{h(k_x + q_x, k_y + q_y) h^*(k_x, k_y)}{|h(k_x + q_x, k_y + q_y) h(k_x, k_y)|} \right],$$

$$(3.8)$$

where only the overlapping integrals on the same atoms are taken into account. The plus and minus signs, respectively, correspond to the Coulomb interactions of intra-Dirac and inter-Dirac cones. To calculate the expectation value in

Eq. (3.8), we express the atomic $2p_z$ orbital in terms of a generalized hydrogen-like wave function, i.e., $\chi(\mathbf{r}) = Cr\cos(\theta)\exp(-Zr/2a_0)$, where C is a normalization factor, θ the azimuthal angle with respect to c axis, a_0 the Bohr radius, and an effective nucleus charge Z chosen as 3.18 according to Ref. [261]. The expectation in the right-hand side of Eq. (3.8) can be reduced to a function of q, defined by $I(q)$, and takes the form of the following integration equation:

$$I(q) = \int \chi^*(\mathbf{r})e^{-i\mathbf{q}\cdot\mathbf{r}}\chi(\mathbf{r})d^3\mathbf{r} = \int |C|^2 r^2 \cos^2(\theta)e^{-Zr/a_0-i\mathbf{q}\cdot\mathbf{r}}d^3\mathbf{r}$$
$$= [1+(qa_0/Z)^2]^{-3}. \tag{3.9}$$

In consequence, the dielectric function $\epsilon(q,\omega)$ is expressed as

$$\epsilon(q,\omega) = \epsilon_0 - 2V_q \sum_{h,h'=c,v} \int_{1stBZ} \frac{dk_x dk_y}{(2\pi)^2} \times \frac{1}{4}I(q)^2$$

$$\times \left|1 \pm \frac{h(k_x+q_x, k_y+q_y)h^*(k_x,k_y)}{|h(k_x+q_x,k_y+q_y)h^*(k_x,k_y)|}\right|^2$$

$$\times \frac{f(E^{h'}(k_x+q_x, k_y+q_y)) - f(E^h(k_x,k_y))}{E^{h'}(k_x+q_x,k_y+q_y) - E^h(k_x,k_y) - (\omega+i\Gamma)}, \tag{3.10}$$

where the Coulomb matrix element $I(q)^2 \simeq \{1+q^2/36\}^{-6}$ is close to one in the long-wavelength limit, and the result is also applicable to layered graphenes. These functions can be calculated numerically by summing overall possible excitations in the first Brillouin zone.

Next, we use analytical expression to analyze the polarization and dielectric functions. By expanding the Hamiltonian near the K point, the low-energy electronic structure measured from the K point are isotropic and described by a conical dispersion relation, i.e., $E^{c,v} \simeq 3\gamma_0 bk/2 = v_f k$, where $v_f (= 3\gamma_0 b/2)$ is the Fermi velocity. Electronic excitations mainly result from the transitions within the Dirac cone. The dielectric function taking into account the spin and valley degeneracy is expressed as

$$\epsilon(q,\omega) = \epsilon_0 - 4V_q P(q,\omega)$$

$$= \epsilon_0 - 4V_q \sum_{\mathbf{k}} \sum_{h,h'=c,v} |\langle \mathbf{k}; h|e^{-i\mathbf{q}\cdot\mathbf{r}}|\mathbf{k}+\mathbf{q}; h'\rangle|^2 \tag{3.11}$$

$$\times \frac{f(E^{h'}(\mathbf{k}+\mathbf{q})) - f(E^h(\mathbf{k}))}{E^{h'}(\mathbf{k}+\mathbf{q}) - E^h(\mathbf{k}) - (\omega+i\Gamma)}.$$

Consider the intraband excitations, the matrix element in the right-hand side of Eq. (3.11) takes the form as follows using the earlier results:

$$\langle \mathbf{k}; h|e^{-i\mathbf{q}\cdot\mathbf{r}}|\mathbf{k}+\mathbf{q}; h'\rangle = \frac{1}{2}I(q)\left[1 + \frac{h(\mathbf{k}+\mathbf{g})h^*(\mathbf{k})}{|h(\mathbf{k}+\mathbf{g})h(\mathbf{k})|}\right]$$

$$= \frac{1}{2}I(q)\left[1 + \frac{\mathbf{k}\cdot(\mathbf{k}+\mathbf{q})e^{-i(\theta_{k+q}-\theta_k)}}{|\mathbf{k}\cdot(\mathbf{k}+\mathbf{q})|}\right], \tag{3.12}$$

where $h(\mathbf{k})$ is given by $3b(k_x + ik_y)/2$, which can be defined as $3bke^{-i\theta_k}/2$ with θ_k being the polar angle of the wave vector. In a similar calculation, we obtain the matrix element for the interband excitations:

$$\langle \mathbf{k}; h | e^{-i\mathbf{q}\cdot\mathbf{r}} | \mathbf{k} + \mathbf{q}; h' \rangle = \frac{1}{2} I(q) \left[1 - \frac{\mathbf{k} \cdot (\mathbf{k} + \mathbf{q}) e^{-i(\theta_{k+q} - \theta_k)}}{|\mathbf{k} \cdot (\mathbf{k} + \mathbf{q})|} \right]. \tag{3.13}$$

In consequence, the dielectric function ϵ in the RPA, including the intraband and interband excitations, is thus expressed as

$$\epsilon(q, \omega) = \epsilon_0 - V_q P(q, \omega)$$

$$= \epsilon_0 - \sum_{h=v, h'=v} 2V_q I(q)^2 \left[1 + \frac{k + q\cos(\phi)}{|\mathbf{k} + \mathbf{q}|} \right]$$

$$\times \frac{f(E^{h'}(\mathbf{k} + \mathbf{q})) - f(E^h(\mathbf{k}))}{E^{h'}(\mathbf{k} + \mathbf{q}) - E^h(\mathbf{k}) - (\omega + i\Gamma)}$$

$$- \sum_{h=v, h'=c} 2V_q I(q)^2 \left[1 - \frac{k + q\cos(\phi)}{|\mathbf{k} + \mathbf{q}|} \right]$$

$$\times \frac{f(E^{h'}(\mathbf{k} + \mathbf{q})) - f(E^h(\mathbf{k}))}{E^{h'}(\mathbf{k} + \mathbf{q}) - E^h(\mathbf{k}) - (\omega + i\Gamma)}, \tag{3.14}$$

where ϕ is the angle between the wave vector $\mathbf{k}$ and transferred momentum q, and the Fermi level and temperature determine whether the channels would be permitted or not.

In an undoped monolayer graphene, the Fermi-momentum state are double-degenerate and located at the Dirac point. The density of free carriers $N(T)$ is described by the Fermi–Dirac function and expressed as

$$N(T) = \frac{2}{\pi} \int \frac{k d\mathbf{k}}{\exp(\frac{E^{c,v}(\mathbf{k}) - \mu(T))}{k_B T}) + 1}. \tag{3.15}$$

At $T = 0$, $N(T) \to 0$, that is, only interband excitations are allowed due to the fully occupied valence bands and fully unoccupied conduction bands. The earlier integral equation can be explicitly calculated at finite temperature. Given by the chemical potential $\mu(T) = 0$ and $v_f(\mathbf{k} - \mathbf{K})$, it gives

$$N(T) = \frac{2}{\pi} \int \frac{k d\mathbf{k}}{\exp(\frac{v_f(\mathbf{k}-\mathbf{K})}{k_B T}) + 1} = \frac{\pi k_B^2 T^2}{6 v_f^2}, \tag{3.16}$$

which shows T^2 dependence. As shown in Figure 3.1d, $N(T)$ is estimated as $\simeq 10^{11}$ e/cm^2 at room temperature $T = 300$ K. With the increasing temperature, thermal populations of electrons and holes increase quickly, giving rise to both intraband and interband excitations. The excitation channels can also be modulated by the Fermi level. The electronic excitation spectra exhibit rich and diversified SPEs and plasmons at low frequencies during the variations of temperature and the Fermi level.

3.2   Numerical and Analytic Results

The dynamic Coulomb response displays SPEs and collective excitations as the transferred q and ω are conserved during electron–electron interactions. These two types of excitations are, respectively, characterized by the bare response function $P_{ll'}^{(1)}(\mathbf{q}, \omega)$ and energy loss function $\text{Im}[-\frac{1}{\epsilon}]$ [237]. The former describes the dynamic charge screening and directly reflects the main features of the band structure. The latter is directly related to the measurements of the intensity of the excitation spectrum by IXS [36] and EELS [35]. For the monolayer graphene, the low-frequency excitations within the Dirac cone described as $E = v_f \mathbf{k}$ give rise to the isotropic effect of the transferred momentum on the excitation spectra. The following calculations of ϵ are based on the $\phi = 0°$ case, in which $\text{Im}\epsilon$ and $\text{Re}\epsilon$ satisfy the Kramers–Kronig (K-K) relation [262]. At various T's and E_F's the q-evolutions of the dielectric functions are shown in Figure 3.2. In the pristine case ($T = 0$ and E_F=0), according to the conservation of transferred momentum and energy, interband excitations are the only available excitation channel. The singularity in the dielectric function appears at $\omega = \omega_{sp}$, which corresponds to the threshold frequency of SPEs related to the Fermi momentum states, i.e., $\omega_{sp} = v_f q$. The former $\text{Im}\epsilon$, representing SPEs, exhibits square-root divergence in the form of $1/\sqrt{\omega - v_f q}$, while $\text{Re}\epsilon$ is in the form of $1/\sqrt{v_f q - \omega}$, a corresponding behavior being preserved up to 0.5 Å^{-1} by the linear dispersion of the Dirac cone [261]. For $T = 0$ and $q = 0.005$ Å^{-1} (Figure 3.2a and b), the singularity is located at $\omega_{sp} = 0.023$ eV, which shift to 0.046 eV and 0.92 eV as q increases to 0.01 and 0.02 Å^{-1}. Temperature plays an important role in the low-frequency electronic excitations. [32,68,263–268] At finite temperatures, both the intraband and interband excitation channels are available in electronic excitations; the former gradually predominates the excitation properties under a sufficiently high free carrier density ($\propto T^2$) [264]. The divergent structures of ϵ at $T \geq 300$ K (Figure 3.2c and d) appear at the same frequencies as those in the zero-temperature case, while the opposite square-root form indicates the fact that they purely come from the intraband excitations. The square-root singularities that $\text{Re}\epsilon$ and $\text{Im}\epsilon$ respectively, display with a negative and positive divergence at the right and left neighborhood of the singular point $\omega = v_f q$ [261]. It should be noted that the T-induced intraband excitations enhance the structures in ϵ, leading to the $\text{Re}\epsilon$ passing through zero. The two zero points of $\text{Re}\epsilon$ indicated by the circles could be interpreted as the collective excitation energies; however, whether the plasmon mode can survive or disappear depends on the magnitude of Landau damping. On the other hand, a higher doping level also induces free carriers in the conduction subband, which give rise to intraband excitations (c-c) and cause a dielectric function similar to the T-induced pattern, as shown in Figure 3.2e and f.

The dimensional energy loss function $\text{Im}[-\frac{1}{\epsilon}]$ is a direct probe of the plasmons and useful in understanding the inelastic scattering probability of the

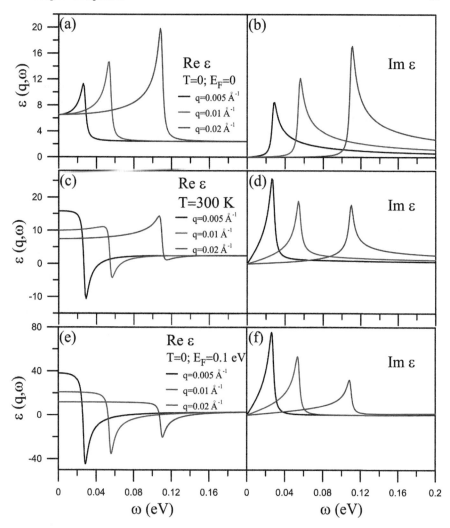

FIGURE 3.2

(a) The real part and (b) imaginary part of the dielectric function ϵ for a pristine monolayer graphene at $q = 0.005$, 0.01, and 0.02 Å^{-1}. (c) and (d) correspond to ϵ at $T = 300$ K and $E_F = 0$. (e) and (f) correspond to ϵ at $T = 0$ and $E_F = 0.1$ eV.

EELS and inelastic light scattering measurements [35, 36]. The measurements are directly related to the dynamic structure factor, which can be deduced in terms of the loss function $\text{Im}[-1/\epsilon]$ in the screened excitation spectrum for a closer study of the temperature-induced and doping-induced SPEs and collective excitations. It is well known that the collective excitations, or plasmons, are referred to the peaks in the loss function and appear at the frequencies where Reϵ passes through zero and Imϵ is nearly vanishing. However, in the opposite example, a finite Imϵ implies that the plasmons might decay into SPEs due to the Landau damping. This statement is used to clarify the plasmons in the screened Coulomb excitation, where the three factors, q, T and E_F, play an important role on the main characteristics of the excitation spectra, as shown in Figure 3.3. In the undoped case, the q-evolutions of the intrinsic plasmons are displayed in the energy loss function for different T's, in Figure 3.3a. It is clearly illustrated that the T^2-dependence of the free carrier density significantly enhances the SPEs and plasmons at higher temperatures. At $T = 0$, there is no obvious peak structure, which is consistent with the fact that the Reϵ is finite without any zero point at low frequencies, as shown in Figure 3.2a. In short, there are no plasmon peaks in the screened excitation spectra for the condition of heavy Landau damping that happens when the interband excitations are much stronger than the intraband ones. On the other hand, an intraband plasmon peak is formed at low frequencies when T is higher than 150 K. Identified as a 2D acoustic plasmon mode, the plasmon frequency ω_p gradually increases with an increase in temperature. The formation of plasmons is manifested by the dielectric function. For example of $T = 300$ K in Figure 3.2c, it is shown that a finite value of Imϵ near the first zero point provides an evidence of the cause of the damped plasmons, while the undamped acoustic plasmon mode corresponds to the frequency of the second zero point, where the small value of Imϵ implies the weak Landau damping. Figure 3.3b illustrates the variation of the acoustic plasmon mode under different transferred momenta at room temperature. The plasmon frequency ω_p increases with q as a result of higher SPE energy ω_{sp}, while the reduced plasmon intensity demonstrates different degrees of Landau dampings of the acoustic plasmon mode. The loss function would not exhibit an obvious plasmon peak under a sufficiently strong Landau damping, e.g., $q \geq 0.02$ Å^{-1} at $T = 300$ K. In addition to T, the Fermi level also enhances the frequencies of the single-particle and collective excitations, as shown in Figure 3.3c; the acoustic plasmon mode with long-q region is clearly presented under heavy doping conditions with negligible Landau dampings.

The E_F- and T-dependent behaviors of the low-frequency acoustic plasmons are important in understanding their characteristics. The strong dispersion relation of ω_p with q is displayed in the (q, ω)-phase diagram in Figure 3.4 at different temperatures. It is shown that as a result of intraband excitations, ω_p of small q is well fitted by $\sqrt{q}$ dispersion relation, which directly reflects the 2D characteristics of the massless Dirac quasiparticles. They are quite different from the optical plasmons in layered graphenes and graphites (detailed in latter chapters). Temperature determines the numbers of intraband and interband excitations. At higher temperature, the former begins to predominate over the latter. The enhancement of free carriers supports the formation of the collective excitations in monolayer graphene; the plasmons could be improved and extended to a relatively long q range at higher temperatures. Moreover, the temperature dependence of the plasmon frequency is

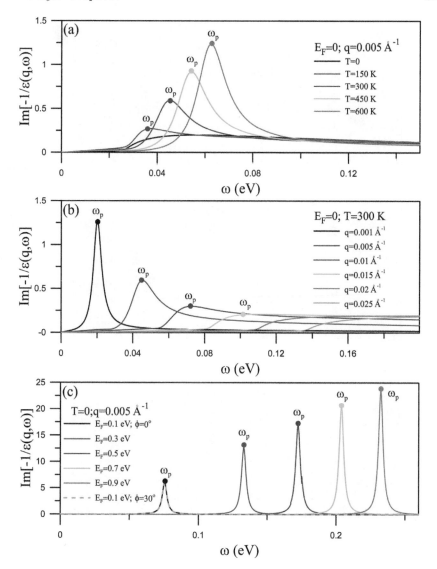

FIGURE 3.3

(a) The intrinsic energy loss function of monolayer graphene for $q = 0.005$ $\mathring{A}^{-1}$ at various temperatures. (b) q-Evolution of the energy loss functions at $T = 300$ K. The low-frequency plasmon peak for $q = 0.005$ $\mathring{A}^{-1}$ and $T = 0$ at various Fermi levels.

approximately fitted by $\omega_p \propto T^2$, and thus $\omega_p^2 \propto T \propto N^2$. The special linear energy dispersion accounts for the novel N or T-dependence of ω_p, which is distinct from the relationship $\omega_p^2 \propto N$ in a 2D electron gas system. At room temperature, the plasmon frequency is about 20 meV for $q = 1$ $\mathring{A}^{-1}$, which is

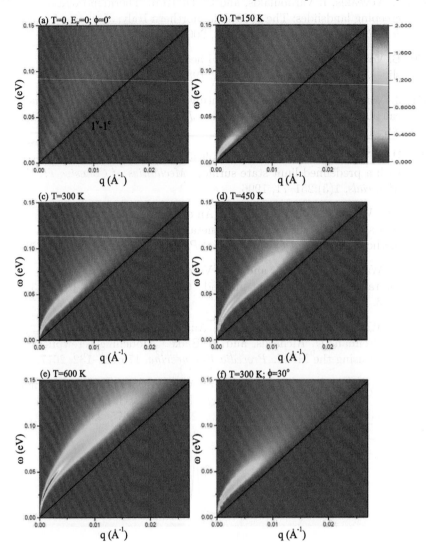

FIGURE 3.4
Temperature-dependent (q, ω)-excitation phase diagrams of monolayer graphene for $\phi = 0°$, and $E_F = 0$ at various temperatures: (a) $T = 0$, (b) $T = 150$ K, (c) $T = 300$ K, (d) $T = 450$ K; (e) $T = 600$ K. Also shown in (f) is that under $\phi = 30°$, and $T = 300$ K.

comparable to that in 2D semiconductor quantum wells or 1D semiconductor quantum wires. While the high-resolution IXS (< 0.1 meV) has been used to successfully identify the acoustic plasmons and electron–hole (e–h) excitations in 2D or 1D electron-gas systems, it could also be used to measure the temperature-induced plasmons in monolayer graphene. In contrast, EELS is unsuitable for measurements, because the energy resolution ~ 10 meV and the momentum resolution ~ 0.05 Å^{-1} are too low to test the infrared plasmons.

In cases of finite electron dopings (Figure 3.5), the (q, ω)-excitation phase diagram can be clearly divided into three domains, indicated as regions I–III; they are enclosed by the excitation boundaries of the intraband and interband channels. With the variation of Fermi level, the (q, ω)-excitation phase diagram of monolayer graphene display fascinating properties of the low-frequency acoustic plasmons, as shown in Figure 3.5a–f. Region I is the transition forbidden region, where neither SPEs nor collective excitations exist below the lower bound of the threshold intraband excitation energies (solid line). The occupied conduction states trigger intense intraband excitations throughout regions II and III. In contrast, the interband excitations are only distributed in region III, which is a Landau damping region. Under screening effect, the collective frequency of the free-carrier excitations must be higher than the corresponding intraband SPE frequency. It is shown that the collective plasmons are favored in region II because of the inhibition of the mentioned interband SPEs that are regarded as the main cause of the Landau damping of the plasmons. The acoustic plasmon mode with 2D $\sqrt{q}$-dependent ω_p apparently predominates over this region, where the plasmon don't overlap with the continuum spectra of e–h pairs. The plasmon wave propagating with a wavelength $2\pi/q$ and group velocity $d\omega_p = dq$ can survive in cases of weak Landau dampings, that is, the prominent acoustic mode are strongly damped when dispersing into the region III of the interband SPEs. The isotropic feature of the massless Dirac plasmons is held by the linear dispersions in energies below 1.0 eV.

In conclusion, the temperature- and momentum-dependent dielectric functions are calculated for a monolayer graphene. The π-band characteristics, the zero gap and the strong wave-vector-dependence, are directly reflected in the electronic excitations. Temperature could generate the free carriers and thus the intraband plasmons. Also noticed that temperature significantly affects the low-frequency plasmons in the doped semiconductors, while it cannot induce them for the undoped semiconductors [227]. A zero-gap graphite sheet is predicted to be the first undoped system that could exhibit temperature-induced plasmons. The pronounced excitation spectra due to these plasmons could be verified from the inelastic light scattering measurements [36]. The intraband plasmons are identified as the 2D plasmons from the momentum dependence of plasmon frequencies. They contrast sharply with the optical plasmons in graphite or of graphite intercalation compounds (GICs) [234].

In general, a doped monolayer graphene sharply contrasts with a 2D electron gas corresponding to a quantum well in Coulomb excitations [269, 270]. The latter has a parabolic energy band accompanied with many conduction electrons. This electronic structure only exhibits the intraband excitations and the 2D-like acoustic plasmon mode according to the Pauli principle and the conservation of transferred momenta and energies. The strong collective excitations survive in the absence of any Landau dampings at long wavelength

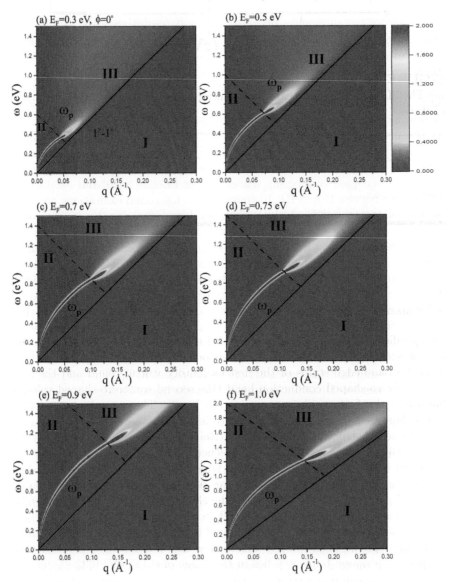

FIGURE 3.5

(q, ω)-excitation phase diagrams of monolayer graphene for $\phi = 0°$ at various Fermi levels: (a) $E_F = 0.3$ eV, (b) $E_F = 0.5$ eV, (c) $E_F = 0.7$ eV, (d) $E_F = 0.75$ eV, (e) $E_F = 0.9$ eV, and $E_F = 1.0$ eV.

limit, while their strength would greatly decline and the q-dependent plasmon frequency dispersion is close to the boundary of the intraband e–h dampings after sufficiently large transferred momenta. Apparently, the intermediate excited states in a 2D electron gas cannot decay by acoustic plasmon modes, but by intraband excitations. The significant differences between two systems cover the distinct boundaries of the intraband excitations dominated by the band structures, the existence/absence of the interband excitations determined by the valence bands, the coexistent regions of the 2D plasmon modes, and the interband excitations/intraband boundaries; the higher/stronger plasmon frequencies the more rich (momentum, frequency)-phase diagrams in graphene system. In addition, the Coulomb decays of quasiparticle states will be thoroughly investigated in Chapter 14.

3.3   Double-Layer Systems: Quasiparticle Lifetimes

Double-layer graphene consists of two unstacked graphene layers without any interlayer atomic interaction between the two layers [271]. Under the influence of Coulomb interactions, the intralayer polarization of electronic transitions keeps the same when compared with that of monolayer graphene, while the interlayer one is vanishing because the probability densities of the two isolated Dirac cones are distributed on different layers. The plasmon modes can be turned by bringing the two monolayer graphene sheets in close proximity to form a double-layer graphene system [272–274]. The interlayer Coulomb interactions result in the coupling of the 2D $\sqrt{q}$-dependent plasmon modes in the two decoupled monolayer graphenes. For a double-layer graphene, the two plasmon modes are determined by the zeros of determinantal equation of the 2×2 dielectric-function matrix, that is,

$$0 = (\epsilon_0 - V_{11}P_{11} - V_{12}P_{11}) \times (\epsilon_0 - V_{11}P_{11} + V_{12}P_{11}), \qquad (3.17)$$

where the solution is given by

$$\begin{aligned} \to \epsilon_0 - (V_{11} + V_{12})P_{11} &= 0; \\ \epsilon_0 - (V_{11} - V_{12})P_{11} &= 0. \end{aligned} \qquad (3.18)$$

The dispersion relations of plasmon frequencies can be obtained by setting the earlier equation to zero. The addition term $(V_{11} + V_{12})$ in the first equation means that the intraband plasmons should belong to a bonding (in-phase) mode [49,50]. The subtraction term $(V_{11} - V_{12})$ in the second equation accounts for the antibonding (out-of phase) mode. The former corresponds to a stronger frequency and intensity than the latter. Displayed by the orange and pink dashed curves in Figure 3.6a, the out-of-phase mode, ω_p^2, appears until $E_F=0.6$ eV at normal distance situation $I_c = 3.5$ Å, and is observed to be much less intense than the in-phase mode, ω_p^1, for $q=0.05$ Å^{-1}, but for $q(=0.1$ Å$^{-1})$ above this momentum, the situation is reversed. The two plasmon modes are significantly enhanced at $E_F=0.8$ eV and $E_F=1.0$ eV, in Figure 3.6b; they can also be modulated by changing the distance between two layers of graphene sheets. The out-of-phase-mode intensity strongly enhanced with a blue shift of up to 0.2 eV with an enlargement of the layer distance. In contrast, the in-phase mode exhibits a relatively small red shift and its intensity is rarely changed.

As shown in Figure 3.7, the phase diagrams of the double-layer graphene indicate a greater impact of the layer distance on the two plasmon modes. At $E_F = 0.2$ eV, the double-layer graphene plasmon resembles a conventional 2D graphene plasmon. However, free carriers are largely enhanced at $E_F = 0.6$ eV so that under the dynamic interlayer Coulomb potentials the charge density fluctuations of the two layers are coupled to form the two plasmon modes, ω_p^1 and ω_p^2, as discussed in the preceding section. The former and the latter are, respectively, described by square-root and linear plasmon dispersion relations, and fall between the two regions of interband and intraband processes. The minimum of the interband excitation energies from the valence bands are increased by the increasing E_F. In this case, the high-energy transitions do not affect the two plasmon modes, being beneficial for the undamped plasmons at low energies. When the layer distance is increased, ω_p^1 mode is rarely changed, while the intensity of ω_p^2 is strongly enhanced with a blue shift of energy ≤ 1.0 eV. The plasmon phenomenon is a straightforward consequence of the change of interlayer Coulomb interactions in double-layer systems.

The double-layer model could be generalized to the superlattices of GICs [261], in which the latter are identified to be successful in fully comprehending the low-frequency Coulomb excitations of the acceptor-/donor-type layered systems. Such systems have sufficiently large interlayer distances. The metal adatoms/molecules could be intercalated between the graphene layers, so that they donate free electrons/holes efficiently. GICs possess a high free carrier density and exhibits an excellent electrical conductivity as best as copper [44–47]. Band structure of each graphitic layer is approximately characterized by the isotropic Dirac cones composed of the linearly intersecting valence and conduction band; furthermore, the Fermi level presents a red/blue shift under the hole/electron doping. When the important interlayer Coulomb interactions are included in the calculations, the dielectric function tensor would become a scalar function by introducing an extra independent parameter of k_z (details in [261]). That is to say, all the calculations are similar to those of monolayer graphene, while there exist the k_z-created Coulomb excitation spectra. The bandlike plasmon structure, which is defined by the various k_z's modes, is formed in the superlattice model [234, 261]. The collective excitations cover the optical and acoustic modes simultaneously, while the former would dominate the energy loss functions according to the detailed theoretical calculations on their strengths. The calculated results are consistent with the measured loss spectra and plasmon modes on GICs [169–172]. Specifically, the q-dependent plasmon frequency belongs to the 3D optical form; that is, $\omega_p(q) \sim \omega_0 + \alpha q^2$, where the coefficient of α depends on the doping free carrier density. For $E_F = 1.0$ eV, the observable plasmon frequency lies in the range of 0.8 eV$\leq \omega_p(q) \leq 2.0$ eV corresponding to $0 \leq q \leq 0.4$ Å^{-1}.

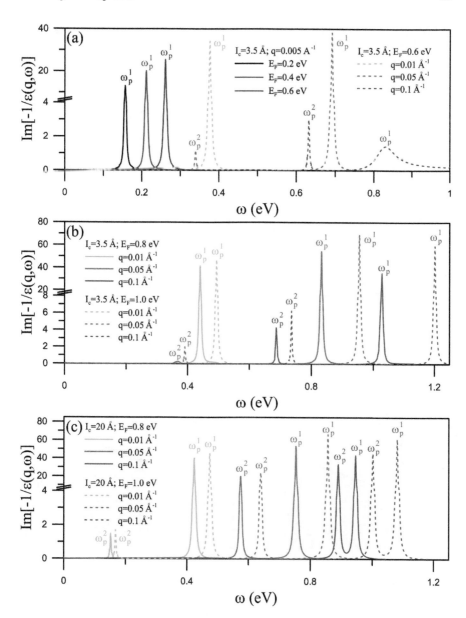

FIGURE 3.6

Energy loss function of a double-layer graphene for $\phi = 0°$ at various Fermi levels and interlayer distances: (a) $I_c = 3.5$ Å and $q = 0.005$ Å^{-1} at $E_F = 0.2$, $E_F = 0.4$, and $E_F = 0.6$; $q = 0.01$, $q = 0.05$, and $q = 0.1$ Å^{-1} at $E_F = 0.6$, (b) $I_c = 3.5$ Å and $q = 0.01$, $q = 0.05$, and $q = 0.1$ Å^{-1} at $E_F = 0.8$ and $E_F = 1.0$ eV, (c) Similar plot in (b), but displayed for $I_c = 20$ Å.

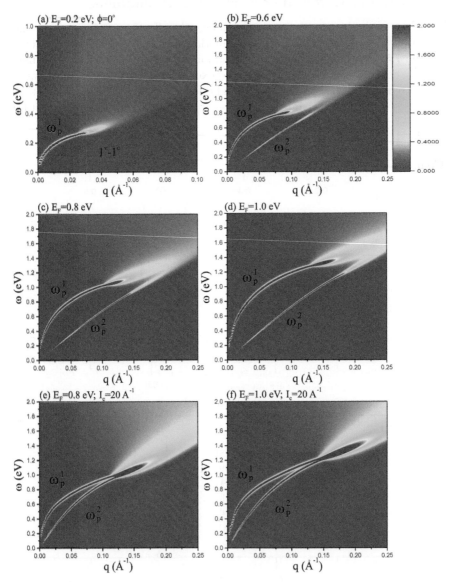

FIGURE 3.7

(q, ω)-excitation phase diagrams of a double layer graphene for $\phi = 0°$ and $I_c = 3.5$ Å at various Fermi levels: (a) $E_F = 0.2$ eV, (b) $E_F = 0.6$ eV, (c) $E_F = 0.8$ eV, and (d) $E_F = 1.0$ eV. Also shown in (e) and (f) are the diagrams for $I_c = 20$ Å at $E_F = 0.8$ and $E_F = 1.0$ eV.

Apparently, GIC is in sharp contrast with monolayer graphene in the collective excitations because of the interlayer Coulomb couplings. The significant plasmon structure further becomes effective Coulomb decay channels, being directly reflected in the optical threshold absorption edges. This has been thoroughly examined in the previous predictions [234, 261].

The main features of free and valence charge screenings, the distance-dependent effective Coulomb potentials and induced charge distributions, have been thoroughly studied using an analytically expressed dielectric function, which is calculated in accordance with the band structure of GICs and the interlayer Coulomb interactions. Due to the unusual energy bands, the screening is predicted to be effective at short distances from an impurity, and a strong but short-period Friedel oscillation also survives at large distances. Compared with a 2D electron gas, there exist certain important differences in static charge screenings, such as the enhanced screening abilities arising from the inter-π-band transitions and the interlayer Coulomb couplings. Also, the linear energy dispersions are found to have important effects on the mobility of electrons. They reduce the effective mass of electrons and even lead to a vanishing backward-scattering amplitude. This practically explains the excellent conductivity of GICs. These features are expected to have important effects on, for example, the resistivity and ordering of impurities in the system. The previous calculations on the residual resistivity are due to the scattering from charged impurities. A direct comparison with the measured results clearly indicates that a small amount of such impurities could explain the observed resistivities in stage-1 C_8M compounds (M for K, Rb, and Cs).

4

AA-Stacked Graphenes

AA-stacking configuration has the highest symmetry, i.e., the carbon atoms of one layer are directly above/below those of another layer [28, 31, 53]. It has been found that AA-stacking bilayer configuration is normally encountered in the synthesis procedure of multilayer graphenes [108, 275] and is very difficult to distinguish from a single-layer graphene by scanning tunneling microscopy [52]. The macroscopically stable configuration has attracted intense interest of theoretical and experimental research even when it is absent in natural graphite. The highly symmetric configuration results in the low-lying multiple Dirac-cone structure that resembles a collection of N pairs of isolated Dirac cones with a rigid shift along the k_y direction [28, 31]. Near the Fermi level, the dispersion is linear and isotropic around the K point in the first Brillouin zone, which implies a high-efficiency carrier transport, even better than in a monolayer graphene [276, 277]. However, the density of free carriers could be efficiently tuned by a perpendicular electric field under a fixed chemical potential [28, 278]. Regardless of external fields, the wave functions are linear symmetric or asymmetric superpositions of the tight-binding functions on different layers [28, 30, 53]. The optical excitations, closely related to the electronic band structure and the dipole transition probability, are not allowed between different Dirac cones due to asymmetry of the wave functions. They will exhibit the unusual optical vertical excitations, mainly owing to the unusual van Hove singularities in density of states (DOS). As a result, the full optical spectrum can be regarded as a combination of several Dirac-cone spectra under different Fermi levels. It displays particular regions of intensity discontinuities and forbidden transitions for certain kinds of excitation channels; these regions highly depend on the number of graphene layers. While AA-stacked graphenes are semimetallic, an optical gap is presented for all the cases of even number of stacking layers. The gap value is determined by the vertical transition energy of the Fermi momentum states within the two identical Dirac cones nearest (above and below) to the Fermi level, or equivalently the same as the energy difference between the two Dirac points [28]. In the presence of a magnetic field, the Landau levels (LLs) are well defined by single-mode Landau wave functions as a result of the Dirac-cone structures. Also the corresponding inter-LL transitions are characterized by a specific selection rule [28, 30, 279]. The aforementioned phenomenon can be realized by the fact that the wave function distributions are responsible for the nonzero and zero probability for the intrapair and interpair optical transitions, regardless of external fields.

In an undoped case, with free electrons and holes within the multiple Dirac cones, semimetallic AA-stacked graphene is expected to display fascinating low-frequency single-particle excitations (SPEs) and collective plasmons [20–23]. The SPEs of intrapair and interpair transition channels give rise to a special 1D structure, the square-root asymmetric divergence, in the imaginary part of the bare response function. A single-peak structure is induced by a major intrapair or interpair channel, while the energy disparity between any two Dirac cones creates a presentation of double-peak structures. The full charge screening due to all the pairs of energy subbands is incorporated in this chapter by the development of the modified random-phase approximation (RPA). An analytic expression for the layer-dependent polarizability is derived based on the modified RPA, which constitutes the basic relationship linking the intrapair and interpair excitation channels to the intralayer and interlayer polarization functions [20, 22]. The two different kinds of polarization functions are similar in structure regarding a specific excitation channel but may differ in phase according to the properties of the layer wave functions. Of interest is the interplay among all excitation channels that give rise to fascinating properties of the Coulomb excitation spectra. The single-particle picture failed to explain the screened Coulomb excitations. In general, the other pristine graphene systems, e.g., the monolayer and AB-stacked graphenes, are semiconducting and only create interband electron–hole (e–h) excitations [23, 264], but not the low-frequency plasmons. This has intrigued us to further investigate the effects due to doping and stacking on the Coulomb excitations and to describe the related diversified phenomena in this book.

4.1   Electronic Properties

The highly symmetric AA-stacked graphene is formed by perpendicularly stacking graphene sheets along the $\hat{z}$ direction. The geometric structure is illustrated on the top right side in Figure 4.1b, in which the four representative atomic interactions taken into account in the tight-binding calculations are indicated: one nearest-neighbor intralayer interaction $\alpha_0 = 2.569$ eV, and three interlayer interactions $\alpha_1 = 0.361$ eV, $\alpha_2 = 0.013$ eV and $\alpha_3 = -0.032$ eV [260]. The electronic band structure of an N-layer AA-stacked graphene comprises N sets of monolayer-like energy bands, with a $2N \times 2N$ Hermitian matrix expressed as

$$\begin{bmatrix} h_1 & h_2 & h_3 & 0 & \cdots & \cdots & 0 \\ h_2 & h_1 & h_2 & h_3 & 0 & \cdots & 0 \\ h_3 & h_2 & h_1 & h_2 & \ddots & \cdots & 0 \\ 0 & h_3 & \ddots & \ddots & \ddots & \cdots & 0 \\ 0 & 0 & \ddots & \ddots & \ddots & \ddots & \ddots \\ 0 & \vdots & 0 & \ddots & h_2 & h_1 & h_2 \\ 0 & \cdots & \cdots & \cdots & h_3 & h_2 & h_1 \end{bmatrix}, \tag{4.1}$$

where h_1, h_2, and h_3 are 2×2 blocks.

The diagonal block h_1 is identical and associated with the intralayer hoppings

$$h_1 = \begin{pmatrix} 0 & \alpha_0 f(k_x, k_y) \\ \alpha_0 f^*(k_x, k_y) & 0 \end{pmatrix}, \tag{4.2}$$

where $f(k_x, k_y) = \sum_{j=1}^3 \exp(i\mathbf{k} \cdot \mathbf{r}_j) = \exp(ibk_x) + \exp(ibk_x/2)\cos(\sqrt{3}bk_y/2)$ represents the phase summation arising from the three nearest neighbors on the same layer.

The off-diagonal blocks, h_2 and h_3, are related to the interlayer interactions

$$h_2 = \begin{pmatrix} \alpha_1 & \alpha_3 f(k_x, k_y) \\ \alpha_3 f^*(k_x, k_y) & \alpha_1 \end{pmatrix}, \tag{4.3}$$

and

$$h_3 = \begin{pmatrix} \alpha_2 & 0 \\ 0 & \alpha_2 \end{pmatrix}. \tag{4.4}$$

By diagonalizing the above Hamiltonian matrix, one can obtain the band structure of AA-stacked graphenes; the low-lying subbands $\pi_j^{c,v}$ ($j = 1, 2, \ldots, N$ counting from low to high energies) resemble a collection of N Dirac cones with rigid energy shifts with respect to the K point in the low-energy region. As shown in Figure 4.1, the multi-Dirac cones of AA-stacked bilayer and trilayer graphenes are chosen as a model study. The energy dispersions and the wave functions are discussed in detail. Based on the low-energy expansion near the K point [280], the Hamiltonian matrix can be reduced into N monolayer-like submatrices, of which each is renormalized due to the interlayer atomic interactions α_1, α_2 and α_3. The energy dispersion of the Dirac-type subband, $\pi_j^{c,v}$ ($j = 1, 2, \ldots, N$), is described as

$$E_j(k_x, k_y) = E_{D,j} + g_j(k_x, k_y) + \mathcal{O}(\alpha_2), \tag{4.5}$$

where the subscript j represents the Dirac-cone number counting from low to high energies. The first term indicates the Dirac-point energy at the K point, the second term describes the conical energy dispersion, and the third term serves as a perturbation energy due to next-neighbor interlayer interactions.

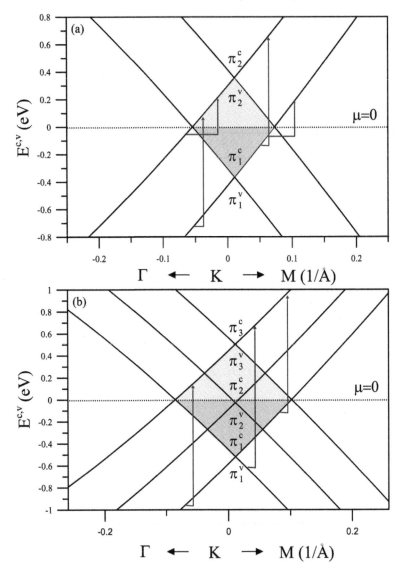

FIGURE 4.1
Band structures of AA-stacked (a) bilayer and (b) trilayer graphenes. The
brown and purple arrows, respectively, point out the intrapair and interpair
excitations. Dashed areas indicate the free carrier dopings of the subbands.

The earlier three quantities are expressed as follows when expanded in the
order of k:

$$E_{D,j} = 2\cos[j\pi/(N+1)]\alpha_1,$$
$$g_j(k_x, k_y) = \pm\,[\alpha_0 + 2\cos(j\pi/(N+1))\alpha_3]|f(k_x, k_y)|. \qquad (4.6)$$

For $N = 2$, $E_{D,1} = -\alpha_1$, $E_{D,2} = \alpha_1$, $g_1(k_x, k_y) = \pm(\alpha_0 - \alpha_3)|f(k_x, k_y)|$ and $g_2(k_x, k_y) = \pm(\alpha_0 + \alpha_3)|f(k_x, k_y)|$. The linear dispersions of the two pairs of Dirac cones, $\pi_1^{c,v}$ and $\pi_2^{c,v}$, in bilayer graphene are expressed as

$$E_1^{c,v}(k_x, k_y) = -\alpha_1 \pm (\alpha_0 - \alpha_3)|f(k_x, k_y)|$$
$$E_2^{c,v}(k_x, k_y) = \alpha_1 \pm (\alpha_0 + \alpha_3)|f(k_x, k_y)|, \qquad (4.7)$$

where the superscript $c(v)$ indicates the conduction (valence) subbands above (below) the corresponding Dirac point. While the two Dirac cones are, respectively, located at the Dirac-point energies, $-\alpha_1$ and α_1, the nonvertical interlayer atomic interaction α_3 causes their slopes to be slightly different. In the undoped case, the Fermi level, evaluated as the average of Dirac point energies, i.e., $E_F = 0$, is higher than $E_{D,1} = -\alpha_1$ and lower than $E_{D,2} = \alpha_1$, implying that there are free electrons in $\pi_1^{c,v}$ Dirac cone and holes in $\pi_2^{c,v}$ Dirac cone, as illustrated by the blue and green shadow areas in Figure 4.1a, respectively. The intersection of these two areas determines the Fermi momentum k_F, which is calculated as $k_F = 2\alpha_1/3\alpha_0 b$. The two Dirac cones have an equal free carrier density given by k_F^2/π, and their respective Fermi velocities are estimated as $v_{F,1} = 3(\alpha_0 - \alpha_3)b/2$ and $v_{F,2} = 3(\alpha_0 + \alpha_3)b/2$. Moreover, the preservation of the Dirac cone structure in the AA-stacking configuration indicates that the wave functions can be described by a linear combination in terms of monolayer eigenfunctions. According to Eqs. (3.8)–(3.14), the wave functions $\psi_i^{c,v}$ (i=1 and 2) corresponding to eigenenergy $E_i^{c,v}$ are expressed as

$$\psi_1^{c,v}(k_x, k_y) = -U_{1A} \pm \frac{f(k_x, k_y)}{|f(k_x, k_y)|}U_{1B} + U_{2A} \mp \frac{f(k_x, k_y)}{|f(k_x, k_y)|}U_{2B}$$

$$\psi_2^{c,v}(k_x, k_y) = U_{1A} \mp \frac{f(k_x, k_y)}{|f(k_x, k_y)|}U_{1B} + U_{2A} \pm \frac{f(k_x, k_y)}{|f(k_x, k_y)|}U_{2B}, \qquad (4.8)$$

where U_{jA} and U_{jB} are the tight-binding functions of sublattices A and B on the jth layer, and $U_{jA} \pm \frac{f}{|f|}U_{jB}$ refers to the wave functions of bonding and antibonding states in monolayer graphene [28, 30, 31]. Equation (3.15) indicates that the probability amplitudes of $\psi_1^{c,v}$ ($\psi_2^{c,v}$) are the same for every A and B sublattice, and the linear symmetric (asymmetric) superpositions of U_{jA} and U_{jB} of different layers construct the total wave function. In this way, the excitation channels resulting from the whole energy band structure can be divided into two kinds, i.e., intrapair and interpair ones. The former possesses two available channels, indicated by the brown arrows in Figure 4.1a, i.e., $\pi_1^v \rightarrow \pi_1^c$ and $\pi_2^v \rightarrow \pi_2^c$. On the other hand, the latter consists of other two channels, $\pi_1^v \rightarrow \pi_1^c$ and $\pi_2^v \rightarrow \pi_2^c$, indicated by the purple arrows. Their main features are determined by the symmetric or asymmetric relationship of the wave functions between the two layers.

In the trilayer case, three sets of monolayer-like subbands, $\pi_1^{c,v}$, $\pi_2^{c,v}$, and $\pi_3^{c,v}$, are displayed by the thin curves in Figure 4.1b; one of the three sets is mapped to the monolayer case, and the other two are oppositely shifted by energy $\sim \alpha_1$. The analytic expression in the low-energy region is given by

$$E_1^{c,v} = (\alpha_2 - \sqrt{8}\alpha_1)/2 \pm (\alpha_0 - \sqrt{8}\alpha_3/2)|f(k_x, k_y)|$$

$$E_2^{c,v} = -\alpha_2 \pm \alpha_0|f(k_x, k_y)|$$

$$E_3^{c,v} = (\alpha_2 + \sqrt{8}\alpha_1)/2 \pm (\alpha_0 + \sqrt{8}\alpha_3/2)|f(k_x, k_y)|. \qquad (4.9)$$

The three low-energy Dirac cones are located at the K point, and the renormalized characteristics depend on the number of inter- and intralayer interactions. Each Dirac cone is affected to different degrees by the interlayer atomic interactions, α_1, α_2, and α_3. It is worth noting that the eigenvalues of $E_2^{c,v}$ shows features similar to that of an undoped monolayer graphene; their only difference is the slight energy shift of $-\alpha_2$ with respect to $E_F = 0$. The associated wave function is $\psi_2^{c,v} = U_{1A} \pm \frac{f}{|f|}U_{1B} + U_{3A} \pm \frac{f}{|f|}U_{3B}$, in which the charge distributions are exclusively located on the two outmost layers. On the other hand, shifted downward and upward by $\sim \sqrt{2}\alpha_1$ with respect to $\pi_2^{c,v}$, the two outermost Dirac cones, $\pi_1^{c,v}$ and $\pi_3^{c,v}$, are described by the associated eigenvectors $|U_{1A}, U_{1B}, \sqrt{2}U_{1A}, \sqrt{2}U_{1B}, U_{3A}, U_{3B}\rangle$ and $|U_{1A}, U_{1B}, -\sqrt{2}U_{1A}, -\sqrt{2}U_{1B}, U_{3A}, U_{3B}\rangle$, respectively. However, a slight modification of the energy dispersions is caused due to the effects of α_3, which is also the main factor causing the particle–hole asymmetry in the electronic band structure. According to the intersection with the Fermi surface, the three subbands have their own Fermi momentums, i.e., $k_{F_2} \simeq 0$, and $k_{F_1} = k_{F_3} \simeq 2\sqrt{2}\alpha_1/3\alpha_0 b$, and free carrier densities, i.e., $D_2 \simeq 0$ and $D_1 = D_3 \simeq k_F^2/\pi$. With the increasing number of layers, the distribution of the vertical multi-Dirac cone structures in 2D case approaches to the K and H points in the $\hat{k}_z$-axis of the 3D simple hexagonal graphite, as a result of the dimensional crossover of the essential electronic properties. The free carrier densities are evaluated from $D_{i'}$ and $k_{F_i'}$. Moreover, the slopes of all the Dirac cones are estimated as $\pm 3\alpha_0 b/2$, the same as that of a monolayer graphene in the low-energy region.

Recently, scanning tunneling spectroscopy (STS) experiments have verified the two Dirac cones in AA-stacked bilayer graphene [281]. The measured slope and location of the Dirac cones are in agreement with theoretical predictions [282, 283]. For multilayer cases, the multiple Dirac-cone structure can be further verified by STS and angle resolved photoemission spectroscopy (ARPES) experiments [54, 55]. Regardless of external fields, the symmetric or asymmetric relationship holds in the AA configuration, while the uniform probability distribution can be modulated by an applied electric field. The characteristics of wave functions dominate the optical dipole transitions and the e–h excitations; the selection rules are derived based on the Fermi golden rule. The optical absorption spectrum consists of several independent Dirac-cone spectra with a particular spectral discontinuity at low frequencies, resulting in an optical gap presented in cases of even number of stacking layers [31].

In the presence of a magnetic field **B**, the N pairs of Dirac cones in N-layer AA-stacked graphene contribute to N groups of LLs in the LL spectrum;

the division is based on the **B**-evolution of LLs and the characteristics of the spatial wave function [30]. Each group exhibits a simple square-root dependence on the field strength B and the quantum number n, corresponding to the individual Dirac cone, and each Landau state is characterized by a single-mode harmonic function. During the variation of **B**, the Fermi distribution evolves with the oscillating Fermi energy, which might change the threshold channel and frequency of each category of intragroup LL transitions and cause a spectral region of forbidden transitions. It is given that each group of LLs displays its own intragroup delta-function-like absorption peaks that satisfy the particular selection rule $\Delta n = \pm 1$. Nevertheless, a zero probability of the intergroup LL transitions is derived from the electric dipole transitions between a symmetric wave function and an asymmetric wave function, which results in a zero inner product [28, 31].

In conclusion, the spatial characteristics of Landau wave functions account for the intragroup optical selection rule and the prohibition transitions between intergroup LLs. Moreover, the generalized Peierls tight-binding model incorporated with the modified RPA is adopted to explore the novel many-particle magnetoexcitation phenomena (details in Chapter 10). The SPEs and plasmons are affected to varying degrees by both the layer numbers and the presence of external fields. In particular, acoustic and optical plasmon modes in the long wavelength limit identify the critical mechanism due to the intra-pair and interpair Dirac-cone excitations in the semimetal system. In the following section, we illustrate the effects due to the electron doping on the bare response function and the energy loss function, and describe the related diversified Coulomb excitation phenomena.

4.2 Stacking- and Doping-Enriched Coulomb Excitations

In layered graphene systems, the polarization function describing dynamic charge screening consists of intralayer P_{ll} and interlayer $P_{ll'}$ ($l \neq l'$) components; the former and the latter, respectively, correspond to the charge correlations on the same layers and between different layers through Coulomb interactions. While the $2 \times N$ subbands contribute to $4N^2$ channels, the only allowed ones are those satisfying the conservation of momentum and energy between the occupied initial state $\pi_i^{c,v}$ and the unoccupied final state $\pi_{i(j)}^{c,v}$. Resulting from intrapair and interpair excitations, the corresponding channels are expressed as $\pi_i^{c,v} \to \pi_i^{c,v}$ and $\pi_i^{c,v} \to \pi_j^{c,v}$, where $i \neq j$ and i (j) represent the band index. Based on the characteristics of layer-dependent wave functions, the polarization functions $P_{ll'}$'s under the electron–electron Coulomb interactions are calculated by the modified RPA and displays layer-dependent properties. The effects due to the intralayer and interlayer charge screening are different in response to the intrapair and interpair excitation channels. The real part and imaginary part of a layer-dependent $P_{ll'}$ obey the Kramers–Kronig relations [262]. The SPEs are characterized by the imaginary part, which reflect the main features of the energy band structure, e.g., energy dispersion and van Hove singularity.

We first see the low-frequency excitation properties of the AA-stacked bilayer graphene. For the four layer-dependent components of the polarization function, i.e., P_{11}, P_{12}, P_{22}, and P_{21}, only the former two are independent, while the relationship, $P_{11} = P_{22}$ and $P_{12} = P_{21}$, holds because of the inversion symmetry of the AA-stacked multilayer graphenes. Given a transferred momentum $\mathbf{q}$, the excitation energy is derived according to the conservation of momentum and energy. The Dirac-cone type excitations exhibit square-root peaks described by $1/\sqrt{\omega^2 - \omega_{sp}^2}$ in $\mathrm{Im}P_{11}$ and $\mathrm{Im}P_{12}$, as shown in Figure 4.2; $\mathrm{Re}P_{11}$ and $\mathrm{Re}P_{12}$ satisfying the Kramers–Kronig relations also show similar structures: $1/\sqrt{\omega_{sp}^2 - \omega^2}$ [261]. The divergent energies ω_{sp}'s are determined by the Fermi velocities and momentums of the Dirac cones. In the case of $q = 0.005$ Å^{-1} and $k_F = 2\alpha_1/3\alpha_0 b \simeq 0.67$ Å^{-1} in Figure 4.2a and b, the SPE spectra exhibit three square-root divergent structures at ω_{sp}^1, ω_{sp}^{2-}, and ω_{sp}^{2+}; the first one and the latter two are, respectively, deduced to be the result of intrapair and interpair excitation channels of the two $\pi_1^{c,v}$ and $\pi_2^{c,v}$ subbands. ω_{sp}^1 is given as follows

$$\omega_{sp}^1 = 3\alpha_0 bq/2 \simeq 0.03\text{eV}, \qquad (4.10)$$

which is the excitation energy of the Fermi momentum states for both the intrapair $\pi_1^c \to \pi_1^c$ and $\pi_2^v \to \pi_2^v$ channels. On the other hand, the latter two peaks at ω_{sp}^{2-} and ω_{sp}^{2+} exhibit double peak structures, due to the disparity of the two $\pi_1^{c,v}$ and $\pi_2^{c,v}$ Dirac cones. The disparity is caused by the interaction α_2 and more obvious at low energies. The excitation energies, ω_{sp}^2 and ω_{sp}^3, respectively, corresponding to the transitions of the Fermi momentum states from the interpair, $\pi_1^v \to \pi_2^v$ and $\pi_1^c \to \pi_2^c$, as indicated by the pink arrows in Figure 4.2a, are expressed as

$$\omega_{sp}^{2-} = \sqrt{2\alpha_1^2} - 3\alpha_0 bq/2 \simeq 0.69\text{eV}$$
$$\omega_{sp}^{2+} = \sqrt{2\alpha_1^2} + 3\alpha_0 bq/2 \simeq 0.76\text{eV}. \qquad (4.11)$$

It should be noticed that the phase relationships between the polarizations P_{11} and P_{12} regarding to the intrapair and interpair channels can be simply checked by the spatial distributions of layer-dependent wave functions. In general, $\mathrm{Im}P_{11}$ must be negative because of the law of causality, while the interlayer tunneling effects cause $\mathrm{Im}P_{12}$ to be positive or negative [23, 262]. As a result of the intrapair channel, these two polarizations are in-phase in the vicinity of ω_{sp}^1. On the other hand, the interpair channel causes them to be out-phase near ω_{sp}^{2-} and ω_{sp}^{2+} because the wave function inner product is

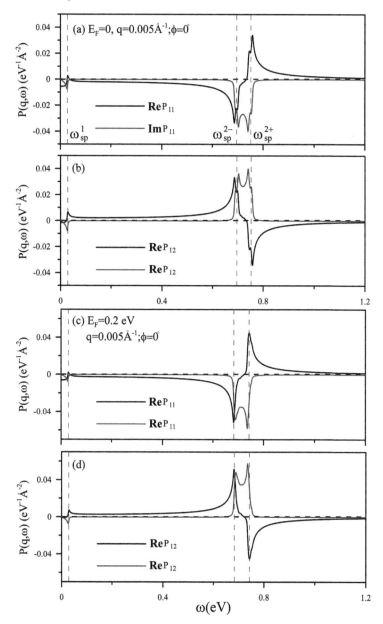

FIGURE 4.2

The two independent bare polarization functions, (a) $\text{Re}[P_{11}]$ and (b) $\text{Im}[P_{12}]$, of the AA-stacked trilayer graphene at $q = 0.005$ Å^{-1}, $E_F = 0$, and $\phi = 0°$; (c) and (d) correspond to those at $q = 0.005$ Å^{-1}, $E_F = 0.2$ eV, and $\phi = 0°$. The blue dashed lines indicate the threshold frequencies of e–h excitations, ω_{sp}^1, and $\omega_{sp}^{2\pm}$.

calculated between the symmetric and antisymmetric components of the associated layers. The aforementioned two kinds of channels gives rise to opposite phases between intralayer and interlayer polarizations. It is expected that two plasmon modes would be generated in the bilayer system; one is acoustic mode with in-phase oscillations and the other is optical with out-of-phase oscillations. The phase relationship between polarizations can provide ideal means of verifying the collective plasmon modes. When the Fermi level is increased to 0.1 eV, the positions where P_{11} and P_{12} diverge are still the same as in the undoped case. This is clearly manifested by the fact that the excitation channels come from the same intrapair or interpair Dirac cone. However, there are only single divergences at the third region, near ω_{sp}^{3-} and ω_{sp}^{3-}, which implies that the disparity of the Dirac cone is reduced with increasing energy.

In the AA-stacked trilayer graphene, a 3×3 bare response function tensor is used to describe the dynamic charge screening under Coulomb interactions. As depicted in Figures 4.3 and 4.4, the real and imaginary parts of the four independent layer-indexed polarization functions, $P_{ll'}$ ($ll' = 11, 22, 12$, and 13), are connected to each other by the Kramers–Kronig relations, and both exhibit several square-root divergent structures. Each divergent structure of $\mathrm{Re}P_{ll'}$ and $\mathrm{Im}P_{ll'}$ results from one kind of major SPE channel, i.e., van Hove singularity of joint DOS. The overall distribution of these divergent structures is spread in three ranges:

$$\omega_{sp}^{1} \simeq 3\alpha_0 bq/2; \ \text{intrapair} \ \pi_i^{c,v} \to \pi_i^{c,v},$$

$$\omega_{sp}^{2\pm} \simeq \sqrt{2\alpha_1^2} \pm \omega_{sp}^{1}; \ \text{interpair} \ \pi_1^{c,v} \to \pi_2^{c,v} \text{and} \pi_2^{c,v} \to \pi_3^{c,v};$$

$$\omega_{sp}^{3\pm} \simeq 2\sqrt{2\alpha_1^2} \pm \omega_{sp}^{1}; \ \text{interpair} \ \pi_1^{c,v} \to \pi_3^{c,v}. \tag{4.12}$$

The earlier equations are derived from the energy dispersions of the $\pi_1^{c,v}$, $\pi_2^{c,v}$ and $\pi_3^{c,v}$ subbands described in Eq. (4.9), and the corresponding channels are presented on the right-hand side. Related to the joint DOS of e–h pairs and the charge distributions, the peak intensity and phase of $\mathrm{Im}P_{ll''}$ are determined by an excitation channel and affected to varying degrees for different polarizations. At $q = 0.005$ Å^{-1} and $\phi = 0°$, three intrapair excitation channels, i.e., $\pi_1^c \to \pi_1^c$, $\pi_2^c \to \pi_2^c$ and $\pi_3^v \to \pi_3^v$, lead to the respective first peak at $\omega_{sp}^{1} \simeq 0.03$ eV for the four polarizations, P_{11}, P_{22}, P_{12}, and P_{13}, with all following the same phase and relatively weak spectral weight. In consideration of the interpair channels between two nearest Dirac cones, $\pi_1^{c,v} \to \pi_2^{c,v}$ and $\pi_2^{c,v} \to \pi_3^{c,v}$, the double-peak divergences are located around $\omega_{sp}^{2-} \simeq 0.47$ eV and $\omega_{sp}^{2+} \simeq 0.55$ eV for the polarizations related to the outermost layers, i.e., P_{11} and P_{13}, but there is absence of prominent structures for those related to the middle layer, i.e., P_{22} and P_{12}. In the vicinity of ω_{sp}^{2-} and ω_{sp}^{2+}, the nearby components are out-of-phase between P_{11} and P_{13}. The energies of these two peaks are determined by the transitions from k_F to $k_F + q$ of the two interpair channels (purple arrows in Figure 4.1b); their disparity causes the spacing in the double-peak structure, which is clearly illustrated in the

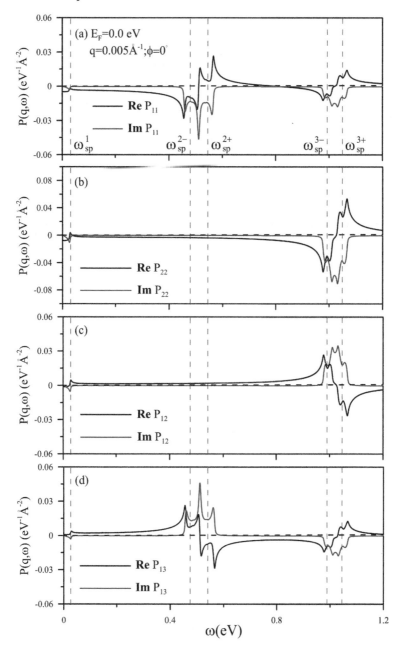

FIGURE 4.3

The four independent bare polarization functions, (a) $\text{Re}[P_{11}]$, (b) $\text{Im}[P_{22}]$, (c) $\text{Re}[P_{12}]$, and (d) $\text{Im}[P_{13}]$ of the AA-stacked trilayer graphene at $q = 0.005$ Å^{-1}, $E_F = 0$, and $\phi = 0°$. The blue dashed lines indicate the threshold frequencies of e–h excitations, ω_{sp}^1, $\omega_{sp}^{2\pm}$, and $\omega_{sp}^{3\pm}$.

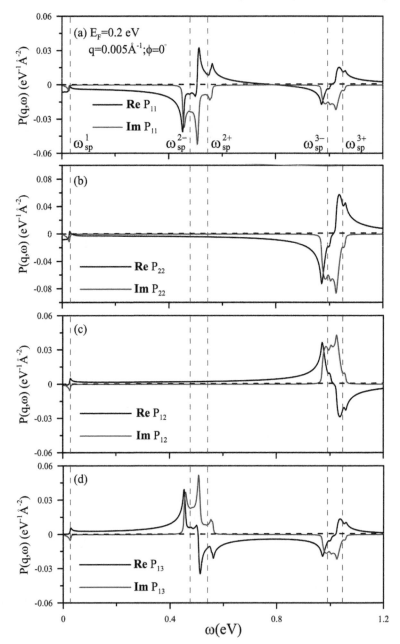

FIGURE 4.4

The four independent bare polarization functions, (a) $\mathrm{Re}[P_{11}]$, (b) $\mathrm{Im}[P_{22}]$, (c) $\mathrm{Re}[P_{12}]$, and (d) $\mathrm{Im}[P_{13}]$ of the AA-stacked trilayer graphene at $q = 0.005\ \mathring{A}^{-1}$, $E_F = 0.2$ eV and $\phi = 0°$. The blue dashed lines indicate the threshold frequencies of e–h excitations: ω_{sp}^1, $\omega_{sp}^{2\pm}$, and $\omega_{sp}^{3\pm}$.

bilayer case. This phenomenon reveals that the spatial charge distribution of the $\pi_2^{c,v}$ subbands are completely constrained to the outermost layers as mentioned in Section 4.1. Moreover, an optical collective mode can be triggered with the out-of-phase excitations between the two outermost layers.

In addition to the aforementioned two kinds of peaks, the third kind consists of two double peaks, which are centered around $\omega_{sp}^{3-} \simeq 0.98$ eV and $\omega_{sp}^{3+} \simeq 1.05$ eV and dominated by the transition of the next-neighbor interpair channel, $\pi_1^{c,v} \to \pi_3^{c,v}$, as illustrated by the brown arrows in Figure 4.1. They are presented with different phases and relatively weak intensities for all types of intralayer and interlayer polarizations, because the wave functions spread over the graphene layers in space. On the other hand, due to the long-range extension of the linear massless Dirac dispersion (up to 0.8 eV), the main features of the polarization functions in AA-stacked graphenes are hardly affected even under the influence of extra free-carrier doping, as shown in Figure 4.4. The effects of doping are similar to those of interlayer atomic interactions, which give rise to a rigid shift of the Dirac cones and induce a number of free carriers in the AA-stacking graphene. In the case of electron doping, the $\pi_1^{c,v}$ and $\pi_2^{c,v}$ Dirac cones get heavier free carrier density, the $\pi_3^{c,v}$ Dirac cone gets weaker, and vice versa for the case of hole doping. The only change worth mentioning is that the unequally occupied free carriers of the two Dirac cones distort the symmetry of the double-peak structures when considering interpair excitation channels. This implies that the plasmon modes are subjected to varying degrees of Landau dampings in cases of different free carrier densities. The channels and features of the low-lying SPEs and Coulomb excitations are diversified when the number of graphene layers is increased, with the estimated one half of Dirac cones getting heavier doping and the other half getting weaker doping. The related diversified phenomena of the Coulomb-excitation spectra are discussed in detail.

The energy loss function, Im$[-\frac{1}{\epsilon}]$ defined in Eq. (2.8), is useful in understanding the collective excitations and the measured excitation spectra from electron energy loss spectroscopy (EELS) and inelastic X-ray scattering (IXS). The peaks in Im$[-\frac{1}{\epsilon}]$ are referred to as plasmons, collective excitations of free carriers, and each plasmon frequency in the screened excitation spectrum is higher than the related SPE frequency. The frequencies of the plasmons can be obtained by solving the eigenfrequencies of the determinantal equation of the $N \times N$ dielectric tensor equation:

$$\epsilon = \prod_l [\epsilon_0 - \sum_m V_{lm}(\mathbf{q}) P_{ml}(\mathbf{q}, \omega)] + \text{(higher order terms)}$$

$$\to \epsilon = \epsilon_0^{N-1} [\epsilon_0 - \sum_{lm} V_{lm}(\mathbf{q}) P_{ml}(\mathbf{q}, \omega)] + \text{(higher order terms)}. \qquad (4.13)$$

Nevertheless, it is necessary to obtain the spectral intensities through the definition of the energy loss function.

AA-stacked graphenes exhibit rich and unique plasmon spectra under the influence of dynamic Coulomb interactions. The energy loss functions of the AA-stacked bilayer graphene are shown in Figure 4.5 for various q's and E_F's. Each panel demonstrates the q-evolutions of plasmons under different E_F's. For $E_F = 0$ and $q = 0.005$ Å^{-1}, there are two intrinsic plasmon peaks, labeled by ω_p^1 and ω_p^2, as illustrated by the black curve in Figure 4.5a. The first peak at ω_p^1 is identified from the two intrapair channels, i.e., $\pi_1^c \to \pi_1^c$ and $\pi_2^v \to \pi_2^v$. By substituting the relationship $P_{11} = P_{12}$ into Eq. (4.13), the plasmon peak frequency ω_p^1 can be derived from the equation $[\epsilon_0^2 - \epsilon_0(V_{11}(\mathbf{q}) + V_{12}(\mathbf{q}))P_{11}(\mathbf{q}, \omega)]|_{\omega = \omega_p^1}$, where the addition in the second term indicates that such plasmons are attributed to an acoustic mode with strong in-phase oscillations between the two layers. In contrast, because of the relationship $P_{11} = -P_{12}$ in the vicinity of ω_p^2, the second peak at ω_p^2 is an optical plasmon mode resulting from the two interpair channels, i.e., $\pi_1^v \to \pi_2^v$ and $\pi_1^c \to \pi_2^c$. The in-phase acoustic plasmon mode has a much stronger intensity than the so-called out-of-phase optical plasmon mode, i.e., Im$[-\frac{1}{\epsilon}(\omega_p^1)]$ is larger than Im$[-\frac{1}{\epsilon}(\omega_p^2)]$ by more than two orders of magnitude. With the increase in q, the two plasmons shift to higher frequencies and their significant intensity difference maintains. When E_F is increased to 0.1 eV (Figure 4.5b), the modes, frequencies, and intensities of the plasmons are nearly the same when compared with the pristine plasmons at $E_F = 0$. The similar loss spectra indicate that the plasmon spectra are identified as the collective excitations of the quasiparticles in massless Dirac cones, and the associated linear energy dispersion is retained at low energies in the semimetallic AA graphenes. That is to say, because of the isotropic low-energy band structures and electronic excitations, the q-evolutions of Im$[-\frac{1}{\epsilon}]$ are almost identical regarding different directions of the transferred momentum under moderate E_F and small q, e.g., the curves in Figure 4.5a–c. In the case of $E_F = 0.3$ eV, the two plasmon intensities become comparable at a sufficiently large q, e.g., $q = 0.02$ Å^{-1} (inset of Figure 4.5d). The acoustic mode largely drops by more than one order of magnitude once $q > 0.02$ Å^{-1}, implying that it suffers significant Landau damping. In particular, the intensity further reduced at $E_F = 0.5$ eV (Figure 4.5e), and the suppressed behavior returns to normal until $E_F = 0.7$ eV (Figure 4.5f). The peculiar behavior directly reflects the interplay between all the intrapair and interpair excitation channels, of which some channels are reduced while the others are enhanced during the variation of the Fermi level. The overall effect leads to that the collective excitations in the multi-Dirac cone are quite different from the 2D Dirac plasmons described in monolayer graphene (Chapter 3). Temperature would alter the carrier occupation number near the Fermi level. The original intrapair excitations at zero temperature are reduced owing to the thermal population of π_1^c and π_2^v subbands above the Fermi level; however, the loss is just compensated by the additional excitations resulting from the unoccupied states below the Fermi level. That the intensities and frequencies of SPEs remain unchanged accounts for the temperature-independent e–h excitations and the intrapair plasmons of the multi-Dirac cones in AA-stacked graphenes.

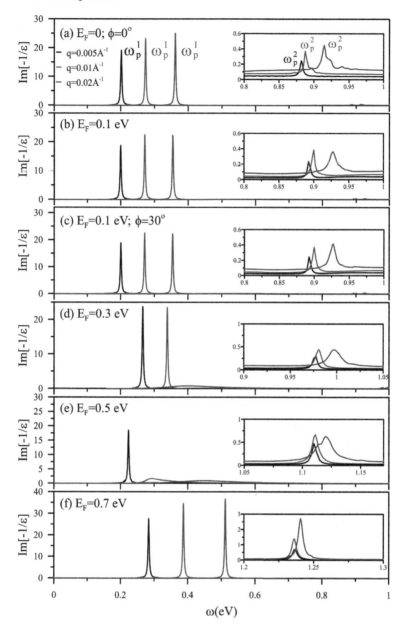

FIGURE 4.5

The energy loss functions of the bilayer AA stacking at $q = 0.005$, $q = 0.01$, and $q = 0.02$ Å^{-1} and under distinct doping levels: (a) $[E_F = 0, \phi = 0°]$, (b) $[E_F = 0.1\,\text{eV}, \phi = 0°]$, (c) $[E_F = 0.1\,\text{eV}, \phi = 30°]$, (d) $[E_F = 0.3\,\text{eV}, \phi = 0°]$, (e) $[E_F = 0.5\,\text{eV}, \phi = 0°]$, and (f) $[E_F = 0.7\,\text{eV}, \phi = 0°]$. The frequencies of the three plasmon modes are indicated by ω_p^1, ω_p^2, and ω_p^3.

For AA trilayer spectra, three plasmon peaks are revealed at low frequencies, as shown in Figure 4.6; the q-evolutions of $\text{Im}[-\frac{1}{\epsilon}]$ are presented under different E_F's in (a)-(f). Similar to the acoustic plasmon in the aforementioned bilayer case, the first peak at ω_p^1 is deduced from the intrapair excitations with in-phase charge oscillations. Besides, there are two optical plasmon modes, ω_p^2 and ω_p^3, as a result of the two kinds of interpair excitation channels. The former mode ω_p^2 is responsible for the channels between two neighboring Dirac cones, i.e., $\pi_1^{c,v} \to \pi_2^{c,v}$ and $\pi_2^{c,v} \to \pi_3^{c,v}$, of which the charge distribution and the polarization relationship $P_{11} = -P_{13}$ near $\omega_{sp}^{2\pm}$ imply that it belongs to an out-of-phase acoustic mode related to the two outermost layers. In the same way, the latter ω_p^3 one is also regarded as an optical mode, while it is associated with every layer and responsible for the collective excitations from the next nearest-neighbor interpair channel, $\pi_1^{c,v} \to \pi_3^{c,v}$.

In the pristine case depicted in Figure 4.6a, all the plasmon modes display positive energy dispersions, while the collective intensities and frequencies are determined by the Landau damping and the strength of intrapair excitation channels (free carrier density within the $\pi_1^{c,v}$, $\pi_2^{c,v}$ and $\pi_3^{c,v}$ subbands). The acoustic plasmon ω_p^1 is apparently broadened and damped with the increasing q. The other two relatively weak optical plasmons are performed with frequencies approximated as $\omega_p^2 \simeq 0.8$ eV and $\omega_p^3 \simeq 1.2$ eV. When the Fermi level is increased, the plasmon modes extend to higher energies owing to an overall increase of free carrier density (Figure 4.6b–f). However, the nonmonotonous behavior of the acoustic plasmon appears in the frequency region 0.2 eV$< \omega <$ 0.6 eV when the Fermi level is located between 0.3 eV (Figure 4.6d) and 0.5 eV (Figure 4.6e), despite the fact that the extra induced intrapair channels enhance the excitation intensity and frequency. The mechanism governing the abnormal frequency region is dominated by the interband Landau damping of the lowest interpair channel. This have been clarified as a consequence of the nearest vertical interlayer atomic interaction α_1 in AA-stacked graphenes. It should be noted that the plasmon modes are highly dependent on the free carrier density and boundaries of SPE channels. For most cases, plasmons and SPEs can coexist in a certain (q, ω) region enclosed by interband SPE boundaries. Nevertheless, the intensities of the plasmons are reduced when their frequencies coincide with those of SPEs, leading to Landau damping of plasmons.

Various plasmon modes are presented in the (q, ω)-excitation phase diagram of AA-stacked graphenes under different E_F's (Figures 4.7 and 4.8). Plasmon modes usually appear in specified domains where the plasmon dispersions don't overlap with the continuum spectra of e–h pairs, i.e., the plasmons can survive in cases of weak Landau dampings. The plasmon frequencies strongly depend on the transferred momentum q. The strong dispersion relation of $\omega_p(q)$ means that the plasmon wave (quantum of the electron density oscillations) propagates with a wavelength $2\pi/q$ and group velocity $d\omega_q/dq$. The (q, ω)-excitation phase diagram exhibits N plasmon branches for an N-layer AA-stacked graphene. The first one is ascribed to the acoustic mode

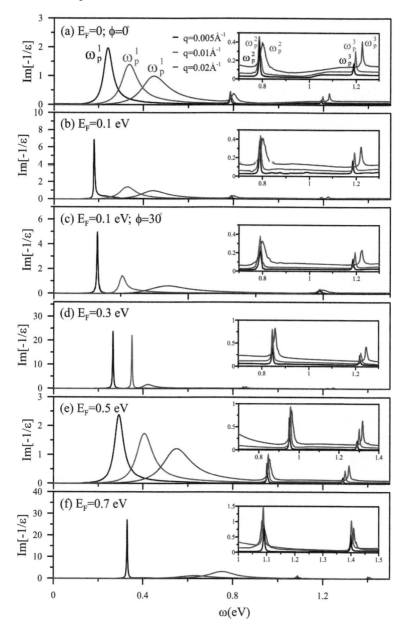

FIGURE 4.6

The energy loss functions of the trilayer AA stacking at $q = 0.005$, $q = 0.01$, and $q = 0.02$ Å^{-1} and under distinct doping levels: (a) $[E_F = 0, \phi = 0°]$, (b) $[E_F = 0.1\,\text{eV}, \phi = 0°]$, (c) $[E_F = 0.1\,\text{eV}, \phi = 30°]$, (d) $[E_F = 0.3\,\text{eV}, \phi = 0°]$, (e) $[E_F = 0.5\,\text{eV}, \phi = 0°]$, and (f) $[E_F = 0.7\,\text{eV}, \phi = 0°]$. The frequencies of the three plasmon modes are indicated by ω_p^1, ω_p^2, and ω_p^3.

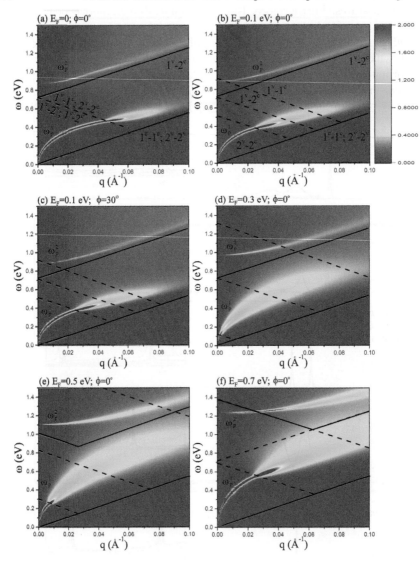

FIGURE 4.7
The (momentum, frequency)-phase diagrams of the bilayer AAA stacking
under various Fermi levels: (a) $[E_F = 0, \phi = 0°]$, (b) $[E_F = 0.1$ eV, $\phi = 0°]$,
(c) $[E_F = 0.1$ eV, $\phi = 30°]$, (d) $[E_F = 0.3$ eV, $\phi = 0°]$, (e) $[E_F = 0.5$ eV,
$\phi = 0°]$, and (f) $[E_F = 0.7$ eV, $\phi = 0°]$.

and mainly caused by the available intraband excitations of the intrapair channels, which have a similar excitation energy under a fixed q value, owing to
the approximately identical linear slopes of the multi-Dirac cones. In addition
to the lowest acoustic branch, the interpair channels result in $N - 1$ optical

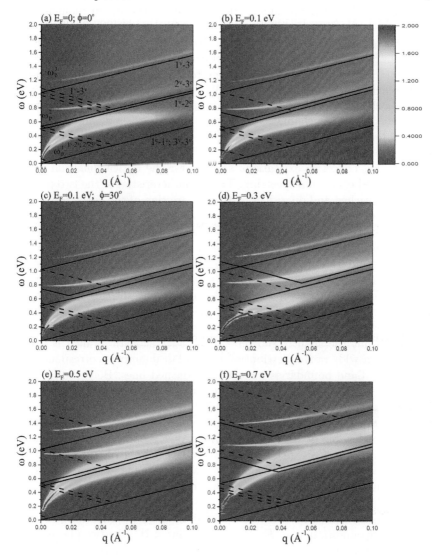

FIGURE 4.8

The (momentum, frequency)-phase diagrams of the trilayer AAA stacking under various Fermi levels: (a) $[E_F = 0, \phi = 0°]$, (b) $[E_F = 0.1$ eV, $\phi = 0°]$, (c) $[E_F = 0.1$ eV, $\phi = 30°]$, (d) $[E_F = 0.3$ eV, $\phi = 0°]$, (e) $[E_F = 0.5$ eV, $\phi = 0°]$, and (f) $[E_F = 0.7$ eV, $\phi = 0°]$.

branches from low- to middle-ω regions. All the plasmon modes are largely damped by about one or two orders when dispersing into the region of SPEs.

In the AA-stacked bilayer graphene, the two plasmon modes are enclosed by SPE boundaries in the low and middle (q, ω) regions, as shown in

Figure 4.7a–f. At $E_F = 0$ (Figure 4.7a), the acoustic plasmon evidently originates from the intrapair channels, $\pi_1^c \to \pi_1^c$ and $\pi_2^v \to \pi_2^v$, and displays a q-dependent behavior of ω_p^1 similar to that of a 2D electron gas, that is, ω_p^1 increases quickly from zero and behaves as a $\sqrt{q}$ dispersion relation. The plasmon intensity is prominent in the region lacking SPEs, while it suffers a significant Landau damping near 0.06 Å^{-1} that matches the interband boundary of the four degenerate channels, $\pi_1^v \to \pi_2^v$, $\pi_1^c \to \pi_2^c$, $\pi_1^v \to \pi_1^c$ and $\pi_2^v \to \pi_2^c$.

As to the second plasmon mode, ω_p^2, its frequency is finite at $q \to 0$ and higher than the SPE energy of interpair channel, $\pi_1^v \to \pi_2^c$. Consequently, the corresponding collective excitations account for the optical plasmon branch in the screened excitation spectrum. These characteristics of the two plasmon modes are enriched and diversified with the variation of the Fermi level. With a variation of E_F, both acoustic and optical branches also exist as a result of excitations related to the two massless Dirac cones, while their dispersion relations and intensities strongly depend on the free carrier densities and the interband boundaries of SPEs. When E_F is increased to 0.1 eV (Figure 4.7b), the two plasmon modes behaves similar to the pristine ones because the SPE boundaries of the main intrapair channels that serve as the cause of these phenomena, i.e., $\pi_1^v \to \pi_2^v$, $\pi_1^c \to \pi_2^c$, and $\pi_1^v \to \pi_2^c$ would remain unchanged once the Fermi level falls between the two Dirac-point energies. However, it should be noticed that the slightly shorter and broader acoustic branch is accompanied with a wide-range distribution of Landau dampings corresponding to the four interband boundaries indicated by dashed lines. Both the SPE and plasmon spectra are isotropic with respect to ϕ due to the linear dispersions within energies between about -1.0 and 1.0 eV, as shown in Figure 4.7c. The isotropic properties of the electronic excitations are preserved for AA-stacked graphenes under low or moderate dopings. As E_F approaches to the upper Dirac-point energy, $E_{D,2}$, (e.g., $E_F = 0.3$ eV in Figure 4.7d and $E_F = 0.5$ eV in Figure 4.7e) the interband channel of $\pi_2^{c,v}$ subbands, $\pi_2^v \to \pi_2^v$, strongly enhances the Landau damping and weakens the acoustic plasmon. On the contrary, the induced free carriers enhance the main SPEs of $\pi_1^v \to \pi_2^v$ at middle energies as well as of the corresponding optical plasmon, ω_p^2. Under a heavy doping condition, e.g., E_F=0.7 eV (Figure 4.7f), the two plasmons, ω_p^1 and ω_p^2, are prominent and distributed throughout the phase diagram.

The frequency dispersion relations of the three plasmon modes of AA-sacked graphene are presented in Figure 4.8. The lowest acoustic branch, ω_p^1, comes from the three intrapair excitation channels, and the other two optical ones, ω_p^2 and ω_p^3, are respectively, from the nearest-neighbor and next nearest-neighbor interpair channels. At E_F=0, the acoustic mode is mainly dominated by the intrinsic intraband plasmons of the two outermost Dirac cones. However, as described in the aforementioned example, the triggered free carriers at finite E_F efficiently modulate the plasmon intensity as a consequence of the competition between SPEs and collective excitations. The effects of a small amount of intrinsic free carriers in the π_2 subbands also induce interband Landau dampings for the acoustic plasmon, in addition to the majority carriers of the two outermost Dirac cones. It is expected that the plasmon intensity of the acoustic mode continuously increases as E_F gradually approaches to the Dirac-point energy $E_{D,2}$. This explains why the intrinsic acoustic plasmons of small q in the pristine case appear to be less intense in comparison to those in other cases, as shown in Figure 4.6. The other two optical plasmon modes are rarely affected by the change of the Fermi level, since the vertical interlayer atomic interaction α_1 dominates the Dirac-point energies and the SPE energies of the main interpair channels, the majority of which keep the same continuum spectrum under different doping levels. The clear demonstration of the various optical plasmon branches is restricted to layered graphenes with AA-stacking configuration in which the sufficient separation between the Dirac points provides enough space in the phase diagram to hold the various plasmon modes.

5

AB-Stacked Graphenes

The Coulomb excitation behaviors are mainly determined by the electronic structures and geometric symmetries. The $2p_z$ orbitals of carbon atoms in graphene-related systems, which built the π valence bands and the $\pi^a st$ conduction bands, are responsible for the electronic excitations lower than the middle frequency (~ 6–10 eV). The low-lying band structures of layered graphenes strongly depend on the intralayer and interlayer hopping integrals of $2p_z$ orbitals. The trilayer ABA stacking (Figure 5.1a) exhibits the unusual energy bands, two pairs of parabolic valence and conduction bands and one pair of distorted Dirac-cone structures (Figure 5.1b); the corresponding van Hove singularities are presented in the density of states (DOS) (Figure 5.1c). As verified from the high-resolution angle-resolved photo emission spectroscopy [64,284,285], the distorted Dirac cone could survive in AB-stacked graphene systems with odd layers [64,284]. Also, there exist the special wave functions arising from the specific superpositions of the six tight-binding functions, being directly reflected in the existence/strength of the Coulomb interactions. Band structure and wave functions are directly included in the current calculations.

There are a lot of theoretical predictions [23,25,49,68,240,261,263–268, 286–296] and experimental measurements [187–197,297–299] on electronic excitation spectra of layered graphenes. Monolayer graphene, with the linear Dirac-cone structure, is predicted to display intraband and interband electron/hole (e–h) excitations, and the low-frequency acoustic plasmon mode (<2 eV) under the extrinsic electron/hole doping [192,261,263,290,292,295, 296] and the temperature effect [32,68,263–268]. The 2D acoustic plasmon, accompanied with the higher-frequency optical plasmon, could survive in the doped bilayer AB stacking [23,49,287,288], but nor in the pristine system. However, the previous theoretical calculations have ignored certain significant interlayer atomic interactions. The phenomenological model studies on AB-stacked systems with layer number higher than three are absent up to now. On the experimental side, the high-resolution electron energy loss spectroscopy (EELS) [188,189,191–198] is successfully used to verify/identify electronic excitations in layered graphenes. The acoustic, π, and $\pi + \sigma$ plasmon modes are identified to appear at low ($\sim 0.1 - 2.0$ eV) [192,195], middle ($\sim 5 - 8$ eV) [191–193,196–198], and high frequencies (> 14 eV) [191,193,198]. Another powerful experimental technique, the inelastic light

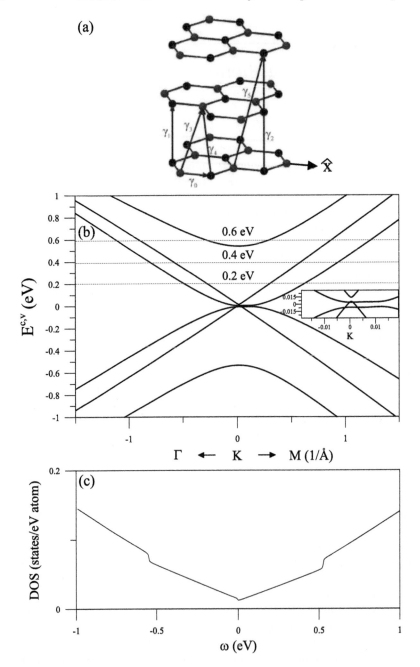

FIGURE 5.1
(a) Geometric structure of AB-stacked trilayer graphene with intralayer and interlayer atomic interactions, and (b) a pristine band structure along the high symmetry points, and (c) the rst Brillouin zone.

scattering spectroscopy, has been used to confirm the low-frequency Coulomb excitations, e.g., ~ 0.01–$0,1$-eV plasmons [165, 222, 225, 300].

For trilayer ABA stacking, the tight-binding model and the modified random-phase approximation (Eqs. 2.4–2.8) are directly combined to thoroughly explore the diverse electronic excitations under intrinsic and extrinsic cases. All the important intrinsic interactions are included in the calculations simultaneously. The special structures of the bare and screened response functions are explored thoroughly, especially for the former with four independent components. Furthermore, the dependence on the doping level is investigated in detail, being useful in comprehending the significant effects arising from the variation of the Fermi–Dirac distribution. A pristine graphene system will be studied whether the interlayer-hopping-induced free electrons and holes could create the lightly damped acoustic plasmon. The dramatic transformations in collective and e–h excitations, which are revealed in the transferred (momentum, frequency)-phase diagrams, are expected to be easily observed as the electron-doping level varies. How many kinds of plasmon modes and the E_F-created single-particle excitation regions is clearly identified from the delicate analysis. The predicted results could be verified by the experimental measurements. In addition, the Coulomb excitation phenomena are also diversified by the electric field and magnetic quantization, being discussed in Chapters 9 and 10, respectively.

5.1 Unusual Essential Properties

The AB-stacked trilayer configuration is clearly shown in Figure 5.1a, in which two neighboring layers shift relative to each other by one C–C bond length (b) along the armchair direction ($\hat{x}$). There are six carbon atoms in a primitive unit cell. The contributions due to the $2p_z$ orbitals of carbon atoms are sufficient for the low-energy energy bands and electronic excitations. The zero-field Hamiltonian is built from the six tight-binding functions associated with the $((A^1, B^1, A^2, B^2, A^3, B^3)$ sublattices. The superscript i represents the i-th layer. The 6×6 Hermitian matrix covers the nonvanishing elements related to the nearest-neighbor intralayer hopping integral ($\gamma_0 = -3.12$ eV), three neighboring-layer hopping integrals ($\gamma_1 = 0.38$ eV, $\gamma_3 = 0.28$ eV; $\gamma_4 = 0.12$ eV), two next-neighboring-layer hopping integrals ($\gamma_2 = -0.021$ eV; $\gamma_5 = -0.003$ eV), and the chemical environment difference between A and B sublattices ($\gamma_6 = -0.0366$ eV). The details of the Hamiltonian matrix could be found in Ref. [301]. All significant atomic interactions are included in the tight-binding model.

The trilayer AB stacking has three pairs of low-lying valence and conduction bands, with a small band overlap [28, 31]. Few free carrier density in this semimetallic pristine system will determine whether the acoustic plasmon could survive. One separated and distorted Dirac-cone structure (the first pair $(\pi_1^{c,v})$ exist near the Fermi level. Their wave functions mainly arise from the first and third layers, so no contributions from the second layer will be reflected in the bare response function (discussed latter in Figures 5.2 and 5.3). Another two pairs of parabolic bands, respectively, appear roughly at E_F (the second pair $(\pi_2^{c,v})$ and $\pm\gamma_1$ (the third pair $(\pi_3^{c,v})$). The valence bands are somewhat asymmetric to the conduction ones about $E_F = 0$, while this property hardly affects the main features of the Coulomb interactions. That is, electron and hole dopings almost lead to similar excitation behaviors. The former doping case is chosen for a model study, in which the Fermi level is located at the conduction band.

Three pairs of energy bands in trilayer ABA stacking could be decomposed into the superposition of bilayer- and monolayer-like ones [28,31]. This perturbation concept is suitable for generalizing to AB-stacked graphene systems [12,302] and even the 3D Bernal graphite [260,303,304]. The N-layer pristine systems, with an odd N, exhibit similar properties, while the other ones only display the bilayer-like band structures and wave functions. Moreover, a natural graphite shows the parabolic and linear Dirac cones, respectively, near the K $[k_z = 0]$ and H (the corner of the first Brillouin zone and $k - z = \pi$) valleys. Such physical picture might be valid for single-particle properties, such as the magnetic quantization of Landau subband spectra [305–309] and optical/magneto-optical absorption spectra [310–314]. The low-lying energy bands of trilayer AB-stacked graphene are magnetically quantized into three pairs of valence and conduction Landau levels (LLs), as identified from experimental measurements [299,315,316] and theoretical predictions [30,302,317,318]. The magnetic wave functions are similar to those in the absence of $B_z\hat{z}$; that is, the layer and sublattice dominance remains unchanged. On the other side, there exist few intergroup LL anticrossing behaviors during the variation of magnetic field [28,30,31,318]. Apparently, the intragroup phenomena frequently appear under the destruction of mirror symmetry by a perpendicular electric field, mainly owing to the dramatic transformation of the well-behaved energy dispersions into the oscillatory ones [28,319–322]; the LL anticrossings have been observed in trilayer sample recently [323].

The optical/magneto-optical excitations due to an electromagnetic wave are worthy of a closer discussion. Under the vertical transitions of the single-particle scheme, three pairs of energy bands only exhibit two shoulder structures (~ 0.4 and 0.8 eV) in the low-frequency absorption spectrum, directly reflecting the forms of van Hove singularities [28,302]. They mainly come from two pairs of parabolic bands, and the interpair transitions between the parabolic and linear ones are forbidden because of the specific symmetric/antisymmetric suppressions of their wave functions. The theoretical predictions are consistent with the experimental measurements [106, 324–326]. The similar physical pictures are revealed in the magneto-optical properties [299,315]. According to the detailed analyses, a lot

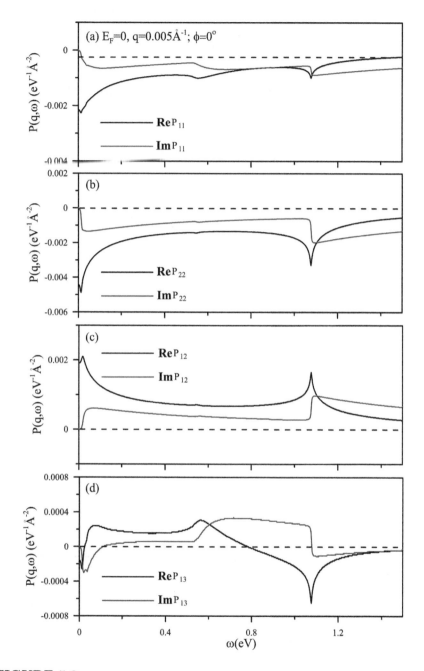

FIGURE 5.2
The independent four bare polarizations, (a) P_{11}, (b) P_{12}, (c) P_{13}, and (d) P_{22} for a pristine trilayer AB stacking at $q = 0.005$ Å^{-1} and $\phi = 0°$.

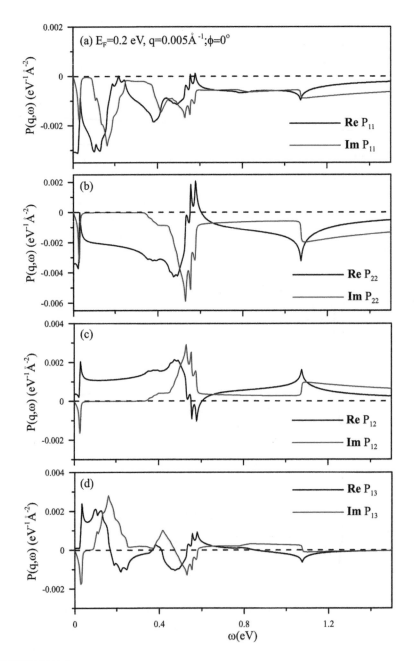

FIGURE 5.3
The Same plot as in Figure 5.1, but shown for an extrinsic system with
$E_F = 0.2$ eV.

of prominent symmetric absorption peaks are classified into five categories of inter-LL excitations, with the specific magneto-optical selection rules. The available channels are only induced by the valence and conduction LLs associated with two pairs of parabolic bands or the distorted Dirac cone. As to the many-particle Coulomb excitations, it needs to check in detail whether the superposition of the bilayer- and monolayer-like behaviors is suitable for the multilayer AB-stacked graphene systems.

5.2 Rich Coulomb Excitations

The single-particle and collective Coulomb excitations are dominated by energy bands and wave functions, being sensitive to the doping level. Electrons are excited from the occupied states to the unoccupied ones under the Fermi–Dirac distribution and the conservation of (q, ϕ, ω). The bare response function, corresponding to the e–h excitations, exhibits the special structure at the specific frequency, if the initial/final state of the allowed transitions comes from the band-edge state with the van Hove singularity or the Fermi-momentum state with the step distribution. It should be noticed that some excitation channels are forbidden because of the symmetric/antisymmetric properties of wave functions. The layer-dependent response functions consist of four independent components: $P_{11}=P_{33}$, P_{22}, $P_{12}=P_{21}=P_{23}=P_{32}$; $P_{13}=P_{31}$.

As for a pristine trilayer ABA stacking (Figure 5.2), the available Coulomb excitation channels include $(\pi_1^v \to \pi_1^c$, $\pi_2^v \to \pi_2^c)$, $(\pi_1^v \to \pi_3^c$, $\pi_3^v \to \pi_1^c)$; $\pi_3^v \to \pi_3^c$, in which the special structures, respectively, appear at very low frequency, ~ 0.56 and ~ 0.92 eV for $q = 0.005$ Å^{-1} and $\phi = 0°$, as shown in Figure 5.2. The imaginary parts of $P_{ll'}$, as shown by the red curves, directly reflect the features of DOS and wave functions, and its special structure relies on the former. The obvious shoulder structures are due to the extreme states (the local maxima/minima). As a result, the symmetric peaks in the logarithmic forms are revealed in the real parts of $P_{ll'}$ (black curves) by the Kramers–Kronig relations. The special structures at $\omega \sim 0.56$ eV strongly rely on the distorted Dirac-cone bands (π_1^v and π_1^c bands) near $E_F = 0$, since they are absent in P_{21} closely related to the significant contribution of the second layer.

Electronic excitations are dramatically changed during the variation of the Fermi level. Parts of them from the valence to conduction bands are suppressed by electron doping, mainly owing to the drastic changes in the Fermi–Dirac distribution. However, there are more free carriers in conduction bands that could build the Fermi surfaces. In addition to the band-edge states, the Fermi-momentum ones, being closely related to the step distribution functions, create the special structures in the bare response functions. Only the pristine $\pi_3^v \to \pi_3^c$ interband excitations are independent of electron doping, if the Fermi level is below the third conduction band, i.e., the special structure above 1 eV remains in a similar form. For a $E_F = 0.2$ eV system, conduction electrons will suppress three valence→conduction excitations ($\pi_1^v \to \pi_1^c$, $\pi_2^v \to \pi_2^c$, $\pi_3^v \to \pi_1^c$), and only the $\pi_1^v \to \pi_3^c$ could survive, leading to the special structure at ~ 0.56 eV, as shown in Figure 5.3. The lower-frequency special structures are generated by the Fermi surfaces. Most importantly, free carriers in conduction bands induce new $\pi_i^c \to \pi_j^c$ excitation channels, covering the intraband and interband transitions simultaneously. Both $\pi_1^c \to \pi_1^c$ and $\pi_2^c \to \pi_2^c$ intraband excitations could create strong responses at almost the same low frequency (<0.1 eV), as indicated by blue arrows. Furthermore, the interband excitations, $\pi_2^c \to \pi_1^c$, $\pi_1^c \to \pi_3^c$ & $\pi_2^c \to \pi_3^c$, respectively, exhibit the special structures near 0.32, 0.42, and 0.5 eV. It should be noticed that the square-root divergent structures are frequently revealed in the imaginary and real parts of $P_{ll'}$ because of the linear excitation energies and the Fermi–Dirac step function [261]. Apparently, the bare response functions will change with a further increase of E_F and the variation of the transferred momentum ((q, ϕ); not shown). The energy loss functions, the screened response spectra, are useful in understanding the plasmon modes and Landau dampings. Furthermore, they directly correspond to the measured excitation spectra. The dimensionless $\text{Im}[-1/\epsilon]$, as clearly shown in Figure 5.4, strongly depends on the doping level and the magnitude (q) of the transferred momentum, but not ϕ. For a pristine system, it is difficult to observe the prominent peak in the loss spectra (the intensity lower than 0.2 in Figure 5.4a at $q = 0.005$ Å^{-1}), indicating the collective excitations fully suppressed by the interband e–h excitations. Too few free carriers are responsible for the absence of strong plasmon modes. Conduction electrons under doping can create two/one prominent peaks in excitation spectra (Figure 5.4b–f and insets), being identified as collective excitations. The lower-frequency plasmon, the first collective mode, has a rather strong intensity, since it is due to intraband excitations of all the conduction carriers. However, the intensity of the higher-frequency plasmon is weaker but easily observable for the sufficiently high E_F ($E_F \geq 0.5$ eV). The second plasmon mode might arise from $\pi_2^c \to \pi_3^c$ interband excitations. The energy loss spectra hardly depend on the direction of the transferred momentum, i.e., they are almost isotropic (Figure 5.4b–e). Coulomb excitations are very sensitive to the magnitude. The plasmon frequencies increases with an increment of q, since the e–h excitation energies behave so, e.g., those of the first and second plasmons at different q's in Figure 5.4b–f.

The $(\mathbf{q}, \omega)$-phase diagrams could provide the full information on the single-particle and collective excitations, as clearly shown in Figure 5.5a–f. Any systems exhibit the vacuum regions that any excitations cannot survive, since electronic states of energy bands (Figure 5.1) do not create some $(\mathbf{q}, \omega)$ Coulomb interactions. For a pristine system, there are no obvious plasmon modes, according to the EELS intensities in the whole $(\mathbf{q}, \omega)$ range (Figure 5.5a). The boundaries of the $\pi_i^c \to \pi_j^v$ interband excitations are characterized by the band-edge states at the K/K' point. The e–h Landau dampings are very

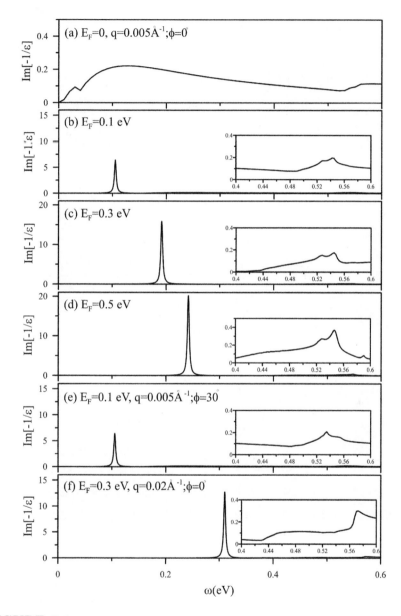

FIGURE 5.4
The energy loss functions at $q = 0.005$ Å^{-1} and $\phi = 0°$ under distinct doping levels: (a) $E_F = 0$, (b) 0.1 eV, (c) 0.3 eV, and (d) 0.5 eV. For a $E_F = 0.1$ system, they change with (e) $q = 0.005$ Å^{-1} & $\phi = 30^0$, and (f) $q = 0.02$ Å^{-1} & $\phi = 0^0$.

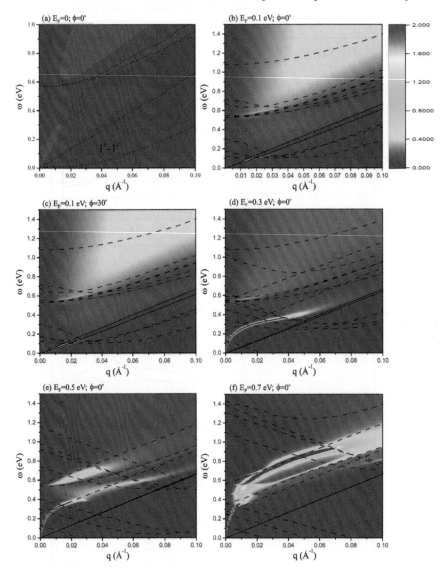

FIGURE 5.5

The transferred momentum-frequency phase diagrams of AB-stacked bilayer graphene at $\phi = 0°$ for (a) $E_F = 0$, (b) 0.1 eV, (c) 0.3 eV, (d) 0.5 eV, and (e) 0.7 eV. Also shown is that (f) under $\phi = 30°$ & $E_F = 0.1$ eV.

strong and effectively suppress the plasmon modes. After the electron/hole doping, all the e–h excitation boundaries are dramatically altered by the distinct Fermi surfaces (Figure 5.5b–f), except that the highest-frequency $\pi_3^v \rightarrow \pi_3^c$ might remain similar under $E_F \leq E^{c_3}(K)$. Apparently, single-particle excitations are enriched by the new $\pi_i^c \rightarrow \pi_j^c$ excitation channels (the solid and dashed curves). Each doped system could display the strongest acoustic plasmon, with the $\sqrt{q}$-dependent frequency at long wavelength limit, as previously verified in a 2D electron gas [261]. The first plasmon mode will gradually decay with the increment of q and disappear at the critical momentum. q_c increases with an increase in doping level (Figure 5.5b–e). The second plasmon-related to $\pi_2^c \rightarrow \pi_3^c$ excitations is identified as an optical mode because of a finite frequency at $q \rightarrow 0$. Its frequency and q do not have a simple relation. The spectral intensity first increases, reaches the maximum, and then declines. This plasmon is easy to observe with an increase in doping level (e.g., 0.7 eV in Figure 5.5f). Moreover, there exists a third plasmon in between the first and second modes under the Fermi level through the π_3^c band. For example, it is revealed in E_F=0.5 eV (Figure 5.5d). Its intensity is lowest among three plasmon modes. The third mode is examined to come from the $\pi_1^c \rightarrow \pi_3^c$ excitations, owing to comparable frequencies. In addition, the phase diagrams almost keep the same as the direction of $\mathbf{q}$ varies ($\phi = 0°$ in Figure 5.5b and $\phi = 30°$ in Figure 5.5c) at $E_F = 0.1$ eV.

The Coulomb excitations are greatly diversified by the stacking configuration and layer number. Monolayer graphene, with the linear Dirac-cone structure, only exhibits interband excitations in the absence of carrier doping. The 2D acoustic plasmon is absent, since this system is a zero-gap semiconductor with a zero DOS at $E_F = 0$. However, the extra intraband excitations and acoustic plasmon could survive under the finite temperature and electron/hole doping, in which the latter experiences serious Landau damping due to interband e–h pairs at large momenta [68, 261, 327]. As to a pristine bilayer AA stacking, there are sufficient free carriers coming from the interlayer atomic interactions, creating two kinds of plasmons, namely, acoustic and optical modes [23]. These two plasmon modes might be changed by the doping effect. On the other hand, the pristine bilayer AB stacking cannot induce acoustic and optical plasmons, mainly owing to very few free carriers associated with rather weak overlap in valence and conduction bands. Both of them exist in doped systems [50]. There are rich and unique excitation behaviors in trilayer AB stacking. The e–h excitation boundaries, being defined by the distinct Fermi surfaces/the band-edge states, become more complicated. One acoustic and two optical plasmon modes are, respectively, related to the intraband and interband excitations of conduction electrons.

The high-resolution EELS could serve as the most powerful experimental technique to investigate the Coulomb excitations in emergent layered systems, such as few-layer graphene, silicene, germanene, tinene, and phosphorene. The EELS measurements on single- and few-layer graphenes have been used to confirm plasmon modes, respectively, arising from free carriers, all the π electrons, and the $\pi + \sigma$ electrons. Specifically, the low-frequency acoustic plasmon (below 1 eV) is identified to experience interband Landau damping at larger momenta [328, 329]. The interband π and $\pi + \sigma$ plasmons are observed at frequencies higher than 4.8 and 14.5 eV, in which their frequencies increase with an increase of layer number [328,329]. However, the experimental identifications on the stacking-enriched electronic excitations are absent up to now. They are very useful in thoroughly understanding the diverse excitation phenomena closely related to the transferred $(\mathbf{q}, \omega)$-phase diagrams. Furthermore, they provide the full information in examining the point of view that all the excitation behaviors are dominated by band structures.

In general, the energy loss functions, the Landau dampings, and the plasmon modes in trilayer ABA stacking do not exhibit the excitation phenomena suitable for the superposition of the monolayer- and bilayer-like ones, as obviously illustrated in Figures 5.4 and 5.5. This is valid under any doping cases, and even for an intrinsic case. The extrinsic systems present the available interband Coulomb excitations between the conduction Dirac cone and parabolic bands, clearly indicating more transition channels. That the bare and screened response functions are drastically changed by the doping effect cannot be understood from a simple concept of linear superposition. On the other hand, only the single-particle excitations in a pristine system are consistent with the perturbation concept, i.e., the interband excitations due to the monolayer- and bilayer-like energy bands are forbidden. The aforementioned physical pictures are easily generalized to N-layer graphene systems.

6

ABC-Stacked Graphenes

To date, the distinct stacking configurations identified in the synthesized graphene systems cover ABC [71,275,330–332], ABA [71,330,333,334], AAB [60,71–73,77,335], and AAA [108,275]. They are the dominant mechanism in determining the essential low-energy properties, e.g., the $2p_z$-orbital-created π-electronic structures. Among them, the ABC stacking, being predicted to have the lowest ground-state energy [53], is frequently revealed in experimental syntheses. This system presents the unusual band structures under the significant vertical and nonvertical interlayer atomic interactions, as verified from angle resolved photoemission spectroscopy (ARPES) measurements [64–66]. Apparently, its electronic energy spectrum cannot be regarded as the superposition bilayer- and monolayer-like ones, and so do the other essential physical properties. For example, trilayer ABC stacking possesses three pairs of energy bands with weakly dispersive, sombrero-shaped, and linear dispersion relations (Figure 6.1a). Such unique wave-vector dependences will be directly reflected in other physical properties, e.g., the optical absorption spectra [28,284,336,337] and the low-frequency plasmon modes [27].

A lot of theoretical [23,25,27,49,240,261,263–268,286–296] and experimental [187–197,297–299] studies have been conducted on the Coulomb excitations of graphene-related systems, clearly indicating that both electron–hole (e–h) excitations and plasmon modes strongly depend on the stacking configurations, number of layers, dimensions, and external fields. For example, the theoretical predictions on intrinsic graphene systems cover the absence (existence) of acoustic plasmon in monolayer graphene at zero (finite) temperature [32,68,263–268], the appearance of acoustic and optical plasmons in few-layer AA stackings even under $T = 0$ [20–22], and the high suppression of collective excitations in layered AB stackings [23]. It might be difficult to observe the low-frequency plasmon mode, as a result of the zero-gap semiconductor, the nonprominent density of states (DOSs) in the low-lying energy bands, or the insufficiently high free carrier density due to the interlayer atomic interactions. As to extrinsic few-layer graphenes, the doping-free carriers can create rich single-particle excitations (SPEs) and plasmon modes [27,49,288–291]. However, most of the previous theoretical predictions only consider the electronic excitations arising from the first pair of valence and conduction bands nearest to the Fermi level. The fully dynamic charge

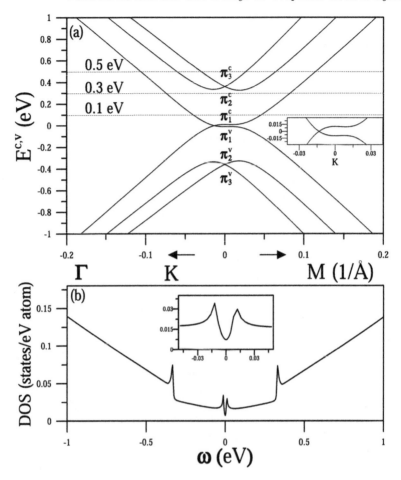

FIGURE 6.1
(a) Geometric structure, (b) low-energy bands, and (c) density of states of
the ABC-stacked trilayer graphene. The Fermi level of the pristine graphene
is set to be zero.

screening closely related to all pairs of energy bands are included in the current calculations, so that diverse Coulomb excitation spectra can be presented
in momentum- and frequency-dependent phase diagrams.

For the ABC-stacked trilayer graphene, electronic properties and Coulomb
interactions are, respectively, evaluated from the tight-binding model and
the modified random-phase approximation (RPA). The van Hove singularities
arising from the band-edge states and the wave functions localized on each
layer are the critical factors for the single-particle and collective excitations.
Specifically, the intralayer & the interlayer atomic interactions and Coulomb
interactions are fully taken into consideration. The layer-based polarization

functions and dielectric functions are built from the sublattice-dependent tight-binding functions, in which there exist four independent components. The special structures in the bare response functions are worthy of a thorough investigation. Many kinds of e–h excitation channels and plasmon modes, which are induced by the composite effects of hopping integrals and electron doping, are explored in detail. Specifically, the strong dependence of electronic excitations on the magnitude of transferred momentum and the Fermi energy will be presented in the diverse (momentum, frequency)-phase diagrams. One of the focus is whether a pristine system exhibits an acoustic plasmon, its critical momentum, and the dramatic transformation into another kind of plasmon mode after the initial electron doping. The predicted results could be verified by the high-resolution electron energy loss spectroscopy [188, 189, 191–193, 195–198] and inelastic X-ray scattering [165, 222, 225, 300].

6.1 Unique Electronic Properties

As shown in Figure 1.1d, the ABC-stacked trilayer graphene has significant interlayer atomic interactions (β_1-β_5) in addition to the intralayer one (β_0) [338]. The former creates the layer-dependent Coulomb excitation behaviors. The π-electronic Hamiltonian is built from six $2p_z$-dependent tight-binding functions. There are three pairs of valence and conduction bands, as shown in Figure 6.1a; their 3D energy dispersions are shown in Figure 6.2, corresponding to the weakly dispersive ($\pi_1^{c,v}$), sombrero-shaped ($\pi_2^{c,v}$), and linear ($\pi_3^{c,v}$) dispersions. These peculiar energy dispersions are reflected in the DOS in Figure 6.1b. The first pair, with a sharp DOSs near the Fermi level, belongs to surface-localized states, since they mainly come from the top and bottom layers. Such pair is revealed in a 2D finite-layer ABC stacking, but not the 3D rhombohedral graphite [31]. The second pair presents the square-root divergent peaks as the van Hove singularities, mainly owing to the constant-energy loops. There are more these energy dispersions when the layer number increases [28, 31]. The electronic structures of ABC-stacked trilayer graphenes have been verified by ARPES [64–66]. Specifically, the presence of the surface-localized states has been clarified for the partially flat subbands centered at the K point. While the DOS in the Fermi level is nearly vanishing, there exists a prominent SPE channel near the Fermi level, which is attributed to the excitations between the van Hove singularities of the conduction and valence partially flatbands, inset in Figure 6.1b. Under the screened Coulomb interaction, its collective excitations account for the intrinsic acoustic plasmon branch. Accordingly, this intrinsic branch is exclusive for multilayer graphene systems with the specific ABC stacking configuration. It is not presented in monolayer graphene because of the rare intrinsic excitations at zero temperature.

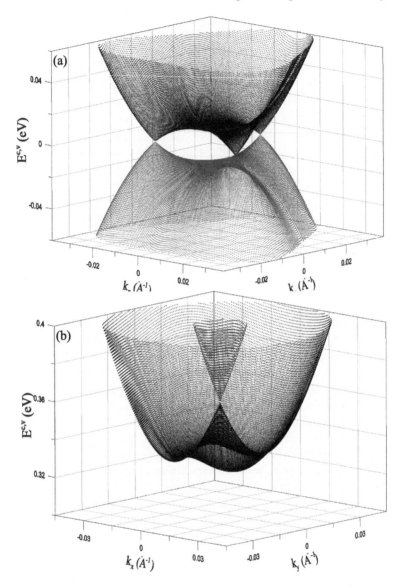

FIGURE 6.2
3D energy band structures of the ABC-stacked trilayer graphene.

Most of the wave functions are composed of six sublattice-based tight-binding functions, indicating the theoretical framework of layer-dependent RPA.

Apparently, the electronic properties of trilayer ABC stacking cannot be decomposed as the superposition of monolayer- and bilayer-like ones, and neither can the magnetic quantization [30, 339]. The electronic states of energy bands are magnetically quantized into three groups of valence and conduction Landau levels (LLs). The simple dispersion relations with the magnetic- and electric-field strengths are absent, in which the abnormal anticrossing phenomena among the intragroup LLs/the intergroup LLs frequently occur in the electronic energy spectra. The well-behaved LLs and the perturbed LLs, which are, respectively, characterized by a specific oscillation mode and the main & side modes, can survive simultaneously [30, 339]. The low-lying LLs of the first group are dominated by the surface localized states. Such LLs are expected to exhibit rich and unique magnetoelectronic Coulomb excitations, which are never investigated up to now. Moreover, the generalized tight-binding mode and the modified RPA are suitable for exploring the novel many-particle excitation phenomena (the related discussions in Chapter 10).

The vertical optical excitations, which correspond to the initial and final states with the same wave vector, mainly arise from the perturbation of an electromagnetic wave. They are closely related to the Coulomb excitations, in which there exist identical excitation spectra at long wavelength limit. The optical absorption spectra could be calculated from the gradient approximation [340, 341]. Three pairs of valence and conduction bands create nine categories of interband transitions. However, only three special structures are revealed in the low-frequency excitation spectrum. They cover the strong symmetric peak, the asymmetric peak of the square-root form, and the shoulder structure, respectively, arising from the first, second, and third pairs of energy bands. This is attributed to the vertical band-edge states. Magnetoelectronic excitations are greatly diversified by the magnetic quantization. Three groups of valence and conduction LLs can create nine categories of inter-LL transitions with three intragroup channels and six intergroup ones. A lot of magnetoabsorption structures display in the single-peak and twin-peak forms, depending on the symmetric LL spectrum about the Fermi level. All the available excitation channels are thoroughly identified by detailed numerical analyses and might be dominated by the specific magneto-optical selection rules. It should be noticed that some extra magneto-optical absorption peaks are related to the anticrossing LLs, leading to the destruction of selection rules [340, 341].

6.2 Dramatic Transformation of Coulomb Excitations under Electron Doping

The dynamic Coulomb response displays SPEs and collective excitations. These two types of excitations are, respectively, described by the bare response function $P_{ll'}(\mathbf{q}, \omega)$ and energy loss function $\mathrm{Im}[-\frac{1}{\epsilon}]$ (Eqs. 2.5 and 2.8) as the transferred momentum and frequency are conserved during the electron–electron (e–e) interactions. The response function describes the dynamic charge screening and directly reflects the main characteristics of electronic properties. $\mathrm{Im}P_{ll'}(\mathbf{q},\omega)$ indicates the intensities of the SPEs and is responsible for Landau dampings of plasmons. On the other hand, $\mathrm{Im}[-\frac{1}{\epsilon}]$ is used to identify the plasmons in the screened Coulomb excitation spectra. Collective excitations or plasmons appear in specified (q,ω) regions. As shown in Figure 6.3, the intrinsic interlayer polarizations $(l = l')$ and intralayer polarizations $(l \neq l')$ show peculiar structures related to the critical points in the energy band structures in Figure 6.1a. The 3×3 polarization function $P_{ll'}(\mathbf{q},\omega)$ depends on the symmetry of the wave function on each layer. The divergent singularities of $\mathrm{Im}[P_{ll'}(\mathbf{q},\omega)]$ correspond to the Van Hove singularities in the DOSs; the structures of $\mathrm{Re}[P_{ll'}(\mathbf{q},\omega)]$ are linked to $\mathrm{Im}[P_{ll'}(\mathbf{q},\omega)]$ via the Kramers–Kronig relations [262]. In response to the same excitation channels, the intralayer and interlayer polarizations with similar structures display phase relationships dependent on the layer-dependent symmetries of wave functions. Due to the inversion symmetry of the ABC stacking configuration, the following relationships are obtained: $P_{11} = P_{33}$, $P_{12} = P_{23}$ and $P_{11} \simeq |P_{13}|$. In the undoped case, the interband excitations contribute to several peaks in $\mathrm{Im}P_{ll'}(\mathbf{q},\omega)$ in the four regions indicated by the dashed gray lines. The square-root divergent peaks are identified as quasi-1D SPE channels; they come from $\pi_1^v \to \pi_1^c$ and $\pi_2^v \to \pi_2^c$ and appear as a result of the nearly isotropic energy dispersions near the K point [28]. In contrast, excitations from $\pi_1^v \to \pi_2^c$ ($\pi_2^v \to \pi_1^c$) and $\pi_2^v \to \pi_3^c$ ($\pi_3^v \to \pi_2^c$) exhibit logarithmic divergences and display a relatively weak response under e–e interactions. It should be noted that the surface-localized states play an important role in the low-energy polarizability. The first prominent square-root divergent structure in $\mathrm{Im}[P_{11}(\mathbf{q},\omega)]$ is attributed to the major low-energy excitations on the two outmost layers, while the empty part in $\mathrm{Im}[P_{22}(\mathbf{q},\omega)]$ reveals the absence of excitations on the middle layer. According to the Kramers–Kronig relations, the square-root and logarithmic peaks in $\mathrm{Re}[P_{ll'}(\mathbf{q},\omega)]$ correspond to the square-root and step discontinuities in $\mathrm{Im}[P_{ll'}(\mathbf{q},\omega)]$.

More electronic excitation channels are triggered by the increasing free carriers under the influence of the Coulomb interactions and the interlayer atomic interactions. At $E_F = 0.2$ eV shown in Figure 6.4, the intralayer and interlayer polarization functions have similar structures; the first logarithmic singularity of the outmost layer polarization $\mathrm{Im}[P_{ll'}^{(1)}(\mathbf{q},\omega)]$ shifts to higher ω, mainly due to the SPEs within the $\pi_1^c \to \pi_1^c$ intraband region. This channel determines the complicated polarization functions in the low-frequency excitation spectrum. The electronic states of the π_1^c subband cause new SPEs reaching up to $\simeq 0.8$ eV. It is claimed that the plasmon intensity is reduced by the Landau damping in the vicinity of the interband SPEs, if the plasmons coincide with the energies of the SPEs [342]. The polarization functions show strong responses at higher E_F. However, the intraband components predominate when compared with the interband ones. This implies that the interplay between interband and intraband excitations enriches and diversifies the electronic excitation spectra, where various plasmon modes are presented with a variation of q and E_F.

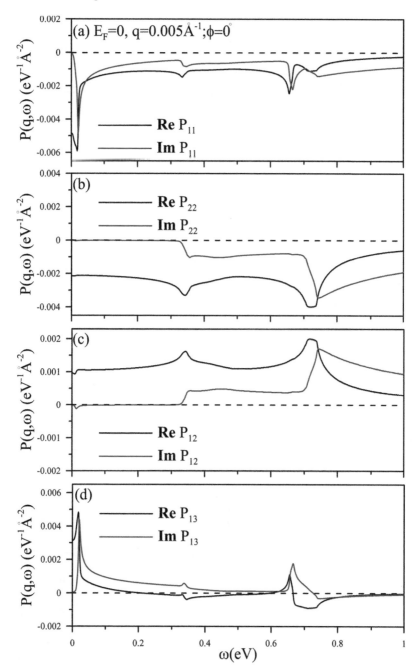

FIGURE 6.3

The four independent bare polarization functions are (a) Re[P_{11}], (b) Im[P_{22}], (c) Re[P_{12}], and (d) Im[P_{13}] of the ABC-stacked trilayer graphene at $q = 0.005$ Å^{-1}, $\phi = 0°$, and $E_F = 0$.

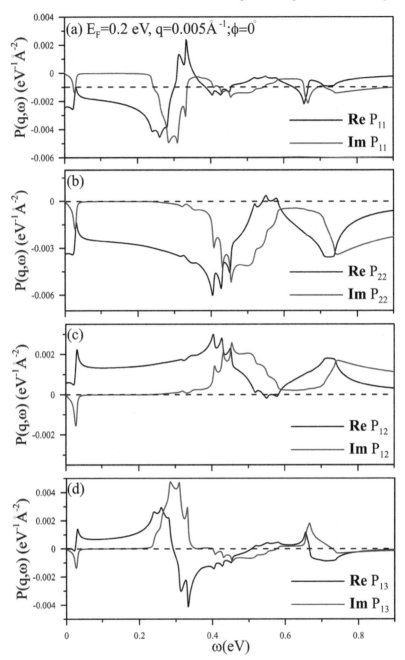

FIGURE 6.4
The four independent bare polarization functions, (a) Re[P_{11}], (b) Im[P_{22}], (c) Re[P_{12}], and (d) Im[P_{13}] of the ABC-stacked trilayer graphene at $q = 0.005$ Å^{-1}, $\phi = 0°$, and $E_F = 0.2$ eV.

The energy loss function $\text{Im}[-\frac{1}{\epsilon}]$ is used to describe the screened Coulomb excitations, as shown in Figure 6.5, where the peaks in $\text{Im}[-\frac{1}{\epsilon}]$ are referred to as plasmons. Due to the Coulomb screening effect, the frequency of the plasmon is higher than the corresponding SPE frequency. For $E_F = 0$, two intrinsic plasmon peaks, labeled by ω_p^1 and ω_p^2, are presented in the pristine ABC-stacked trilayer graphene (Figure 6.5a). The plasmon energies, resulting from the specified interband channel, i.e., $\pi_1^v \to \pi_1^c$, correspond to the weak Landau damping described in Figure 6.3. Responsible for the high DOS of the surface-localized states [31, 343–345], the interband plasmon mode is categorized as the first kind of plasmons, ω_p^1. The ω_p^1 mode extends up to $\simeq 0.25$ eV, while the intensity decrease of the energy-loss function for $\omega \simeq 0.32$ eV is attributed to the Landau damping that matches the energies of $\pi_1^v \to \pi_2^c$ SPEs. Modulated by the electron doping level, the loss function is enhanced for $E_F = 0.1$ and 0.3 eV by both intraband and interband excitations (Figure 6.5b and c). There are three extrinsic plasmon modes, ω_p^1, ω_p^2, and ω_p^3. The first plasmon mode $\omega_p^1 (\simeq 0.1)$ eV comes from the $\pi_1^c \to \pi_1^c$ intraband excitation channel, leading to a relatively prominent plasmon intensity. The latter two modes, ω_p^2 and ω_p^3, near 0.3 eV mainly originate from the $\pi_1^c \to \pi_2^c$ and $\pi_1^v \to \pi_1^c$ interband excitations, respectively; however, the higher excitations also make considerable contributions. The free-carrier excitations lead to a dramatic change of plasmon modes, as the doping level is higher than the critical point of the subbands $\pi_2^c \to \pi_3^c$. At $E_F = 0.5$ eV (Figure 6.1d), the large suppression of the low-frequency plasmons implies the significant Landau dampings of the interband SPEs related to the surface-localized states. Moreover, the low-frequency plasmons are ascribed to the multimode excitations of various intraband and interband channels. The effect of the q-direction on plasmons is negligible in the electronic excitations (Figure 6.5e). With an increment of q, more available SPE channels are triggered and the enhanced Landau damping quickly reduces the plasmons, as shown in Figure 6.5f. Under a sufficiently large E_F, the ω_p^1 is prominent in the energy-loss spectra; the intensity and frequency depend on the free carrier densities. It should be noted that the dispersion of each plasmon mode is highly dependent on the boundaries of SPE channels and, moreover, for most interband excitations, plasmons and SPEs can coexist in a certain (q, ω) region.

Trilayer ABC-stacked graphene displays rich and unique plasmon spectra under the influence of dynamic Coulomb interactions. Various plasmon modes are presented in the (q, ω)-excitation phase diagram. The plasmon mode usually appears in several specified domains of the phase diagram, because the Landau dampings occur in the region where the plasmon dispersion overlaps with the continuum spectrum of e–h pairs [342]. In the undoped case, there exists a strong SPE channel resulting from the excitations between $\pi_1^v \to \pi_1^c$ partially flat subbands. Under the screening effect, the corresponding collective excitations account for the low-ω plasmon branch. The dispersion relations of the intrinsic plasmons ω_p^1 and ω_p^2 are shown in Figure 6.6a. The interband SPEs create strong Landau dampings near $\omega \sim 0.35$ and ~ 0.65 eV.

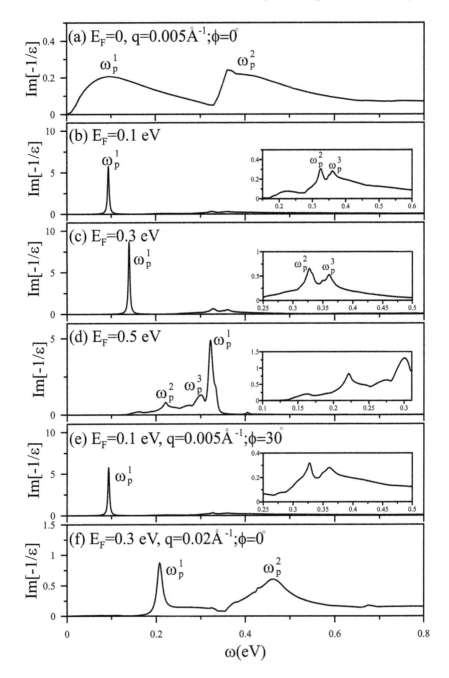

FIGURE 6.5
Energy loss spectra of ABC-stacked trilayer graphene at various q's, ϕ's, and
E_Fs.

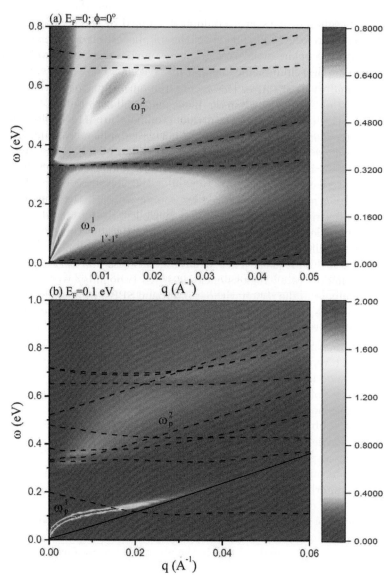

FIGURE 6.6

(q, ω)-excitation phase diagrams of ABC-stacked trilayer graphene at $\phi = 0°$ and (a) $E_F = 0$ and (b) $E_F = 0.1$ eV.

In particular, the first plasmon ω_p^1 is assigned to an acoustic mode, of which the frequency approaches to zero as q$\longrightarrow 0$ [346], and behaves a linear dependence on q as a consequence of the collective excitation mode of the surface-localized states. The intrinsic acoustic mode at zero temperature is exclusive for multilayer graphene systems with the specific ABC stacking configuration.

The linear plasmon dispersion, well defined up to 0.25 eV, is describable by the band-structure effect. Distinct from the $\sqrt{q}$ dispersion of the 2D electron gas and from that of the monolayer graphene, such a plasmon mode displays strong damping and disappears at small $q \simeq 0.01$ Å^{-1} (the SPE boundaries of $\pi_1^v \to \pi_2^c$ and $\pi_1^v \to \pi_3^c$. After this region, the optical plasmon ω_p^2 is formed near $\omega \simeq 0.32$ eV, with the plasmon dispersion similar to the ω_p^1 one. These two modes have a similar dispersion that is mainly attributed to the same $\pi_1^v \to \pi_1^c$ interband excitation channel. Another prominent characteristic of the ω_p^2 mode is that its frequency reaches up to 0.6 eV. This can be manifested by the fact that the high DOS of the $\pi_1^{c,v}$ subbands prevents the coupling from other interband excitations.

The plasmon modes are improved by doping to increase the free charge density in the extrinsic condition. As E_F is increased, the interband and intraband excitations lead to new plasmon modes and diversified phase diagrams. Plasmons with different dispersion relationships are revealed at $E_F = 0.1$ eV, as shown in Figure 6.6b. They behave as acoustic and optical modes in the low and middle (q, ω) regions enclosed by SPE boundaries. The acoustic mode is prominent in the region without interband SPEs, while showing strong damping when dispersing into the region of the $\pi_1^v \to \pi_1^c$ interband SPEs. Its intensity quickly drops by more than one order of magnitude at $q \simeq 0.017$ Å^{-1} and disappears beyond $q \simeq 0.05$ Å^{-1}, a characteristic being dominated by the nearest vertical interlayer atomic interaction γ_1. On the other hand, the optical mode is separated into several parts, each of which appears with different degrees of Landau damping in a specified domain. For 0.3 eV $\leq \omega \leq$ 0.4 eV, the plasmon dispersion is approximately flat, reflecting the particular partially flat subbands. In addition to the original interband channels, the induced free carriers also contribute to the optical plasmon.

The acoustic plasmon deserves a closer examination in the low-energy region. With an increment of E_F, the collective excitation channel is transformed from interband $(\pi_1^v \to \pi_1^c)$ to intraband $(\pi_1^c \to \pi_1^c)$. Accordingly, the acoustic plasmon deviates from the linear dispersion of the pristine graphene even in the case of weak doping, in Figure 6.7. The dispersion and intensity of the acoustic plasmon are enhanced, because the intraband collective excitations gradually become predominant in the plasmon spectra. Furthermore, the acoustic mode extends over a wider (q, ω) range than in the case of zero doping as the SPE boundaries shift to higher q and ω. The existence of acoustic plasmons with different dispersion relationships indicates the effects of band structure and doping carrier densities.

With a variation of E_F, phase diagrams are dramatically changed due to the conservation of the transferred momentum q and the energy ω, as shown in Figure 6.8. Figure 6.8a indicates nearly isotropic effects of q on the electronic excitations. At E_F=0.3 eV (Figure 6.8b), the plasmon modes extend to higher energy due to the increasing free carriers. The most striking behavior of the ω_p^1 acoustic mode is its enhanced intensity and square-root dispersion, which are in sharp contrast to the acoustic plasmon in cases of zero and

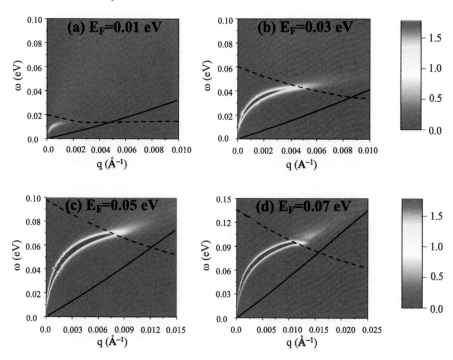

FIGURE 6.7

The acoustic plasmon of ABC-stacked trilayer graphene at various E_F's.

low dopings. Nevertheless, if the subbands π_2^c and π_3^c are partially occupied, the plasmon modes are drastically changed. At E_F=0.5 eV (Figure 6.8c), the acoustic plasmon arises from the three kinds of intraband excitations. i.e., $\pi_i^c \to \pi_i^c$ (i = 1, 2, and 3). In addition, the interplay between interband and intraband excitations also gives rise to new plasmon modes and diversified phase diagrams. According to the band effects, the Landau damping is strong for the induced interband SPEs, e.g., $\pi_1^c \to \pi_2^c$ and $\pi_1^c \to \pi_3^c$ in the region of 0.2 eV $\leq \omega \leq$ 0.3 eV. This leads to a weak plasmon mode with a concave and convex dispersion, which indicates the robust Landau dampings associated with the particular partially flat and sombrero subbands. On the other hand, the acoustic mode is enhanced and shifted to higher ω by the induced collective excitation channels. With a further increase of E_F, the plasmon is hardly affected by Landau dampings associated with the induced interband SPEs. At E_F=0.7 eV (Figure 6.8d), the various plasmons gradually merge into a long-range single acoustic mode in the (q, ω)-excitation phase diagram. It is claimed that the 2D $\sqrt{q}$-dependent plasmon appears in layered graphenes under a heavy doping condition.

Trilayer ABC-stacked graphene is predicted to exhibit rich and unique Coulomb excitations. There are many SPE channels and five kinds of plasmon modes, mainly arising from three pairs of energy bands and doping carrier

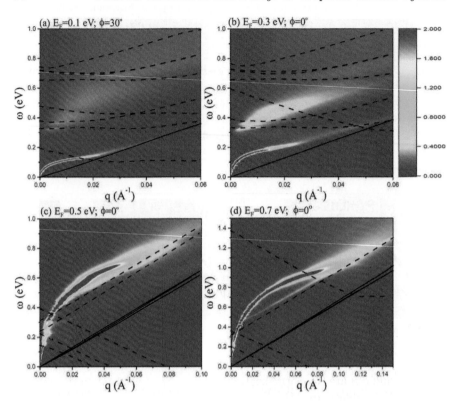

FIGURE 6.8

(q, ω)-excitation phase diagrams of ABC-stacked trilayer graphene at (a) $\phi = 30°$ and $E_F = 0.1$ eV; $\phi = 0°$ and (b) $E_F = 0.3$ eV, (c) $E_F = 0.5$ eV, and (d) $E_F = 0.7$ eV.

densities. Their complicated relations create the diverse (q, ω)-excitation phase diagrams. The plasmon peaks in the energy loss spectra might decline and even disappear under various Landau dampings. The linear acoustic plasmon is related to the surface states in pristine system, while it becomes square-root acoustic mode at any doping. Specifically, all the layer-dependent atomic interactions and Coulomb interactions have been included in polarization function and dielectric function. The theoretical framework of the layer-based RPA could be further generalized to study the e–e interactions in emergent 2D materials, e.g., silicene [34] and germanene [238].

7

AAB-Stacked Graphene

For AAB-stacked few-layer graphene systems, there are some experimental and theoretical studies on the growth [60, 70–79, 83] band structure, [81, 82], density of states (DOSs) [81, 82], magnetic quantization [82], and magneto-optical properties [81]. However, significant researches about the many-particle carrier excitations and deexcitations are absent up to now. This system owns a narrow bandgap and unusual energy dispersions; therefore, the electron–electron (e–e) Coulomb interactions are expected to create the rich and unique excitation phenomena, such as the temperature- and doping-created low-frequency plasmons. The former characteristics have been identified from scanning tunneling spectroscopy (STS) measurements [121]. It is also noticed that most of the pristine-layered graphenes exhibit semimetallic behaviors, e.g., the AAA, ABA, and ABC stacking, and the bilayer twisted and sliding systems [28, 347]. From the tight-binding model [82] and the first-principles method [53], there exist three pairs of low-lying valence and conduction bands, with the oscillatory, sombrero-shaped, and parabolic dispersions. They will exhibit the unusual optical vertical excitations, mainly owing to the unusual van Hove singularities in DOS [82]. Also, such energy bands induce magnetically quantized Landau levels, with the noncrossing, crossing, and anti-crossing B_z-dependent energy spectra [82], being quite different from those of monolayer, bilayer, and AAA/ABA/ABC trilayer graphene systems [28]. The lower-symmetry AAB stacking is predicted to have more complicated magneto-optical absorption spectra [81].

The electronic properties and Coulomb excitations of the lower-symmetry trilayer AAB are, respectively, explored by the tight-binding model and the modified RPA in detail. Specifically, the intralayer & interlayer hopping integrals and the intralayer & interlayer e–e interactions are taken into consideration simultaneously through the layer-dependent polarization functions. The low-lying energy bands mainly come from the $2p_z$ orbitals of six carbon atoms in a primitive unit cell, being examined with the first-principles calculations [53]. Three pairs of valence and conduction bands are identified to present the unusual energy dispersions: the oscillatory, sombrero-shaped, and parabolic ones, as measured from the Fermi level. The former two kinds of energy bands are predicted to present a strong van Hove in DOSs as the square-root asymmetric peaks; that is, they could be regarded as quasi-1D parabolic bands. Moreover, there exists a very narrow energy gap (<10 meV). The single-particle excitations are expected to exhibit a lot of special structures, since the nine categories of interband transitions, being related to the low-symmetry wave functions, are effective under nonvertical Coulomb excitations. This study clearly shows that a prominent plasmon peak, with the low frequency (<0.2 eV), appears in the energy loss function, and it is deduced to originate from the interband excitations of the first pair of energy bands. The dependence on the transferred momentum and temperature will be explored thoroughly. Furthermore, the similarities between 2D AAB trilayer stacking and 1D narrow-gap carbon nanotubes are fully discussed. How to dramatically change the low-frequency plasmon modes by the electron/hole (e–h) doping is another focus of Chapter 7. The number, intensity, frequency, and optical/acoustic mode of the Fermi-level-related plasmons and the various regions of intraband and interband e–h excitations are worthy of a systematic investigation, the diverse (momentum, frequency)-phase diagrams with the specific plasmon modes, and the distinct e–h boundaries. Finally, a detailed comparison among AAA, ABC, ABA, and AAA stacking is made in the single- and many-particle properties.

7.1   Stacking- and Temperature-Enriched Coulomb Excitations

The low-lying energy bands of the trilayer AAB stacking, which principally originate from the $2p_z$ orbitals, is calculated with the tight-binding model. They are almost consistent with those using the first-principles method [53]. The first two layers (the second and third layers), as clearly shown in Figure 7.1, are arranged in the AA (AB) stacking. In this system, the A atoms (black balls) possess the same (x, y) coordinates, while the B atoms (red balls) on the third layer are projected into the hexagonal centers of the other two layers. The interlayer distance and the CVC bond length are, respectively, $d = 3.37$ Å and $b = 1.42$ Å. There are six carbon atoms in a primitive unit cell, being similar to those in the trilayer AAA, ABA, and ABC stackings. The low-energy electronic properties are characterized by the complicated atomic interactions of the carbon $2p_z$ orbitals. The zero-field Hamiltonian, being built from the six tight-binding functions associated with the periodical $2p_z$ orbitals, is dominated by the intralayer and interlayer atomic interactions, γ_i's [82]. As a result of the lower-symmetry stacking configuration, there exist ten kinds of hopping integrals in the Hamiltonian matrix elements. $\gamma_0 = -2.569$ eV represents the nearest-neighbor intralayer atomic interaction; $\gamma_1 = -0.263$ eV & $\gamma_2 = 0.32$ eV, respectively, present the interlayer atomic interactions between the first and second layer; $\gamma_3 = -0.413$ eV$/\gamma_4 = -0.177$ eV$/\gamma_5 = -0.319$ eV are associated with the interlayer atomic interactions between the second and third layer; $\gamma_6 = -0.013$ eV, $\gamma_7 = -0.0177$ eV, & $\gamma_8 = -0.0319$ eV relate to the interlayer atomic interactions between the first

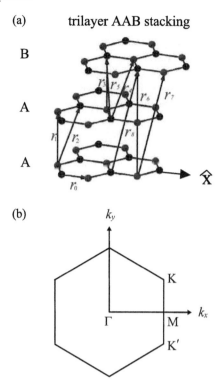

(a) trilayer AAB stacking

(b)

FIGURE 7.1
The geometric structure of the trilayer AAB-stacked graphene in the presence
of various intralayer and interlayer hopping integrals and its first Brillouin
zone.

and third layers; $\gamma = -0.012$ eV accounts for the difference of the chemical
environment in A and B atoms. The specific hopping integrals, γ_1, γ_3, &
γ_5, belong to the vertical interlayer hopping integrals, while the others are
nonvertical ones. It is also noted that such tight-binding parameters are thoroughly examined to characterize the complex and unique energy bands of the
AAB-stacked trilayer graphene (Figure 7.2a). In addition, their magnitudes
are comparable with those used in the other stacking systems, e.g., AAA,
ABA, and ABC stackings [28].

The first, second, and third pairs of valence and conduction bands
in trilayer AAB stacking, being measured from the Fermi level, exhibit
unique energy dispersions: the oscillatory ($S_1^{c,v}$), sombrero-shaped ($S_2^{c,v}$), and
parabolic ones, as clearly shown in Figure 7.2a and b. The first conduction
band begins to grow from a local minimum value of $\sim$4 meV at the K point,
along the KΓ and KM directions. After reaching a local maximum value of
$\sim$58 meV, it declines, recovers to a local minimum energy again ($\sim$4 meV),
and then increases steadily. The first conduction band is almost symmetric to

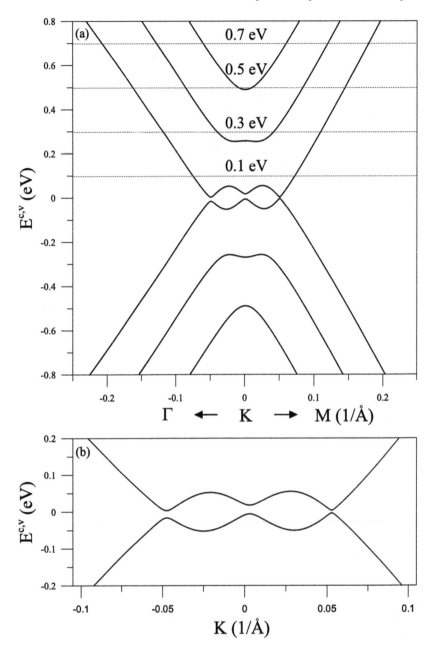

FIGURE 7.2
(a) The low-lying valence and conduction bands of the trilayer AAB stacking with (b) a narrow gap of $E_g \sim 8$ meV arising from the oscillatory energy bands. Also shown by the dashed curves are those from the first-principles method.

the first valence band about $E_F = 0$, corresponding to the opposite curvature. As a result, there exist a narrow energy gap of $E_g \sim 8$ meV and four constant-energy contours at ± 4 & ± 58 meV. Specifically, such loops in the 3D energy-wave-vector space could be effectively regarded as the 1D parabolic bands (Figure 7.3a). The multifold degenerate states are not suitable in doing the

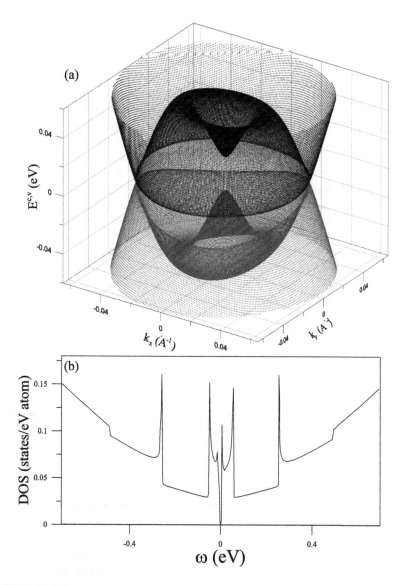

FIGURE 7.3
(a) The 3D energy bands near the Fermi level and (b) the low-energy density of states.

low-energy expansion; that is, the effective-mass model cannot deal with the low-energy essential properties. As to the $S_2^{c,v}$ energy bands, their conduction-band (valence-band) states have a sombrero-shaped energy dispersion with the local energy minimum (maximum) and maximum (minimum), being situated at around 0.24 eV (-0.24 eV) and 0.26 eV (-0.26 eV), respectively. The narrow energy difference between two extreme points is ~ 4 meV. Specifically, the third pair of $S_3^{c,v}$ consists of monotonic parabolic bands, in which the minimum (maximum) conduction (valence) state energy minimum is about 0.49 eV (-0.49 eV). The aforementioned features of the low-lying energy bands agree with those from the first-principle calculations (the dashed curves in Figure 7.2a and b), clearly indicating that the complicated interlayer hopping integral used in the tight-binding model are suitable and reliable [28]. Such energy dispersions also create the unusual magnetic quantization and magneto-optical properties [28], such as the frequent intragroup and intergroup LL anticrossings, and the complicated and rich magneto-optical absorption spectra, being never revealed in other stacking systems.

The unusual energy dispersions create two kinds of van Hove singularities in trilayer AAB stacking. The asymmetric square-root peaks and the shoulder structures, respectively, arise from the constant-energy contours (Figure 7.2a) and the band-edge states of the parabolic dispersions. Two pairs of asymmetric peak structures (Taiwanese temple structures), which are related to the first pair of oscillatory energy bands, are centered about the Fermi level. The higher-energy/deeper-energy asymmetric peak only presents at the right-hand/left-hand side, since the sombrero-shaped conduction/valence band is too shallow. The rich DOS structures have led to very complicated optical excitations, in which there are nine kinds of available vertical transitions arising from three pairs of valence and conduction bands. The number, frequency, intensity, and form of the optical absorption structures are predicted to be very sensitive to the strength of an external electric field [28].

The unusual energy dispersions and van Hove singularities lead to the special structures in the single-particle excitation spectra, There exist six independent bare polarization functions: P_{11}, P_{22} P_{33}, P_{12}, P_{13}, and P_{23}. This number is larger than from that (four) in AAA (four), ABA, and ABC stackings. For the pristine trilayer AAB stacking ($E_F = 0$ in Figure 7.2a), three pairs of energy bands can create nine categories of interband transition channels, being expressed as $\pi_i^v \rightarrow \pi_j^c$ for i & $j = 1, 2, 3$. The first pair of valence and conduction bands presents two asymmetric peak structures in the lower-frequency polarization functions at small transferred momenta, e.g., P_{ij} at $\omega = 0.01$ and 0.12 eV's/$\omega = 0.015$ and 0.12 eV's for $q = 0.005/q = 0.01$ Å^{-1} (the black/red curves in Figures 7.4 and 7.5). The real and imaginary parts, respectively, the square-root divergent peak and a pair asymmetric peak structures, directly reflect the Kramer–Kronig relations. It should be noted that the Coulomb excitations belong to the nonvertical transitions. However, the valence to conduction band-edge state transitions could survive under small q's. With the increasing excitation frequency, the special structures, without

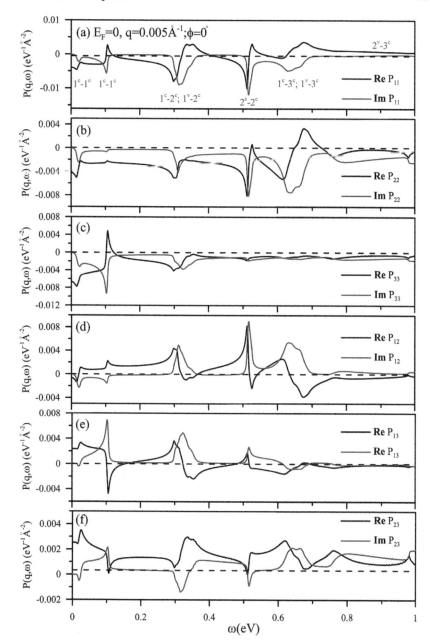

FIGURE 7.4

The bare polarization functions of the trilayer AAB stacking under different Fermi energies and transferred momenta: (a) Re[P_{11}], (b) Im[P_{11}], (c) Re[P_{22}], (d) Im[P_{22}], (e) Re[P_{33}], and (f) Im[P_{33}].

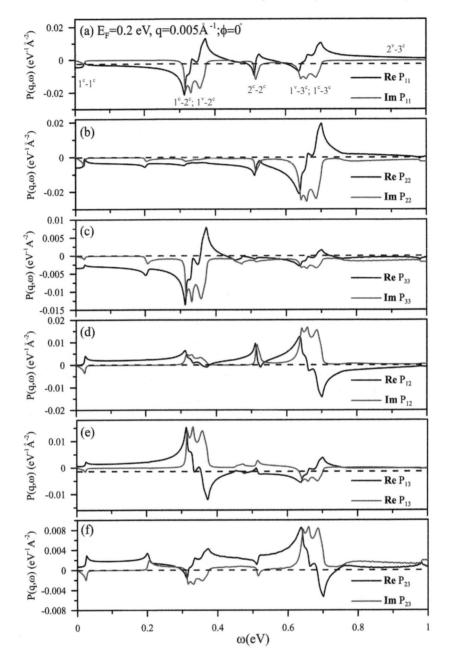

FIGURE 7.5
Similar plots as Figure 7.4a–f, but shown for (a) Re[P_{12}], (b) Im[P_{12}], (c) Re[P_{13}], (d) Im[P_{13}], (e) Re[P_{23}], and (f) Im[P_{23}].

the specific form, occur in the range of 0.2 eV$< \omega <$0.4 eV. The main reason is that such composite structures are associated with the large joint DOSs from the band-edge states of the first oscillatory valence band and the second sombrero-shaped conduction band (the second sombrero-shaped valence band and the first oscillatory conduction band). The second pair of shallow sombrero-shaped energy bands only induce a square-root peak (a pair of asymmetric structures) in Im$[P_{ij}]$'s (Re$[P_{ij}]$'s within the range of 0.4 eV $< \omega < 0.6$ eV. The second (third) valence states could also be excited to the third (second) conduction ones, in which the band-edge states near the K point might lead to special structures at 0.6 eV $< \omega < 0.8$ eV. Finally, the third pair of parabolic band show the shoulder structure in Im$[P_{ij}]$'s/the logarithmically symmetric peak in Im$[P_{ij}]$'s.

The energy loss spectra of the undoped trilayer AAB stacking can exhibit a very prominent peak, clearly illustrating the dominating band-structure effect. The plasmon frequency lies in the range of 0.145 eV$-$0.22 eV for the momentum range of $q = 0.005 - 0.02$ Å^{-1}, as shown in Figure 7.6a by the distinct solid curves. Its intensity and frequency, respectively, declines and increases with the increment of q. Apparently, this strong plasmon mode is related to the interband transitions of the first pair of oscillatory energy bands (Figure 7.2a; Figures 7.3 and 7.4), in which it directly reflects the significant van Hove singularities nearest to the Fermi level. The low-frequency plasmon peak is prominent, since it is far away from Landau dampings due to the higher-frequency interband excitations of $\pi_1^v \to \pi_2^c$ & $\pi_2^v \to \pi_1^c$. Also, other low-intensity plasmon peaks could be observed in the screened response functions, e.g., the $0.32 - 0.38$ eV peaks arising from the first/second valence band and the second/first conduction band (the inset of Figure 7.6a). Specifically, the low-frequency collective excitations, with the significant peak, are revealed in the narrow-gap carbon nanotubes (the detailed discussion in Section 12.2). Free electronic and holes cannot survive in these two systems, while the creation of the low-frequency prominent plasmon modes is attributed to the 2D oscillatory bands/the 1D parabolic bands with the large DOS; the dimension- and wave-function-dependent strong Coulomb interactions under the long wavelength limit.

7.2   Doping-Diversified Excitation Phenomena

The electron doping can dramatically change the effective excitations and the special structures in the bare response functions, as clearly indicated in Figures 7.4 and 7.5. For example, as to the low-doping case of $E_F = 0.1$ eV (the blue curves), certain intraband and interband excitations are, respectively, created and forbidden by the Pauli principle. The low-frequency interband $\pi_1^v \to \pi_1^c$ channel is replaced by the intraband $\pi_1^c \to \pi_1^c$ so that the only special structure at $\omega \sim 0.022$ meV mainly comes from the Fermi-momentum states, but not the band-edge ones. The linear energy dispersions near E_F also lead to the asymmetric square-root peak [261]. In the range of 0.2 eV $< \omega < 0.4$ eV, the special structures are due to the extra $\pi_1^c \rightarrow \pi_2^c$ excitations and the original interband $\pi_1^v \rightarrow \pi_2^c$ excitations simultaneously, in which they are related to the band-edge states. With the further increase of frequency, the identical structure at $\omega \sim 0.522$ eV is induced by the same interband $\pi_2^v \rightarrow \pi_2^c$. The $\pi_3^v \rightarrow \pi_1^c$ channel is changed into the $\pi_1^c \rightarrow \pi_3^c$ one at higher frequencies, and then the other keeps it similar, e.g., the $\pi_1^v \rightarrow \pi_3^c$, $\pi_2^v \rightarrow \pi_3^c$, and $\pi_3^v \rightarrow \pi_c^c$ channels. Apparently, the variation of the Fermi level will modify the single-particle excitations and even results in the creation or destruction of different plasmon modes. It is also noticed that the threshold asymmetric structure at the lowest frequency could survive under any Fermi levels, and so does the prominent acoustic plasmon peak.

After the electron doping, the screened response functions, revealing the distinct collective excitations and the various Landau dampings, present a rich and unique phenomena (Figure 7.6a–f). For the lower doping of $E_F = 0.1$ (Figure 7.6b), the frequency and intensity of the first plasmon peak are slightly reduced under the same transferred momenta, compared with those of $E_F = 0$ (Figure 7.6a). The critical mechanisms are in sharp contrast to each other, in which the former and latter cases, respectively, correspond to the Fermi-momentum states and the valence band-edge states; that is, the collective oscillations of charge carriers mainly come from the free conduction electrons and the valence ones in the first pair of oscillatory energy bands. The plasmon is very sensitive to the angle of the transferred momentum, e.g., the great reduction in the strength and frequency of the low-frequency plasmon as a result of the significant interband Landau dampings (Figure 7.6c). Furthermore, the observable plasmons at higher frequencies, which are located in the range of 0.2 eV $< \omega < 0.4$ eV (the insets in Figure 7.6b and c) are two distinct modes due to the $\pi_1^v \rightarrow \pi_2^c$ and $\pi_1^c \rightarrow \pi_2^c$ channels. However, there is one interband plasmon peak for the undoped case (the inset in Figure 7.6a) because of the approximately same excitation frequency of the $\pi_1^v \rightarrow \pi_2^c$ and $\pi_2^v \rightarrow \pi_1^c$ channels. With the increase of the Fermi level (e.g., $E_F = 0.3$ eV in Figure 7.6d), there are more occupied states (free electrons) arising from the first and second conduction bands, so that two intraband excitation channels, the $\pi_1^c \rightarrow \pi_1^c$ and $\pi_2^c \rightarrow \pi_2^c$ ones, would enhance the bare and screened response functions. Consequently, the first plasmon mode presents the enhanced frequency and strength. The similar excitation phenomena appear at the higher-E_F cases, such as $E_F = 0.5$ and 0.7 eV (Figure 7.6e and f) crossing the third conduction band (Figure 7.2a). Also, the energy loss spectra clearly display the drastic changes in the number, frequency, and intensity of the higher-frequency plasmon peaks during the variations of E_F (the insets in Figure 7.6a–f).

Obviously, the (momentum, frequency)-phase diagrams are diversified by the increasing doping density, as clearly indicated in Figure 7.7a–f. The single-particle regions are enriched by the doping effects. For the pristine system at

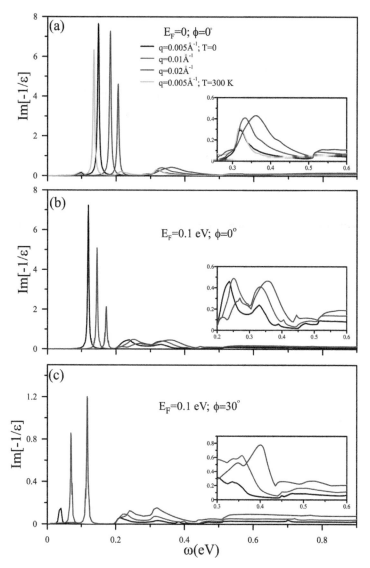

FIGURE 7.6
The energy loss functions under the various transferred momenta and $\phi = 0°$ for (a) $E_F = 0$, and (b) 0.1 eV. Also shown in (c) are those at $\phi = 30°$ and $E_F = 0.1$ eV. The insets show the higher-frequency loss spectra covering one or two optical plasmon modes.

zero temperature, nine available interband excitations present the distinct boundaries (Figure 7.7a), being characterized by the band-edge states (Figure 7.2), but not the Fermi-momentum states. In general, the whole phase space is almost full of e–h excitations; furthermore, the specific region of the small q's

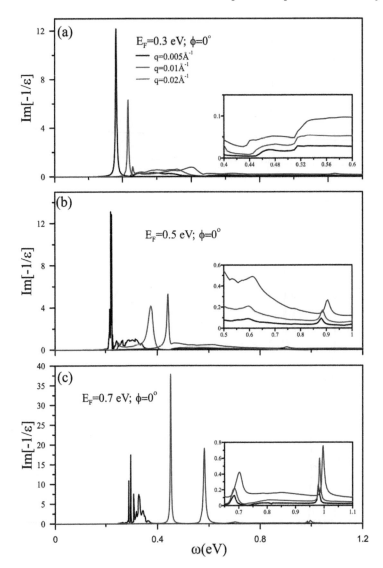

FIGURE 7.7
The energy loss functions under various transferred momenta and $\phi = 0°$ for
(a) $E_F = 0.3$. (b) 0.5 eV, and (c) 0.7 eV. The insets show the higher-frequency
loss spectra covering one or two optical plasmon modes.

and ω's exhibit very weak e–h excitations (the left-hand down corner in Figure
7.7a). A similar behavior is revealed in the doped case, without the Landau
dampings in the specific region (the crossed regions under $E_F \neq 0$ in Figure
7.7b–f). The original boundaries are drastically modified by the Fermi level if
the final Coulomb scattering states are associated with the partially occupied

conduction energy bands, e.g., $E_F = 0.1$ eV (Figure 7.7b), 0.3 eV (Figure 7.7c and d), 0.5 eV (Figure 7.7e), and 0.7 eV (Figure 7.7f). The interband valence-state excitation regions are greatly reduced by the increasing E_F, being more suitable for the existence of the low-frequency acoustic plasmon (Figure 7.7f). Furthermore, the new excitation boundaries are created by the doping effects. For example, three kinds of extra e–h boundaries principally come from the $\pi_1^c \to \pi_1^c$, $\pi_1^c \to \pi_1^c$ & $\pi_1^c \to \pi_1^c$ channels if the Fermi level is between the first and second conduction energy bands, such as the single-particle excitation regions under $E_F = 0.1$ eV by the red notations (Figure 7.8).

There exists one acoustic plasmon in any doping, in which its frequency approximately present the similar $\sqrt{q}$-dependence at long wavelength limit. This mode in the pristine system is created by the valence electrons associated with a narrow energy gap, while it arises from the conduction electrons related to intraband excitations. The low-frequency plasmon in trilayer AAB stacking, which resembles the acoustic mode in 2D electron gas [269], further illustrates the critical mechanism due to the large valence DOS of oscillatory band or the free conduction electrons (Figure 7.2a). However, the critical momentum (q_c) strongly depends on the Fermi level, since the higher-energy interband excitations are mainly determined by it. The low-frequency plasmon would disappear (Figure 7.7a–d) or merge with the optical plasmon mode (Figure 7.7e and f) after entering into the interband e–h excitations for $q > q_c$. A simple relation between q_c and E_F is absent because of the complicated nonvertical interband Coulomb excitations, such as $q_c = 0.054$, 0.03, 0.02, 0.04, and ~ 0.1 Å^{-1}, respectively, corresponding to $E_F = 0$, 0.1, 0.3, 0.5, and 0.7 eV. This plasmon mode is prominent under the pristine and high-E_F cases (Figure 7.7a–f); that is, it is relatively easy to measure the low-frequency collective excitations in the absence (presence) of doping (high doping). As to the higher-frequency plasmons, the energy loss spectra might present one or two observable modes, being sensitive to the Fermi level. One or two optical modes, respectively, correspond to $[E_F = 0; 0.3$ eV$]$ and $[E_F - 0.1, 0.5, 0.7]$. In addition, the strongly hybridized optical plasmon modes might survive under a certain range of Fermi level, e.g., those within $E_F \sim 0.3 - 0.5$ eV (Figure 7.7c–e). The optical plasmon in the pristine system is due to the interband $\pi_1^v \to \pi_2^c$ & $\pi_1^v \to \pi_2^c$ excitations. On the other hand, the diverse critical mechanisms, being sensitive to E_F, are revealed for two distinct optical modes under the doped cases. For example, at $E_F = 0.1$ $[E_F = 0.7$ eV$]$, the second and third plasmons, with frequencies higher than 0.2 and 0.3 eV (0.65 and 0.95 eV), dominated the interband $\pi_1^c \to \pi_2^c$ & $\pi_1^v \to \pi_2^c$ excitations at small transferred momenta. In general, two optical plasmons should be closely related to the multi-interband excitation channels (Figure 7.7b–f).

The significant differences among distinct trilayer stackings are worthy of a detailed comparison. Obviously, the AAB, ABC, AAA, and BAA stackings exhibit the rich and unique Coulomb excitations, clearly illustrating the geometry-diversified phenomena. According to the calculated bare and

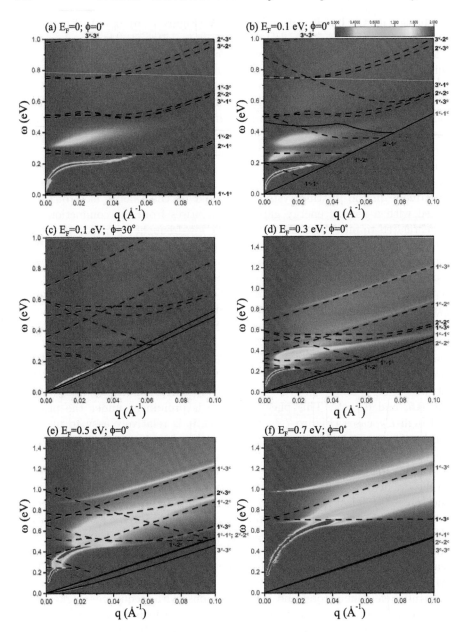

FIGURE 7.8

The (momentum, frequency)-phase diagrams of the trilayer AAB stacking under various Fermi levels: (a) $[E_F = 0, \phi = 0°]$, (b) $[E_F = 0.1 \text{ eV}, \phi = 0°]$, (c) $[E_F = 0.1 \text{ eV}, \phi = 30°]$, (d) $[E_F = 0.3 \text{ eV}, \phi = 0°]$, (e) $[E_F = 0.5 \text{ eV}, \phi = 0°]$, and (f) $[E_F = 0.7 \text{ eV}, \phi = 0°]$. The region without e–h excitations is indicated by crosses.

screened response function, the geometric symmetries, which principally determine the various interlayer hopping integrals, fully dominate the low-lying energy bands and thus the single-particle & collective excitations in the frequency range below ~ 1 eV. Obviously, the e–h excitation boundaries/regions are related to the band-edge states, with the van Hove singularities and the Fermi-moment states being sensitive to the stacking-generated energy dispersions. As a result, they are quite different among four kinds of stacking configurations. Concerning the pristine AAB and ABC (details in Section 6.1) stackings (the AAA stacking; Section 4.1), the first pair of valence and con duction bands nearest to the Fermi level (the two pairs of Dirac cones below and above the Fermi level, but not the specific one through E_F) can create a very strong special structure in the bare polarization and a prominent peak in the energy loss spectrum at low frequency. That is, the low-frequency acoustic plasmon is identified to come from the valence states with the large DOSs in a narrow-gap (AAB) or gapless (ABC) system, or the free electrons and holes (AAA) due to the strong interlayer atomic interactions. The critical momenta are, respectively, ~ 0.04, ~ 0.02, and ~ 0.1 $\mathring{A}^{-1}$ for the AAB, ABC, and AAA stackings. The largest q_c in the AAA stacking clearly indicates that this system possesses sufficiently high free carrier density. On the other hand, it is very difficult to observe this plasmon in the ABA stacking, mainly owing to the low-DOS valence band-edge states (Chapter 5). Of course, such low-frequency plasmon could survive in any doped graphenes, since it is induced by all the free carriers. There exist dramatic transformations in the available low-frequency excitation channels as the Fermi gradually increases from $E_F = 0$. In general, a simple relation between the main features of the acoustic plasmon and E_F is absent under low doping, except for the ABA stacking. That is, the frequency and intensity does present a monotonous dependence on E_F. Such result is attributed to the strong competition in AAB and ABC stackings between the low-frequency interband excitations related to the valence band-edge states and the intraband excitations associated with the conduction electrons (the opposite variations of free electrons and holes in AAA stacking). However, the opposite is true for the ABA stacking. As to the higher-frequency collective excitations, all the pristine AAB, ABC, and ABC stackings show an observable optical mode arising from the most significant interband excitations. No optical plasmons are revealed in the undoped ABA stacking. Specifically, there are one or two optical modes in any doped systems, being enriched by a lot of interband excitations due to the valence and conduction electrons.

8

Sliding Bilayer Graphene

In addition to the normal AA and AB stackings, the geometric structures of bilayer graphene could be manipulated by the relative shift along the specific direction [84, 86–88, 91, 95–97], the twisted angle between two layers [98–100], and the modulated period due to the gradual transformation of stacking configuration [89, 90] (e.g., AB/domain wall/BA/domain wall; details in [89]). These three types of typical bilayer systems have been successfully synthesized under various experimental methods [60, 70–79, 83] and band structure [81, 82]. Apparently, they present the different unit cells, corresponding to the original carbon atoms in a regular hexagon, many atoms in a Morie superlattice cell, and more than one thousand atoms in a geometry-modulated periodical cell. As a result, the sliding and twisted bilayer graphenes, respectively, possess two pairs and many pairs of 2D energy bands due to the $2p_z$ orbitals, as clearly displayed along the high symmetry points in the first Brillouin zone [91]. However, the geometry-modulated bilayer graphene possesses a lot of 1D energy subbands. The first and second systems exhibit 2D behaviors, while the third one shows 1D phenomena. For example, the van Hove singularities belong to the dimension-dependent special structures; furthermore, the 2D and 1D parabolic bands in density of state (DOS)/optical absorption spectra/imaginary parts of bare polarization functions display the shoulder and square-root asymmetric peaks, respectively. In Chapter 10, the sliding bilayer graphene is chosen for a model study, so that the asymmetry-enriched Coulomb excitations could be explored in detail.

There are certain significant studies on the diversified essential properties of the sliding bilayer graphene systems. The previous theoretical works show that the intermediate configurations between AA and AB stackings exhibit an anomalous optical phonon splitting [97] and an unusual electronic transmission [91]. By the first-principle calculations, the dependence of phonon frequency on the sliding stackings affects the polarized Raman scattering intensity, which is deduced to be available in identifying tiny misalignments in bilayer graphene [97]. Moreover, according to the combination of the generalized tight-binding model and the gradient approximation, electronic and optical properties under a uniform perpendicular magnetic field could be explored in detail [28, 91]. The well-behaved, perturbed, and undefined Landau levels (LLs) are, respectively, associated with the normal, deviated, and fully random stacking configurations (the high, middle, and low symmetries), corresponding to the Dirac-cone structures/parabolic energy bands, the distorted band structures, and the thorough destruction of Dirac points. Also, the magnetoabsorption spectra are characterized by a specific selection, the extra selection rules, and the absence of selection rule. However, a full comprehension of the Coulomb excitation phenomena under various stacking configurations has not been previously achieved.

The essential electronic properties, band structures and DOSs, of the sliding bilayer graphene systems are thoroughly explored by the tight-binding model, in which the position-related interlayer hopping integrals under the empirical formula are more convenient for the numerical calculations [91]. The shift-created features, the vertical/nonvertical Dirac-cone structures, the well-behaved parabolic dispersions, the highly distorted energy bands, the Fermi-momentum states, the extreme points, the saddle points, and the simple/complicated and linear/nonlinear superposition of the subenvelope functions on four sublattices will be investigated in detail. The work is focused on illustrating the close relations between the main characteristics of energy bands and wave functions and the diverse Coulomb excitation phenomena. According to two pairs of energy bands in bilayer graphenes, whether the single-particle excitation channels, such as the intrapair intraband, intrapair interband, and interpair transitions, present an unusual dependence on the relative shift of two graphene layers and the doping level is worthy of study. Furthermore, the form, number, frequency, and intensity of special structures in bare response functions are included in the investigations. Most important, the stacking-configuration- and doping-dependent energy loss spectra are expected to exhibit diverse/various electron–hole boundaries, the acoustic & optical plasmon modes, significant Landau dampings, and the critical transferred momenta. Specifically, the (momentum, frequency)-phase diagrams are evaluated for the distinct Fermi energies, where they display the screened response functions in the whole region.

8.1   Rich and Unique Electronic Properties

The continuous stacking configurations could be achieved by a relative shift between two graphene layers. The six ones, chosen later, clearly illustrate the diversified essential properties. One layer is gradually shifted along the armchair direction ($\hat{x}$) according to the following path (Figure 8.1): (a) $\delta = 0$ (AA), (b) $\delta = 1/8$, (c) $\delta = 6/8$, (d) $\delta = 1$ (AB), (e) $\delta = 11/8$, (f) $\delta = 12/8$ (AA') (shift in the unit of C–C bond length). The interlayer distance is assumed to remain unchanged within the tight-binding model, while it weakly changes with the variation of stacking configuration from the Vienna Ab initio simulation package (VASP) calculations [85]. Also, the numerical method shows that the AB stacking is most stable among all stackings. The low-energy Hamiltonian, which mainly comes with four $2p_z$ orbitals in a primitive unit cell, is expressed as

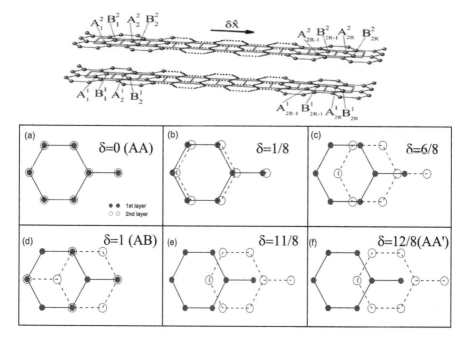

FIGURE 8.1
Geometric structures of bilayer graphene a shift relative along the armchair direction: (a) $\delta = 0$ (AA), (b) $\delta = 1/8$, (c) $\delta = 6/8$, (d) $\delta = 1$ (AB), (e) $\delta = 11/8$, and (f) $\delta = 12/8$ (AA').

$$H = -\sum_{i,j} \gamma_{ij} c_i^\dagger c_j. \tag{8.1}$$

where γ_{ij} is the intralayer or interlayer hopping integral from the i- and j-th lattice sites. Using the well-fitting parameters in Ref. [91], the distance- and angle-dependent $2p_z$-orbital interactions have the analytic form

$$\gamma_{ij} = \gamma_0 e^{-\frac{d-b_0}{\rho}} \left[1 - \left(\frac{\mathbf{d} \cdot \mathbf{e_z}}{d} \right)^2 \right] + \gamma_1 e^{-\frac{d-d_0}{\rho}} \left(\frac{\mathbf{d} \cdot \mathbf{e_z}}{d} \right)^2. \tag{8.2}$$

$\gamma_0 = -2.7$ eV is the intralayer nearest-neighbor hopping integral, $\gamma_1 = 0.48$ eV the interlayer vertical hopping integral, $\mathbf{d}$ the position vector connecting two lattice sites, $d_0 = 3.35$ Å the interlayer distance, and $\rho = 0.184b$ the decay length. This model is convenient and reliable in describing the various layer–layer interactions in graphene-related sp^2 systems, such as multiwalled carbon nanotubes [348] and multilayer graphenes [30]. It has been successfully generalized to electronic properties under a uniform magnetic field, directly revealing the sliding-induced three kinds of LLs with the special optical selection rules [91].

The low-lying band structures directly reflect the various stacking symmetries in the sliding bilayer graphene. They are equivalent about two valleys K and K', since the inversion symmetry/the equivalent A and B sublattices exist in all systems. The AA bilayer stacking, as shown in Figure 8.2a, the first (second) pair of the Dirac cones in the AA-stacked bilayer graphene is centered at the K(K') point, and has initial energies $E^c = 0.32$ eV ($E^v = -0.36$ eV). A strong overlap of two vertical Dirac cones indicates the existence of free electrons and holes. When a relative shift gradually occurs, e.g., $\delta = 1/8$ in Figure 8.2b, a strong hybridization appears between the conduction states of the lower cone and the valence states of the upper cone. This leads to distorted Dirac-cone structures along $\hat{k}_x$ with a created arc-shaped stateless region near E_F, indicating the reduction of free electron/hole density and holes. Apparently, the conduction and valence bands only touch each other at two points on the edge of the arc-shaped region; furthermore, these points belong to the Fermi-momentum states (k_F's) within the sliding of AA $\rightarrow$ AB. Moreover, there are two pairs of saddle points on the top and bottom of the arc-shaped stateless region within the range $0 < \delta < 5/8$, where the critical displacement is $\delta_c \sim 5/8$. The conservation of the number of electronic states means that the low-energy states near the K point are transferred to its neighbor regions, especially at the induced saddle points. However, the energies of two pairs of saddle points are nearly the same for small shifts (energy difference of less than 0.02 eV). With the further increase of shift, the arc-shaped region expands quickly, and energy dispersions are highly distorted at $\delta = 6/8$ (Figure 8.3c), where the saddle-point energies are completely split. The complete separations of the upper and the lower Dirac cones clearly illustrate that two pairs of energy bands are reformed at different initial energies. The band-edge states, corresponding to the first and second pairs of energy bands, have $E^{c,v} \sim 0$ and $E^c = 0.32$ eV & $E^v = -0.36$ eV, respectively.

Also, a further change from $\delta = 1$ to $12/8$ leads to dramatic transformations of low-lying band structures. As is clear from Figure 8.2d, the AB stacking possesses two pairs of parabolic dispersions, being characterized by a weak band overlap near E_F. With increments of relative shifts, the parabolic bands of the first pair are strongly distorted along $\hat{k}_y$ and $-\hat{k}_y$ simultaneously, as apparently shown at $\delta = 11/8$ in Figure 8.2e. The region outside the created eye-shape region, with two new Dirac points at distinct energies, grows rapidly. Furthermore, two neighboring conduction (valence) bands form strong hybrids. Finally, two pairs of isotropic Dirac cones are formed in the AA' stacking ($\delta = 12/8$ in Figure 8.2f), where the linearly intersecting Dirac points are situated at different wave vectors with $E^{c,v} = -0.11$ and 0.1 eV. The tilted cone axes lie in the opposite directions for the conduction and valence bands, being clearly identified from the distinct loops with a constant energy measured from the current Dirac point [91].

The diverse electronic structures are revealed as various van Hove singularities (vHSs) and DOSs, as clearly shown in Figure 8.3. The low-energy DOSs exhibit two types of vHSs near the Fermi level, namely, plateau and

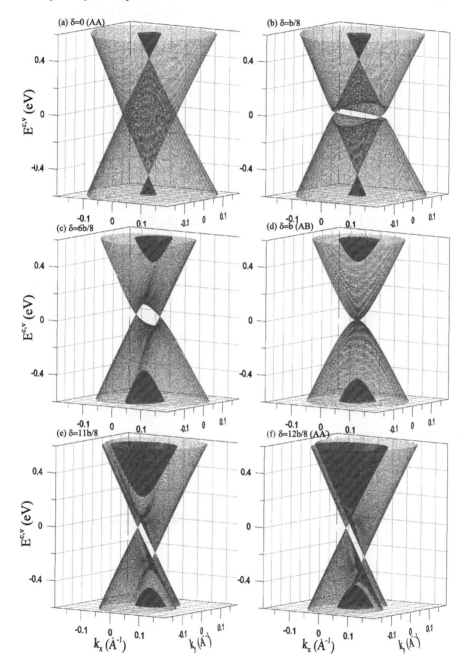

FIGURE 8.2
The 3D low-lying energy bands of the sliding bilayer graphene systems under
(a) $\delta = 0$, (b) $\delta = 1/8$, (c) $\delta = 6/8$ (d) $\delta = 1$ (e) $\delta = 11/8$, and (f) $\delta = 12/8$.

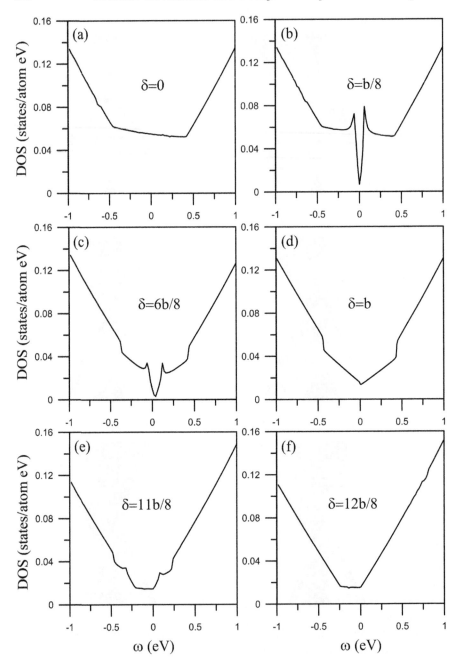

FIGURE 8.3
The low-energy densities of states under various shifts: (a) $\delta = 0$, (b) $\delta = 1/8$, (c) $\delta = 6/8$, (d) $\delta = 1$, (e) $\delta = 11/8$, and (f) $\delta = 12/8$ (AA′).

dip structures. The former is due to the linear energy dispersions across E_F, corresponding to the AA stacking and the configurations close to AA stacking (Figure 8.2a, e, and f). However, the latter comes from the band-edge states of parabolic energy bands (Figure 8.2b–d). Such DOSs are proportional to the free carrier densities created by interlayer atomic interactions, so that they dominate the low-frequency plasmons in the sliding bilayer graphene systems. Except for the well-behaved AA, AB, and AA′ stackings, the other stackings display symmetric peak structures at lower energies ($\sim \pm 0.2$ eV), being induced by the new saddle points at the top and bottom of the arc shaped stateless region (Figure 8.2b, c, and e). The number and energy of significant peaks are sensitive to the relative shift (details in Ref. [85]).

8.2 Diverse Coulomb Excitations in Pristine Systems and Doping-Enriched Excitation Phenomena

The unusual band structures of the sliding bilayer systems will create diverse Coulomb excitation phenomena. These systems possess four atoms in a primitive unit cell under the theoretical calculations, so that the single-particle layer-dependent polarization functions have four components or two independent elements ($P_{11} = P_{22}$ & $P_{12} = P_{21}$). The special structures in P_{11} and P_{12} directly originate from the Fermi-momentum states and the band-edge states/the critical points in the energy-wave-vector space, in which their forms could be examined/identified from the energy dispersions, dimensionalities, and Kramers–Kronig relations. They are very useful in determining the distinct/composite electron–hole excitation regions and the Landau dampings of the plasmon modes.

The pristine bilayer AA stacking exhibits rich special structures in the bare polarization functions, as clearly shown on Figure 8.4a and b by the black curves. The first square-root asymmetric peaks in $\mathrm{Im}[P_{11}]$ and $\mathrm{Im}[P_{12}]$ (the black curves), corresponding to the form of $1/\sqrt{(\omega_- - \omega)}$, is located at $\omega_- \sim 3.0\gamma_0 bq$. By the Kramers–Kronig relations, the real parts are square-root divergent in the opposite form. Such prominent structures are due to the intrapair intraband transitions, namely $\pi_1^v \to \pi_1^v$ and $\pi_2^c \to \pi_2^c$. Furthermore, they originate from the Fermi-moment states in the linear dispersions [23]. Such excitations could only survive at small transferred momenta (<0.1 Å^{-1}) comparable to the Fermi momenta. And then, the intrapair electronic excitations are forbidden within a sufficient wide frequency range of $\omega_- < \omega < \omega_+$ ($\sim 0.1 - 0.8$ eV). They are obviously indicated by the vanishing single-particle electronic polarizations ($\mathrm{Im}[P_{11}] = \mathrm{Im}[P_{12}] = 0$). The absence of intrapair interband transitions will create significant 2D plasmon modes due to the free electrons and holes (Figure 8.7). Such excitations, $\pi_1^v \to \pi_1^c$ & $\pi_2^v \to \pi_2^c$, will exist only for high enough transferred frequency of $\omega > \omega_+$, e.g., four

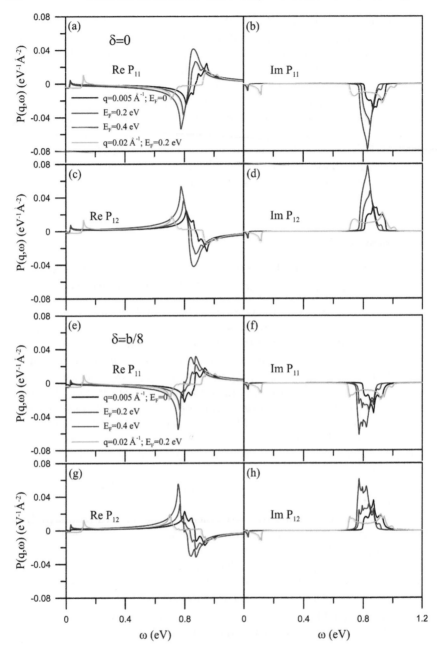

FIGURE 8.4
The two independent bare polarization functions under [$q = 0.005$ Å^{-1},
$\phi = 0°$; $E_F = 0, 2$ & 0.4 eV's] and [$q = 0.02$ Å^{-1}; $E_F = 0, 2$ eV] for the
$\delta = 0$ bilayer stacking: (a) Re[P_{11}], (b) Im[P_{11}], (c) Re[P_{12}], and (d) Im[P_{11}].
Similar plots of the $\delta = 1/8$ system in (e)–(h).

$1/\sqrt{(\omega - \omega_+)}$-form asymmetric peaks at $\omega > 0.8$ eV. It should be noticed that the specific intrapair interband excitation channels might induce two close peaks as a result of asymmetric Dirac cones/the distinct linear energy dispersions about the K/K′ point. However, only one peak is revealed in the previous investigation on the same system (Figure 4.3 in Section 4.1). The minor differences lie in the theoretical model, since the current interlayer interactions include more nonvertical hopping integrals.

The small shift deviated from the AA stacking might induce similar excitation behaviors and significant differences simultaneously. The $\delta = 1/8$ bilayer system also displays one-/four-peak structure associated with the intrapair intraband/interband excitations in the single-particle polarization functions, as clearly illustrated by the black curves in Figure 8.4e–h under small transferred momenta. Specifically, the intensity of the lowest-frequency asymmetric peak obviously presents a great decrease, directly reflecting the reduced screening ability. The decline of free carrier density (the decrease of DOS at the Fermi level) and the creation of the stateless eye-shape region are responsible for the weakened Coulomb response. Moreover, the higher-frequency electronic excitations, which arise from the intrapair interband transitions, are available in a more wide ω-range, mainly owing to the enhanced asymmetry of energy spectrum. It should be noticed that the saddle points in the low-lying energy bands, with a very high DOS, do not present any special structures in bare polarization functions. However, the valence and conduction could induce a prominent peak structure in the optical absorption spectrum [28]. Apparently, this important difference between the Coulomb and electromagnetic (EM)-wave excitations mainly lie in the nonvertical and vertical transitions, respectively.

There exists the dramatic transformation in low-energy electronic properties under a sufficiently large shift, and so do the Coulomb excitation phenomena. The $\delta = 6/8$ bilayer graphene, as clearly shown in Figure 8.5a–d by the black solid curves, exhibits single-particle excitations in the overall region. This result obviously indicates more available excitation channels due to the abnormal superposition of four subenvelope functions on the first and second layers. The intra- and interpair interband transitions, $\pi_1^v \to \pi_1^c$, $\pi_1^v \to \pi_2^c$, $\pi_2^v \to \pi_1^c$ & $\pi_2^v \to \pi_2^c$, are very efficient during the nonvertical Coulomb excitations. Moreover, three kinds of special structures appear the at the imaginary and real parts of polarization functions, namely, the square-root asymmetric peaks, the logarithmically symmetric peaks, and the shoulders, in which the latter two structures are consistent with each other in the principle-value integrations. Such structures, respectively, originate from the Fermi-momentum states with the linear energy dispersions, the saddle points, and the band-edge states of the parabolic bands. The latter two structures are consistent with each other in the principle-value integrations. For example, under a specific $q = 0.005$ Å^{-1}, the strong threshold response appears at ~ 0.022, being due to the Fermi-momentum initial/final states in the first pair of valence and conduction bands. The second symmetric/shoulder structure at ~ 0.24 eV

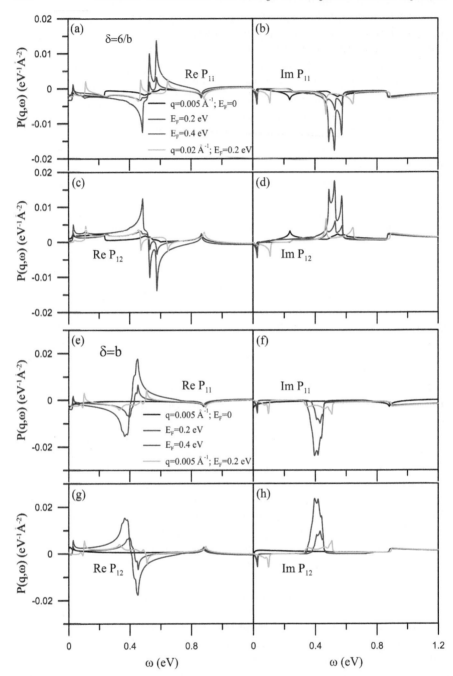

FIGURE 8.5
Similar bare response functions in Figure 8.4, but displayed for the $\delta = 6/8$ and $\delta = 1$ bilayer graphenes in (a–d) and (e–h), respectively.

presents the logarithmic/discontinuous form in the imaginary/real parts of polarization functions (Figure 8.5b and d/Figure 8.5a and c). This structure is closely related to the occupied and unoccupied saddle points in the first pair of interband transitions. With the increase in frequency, the interpair interband excitations, $\pi_1^v \to \pi_2^c$ & $\pi_2^v \to \pi_1^c$, exist within the range of 0.45 eV$< \omega <$ 0.62 eV, clearly revealing three special structures. It is very difficult to identify their specific forms, since energy bands display a strong asymmetry about the Fermi level (Figure 8.2c). Finally, the band-edge states of the second parabolic valence/conduction bands could create an obvious shoulder/logarithmic peak at ~ 0.88 eV in the imaginary-/real-part bare response functions. The active single-particle excitations will become a very high barrier in generating the undamped plasmon modes.

The polarization functions become more concise under the normal AB stacking of $\delta = 1$, compared with those of the other stackings. They are clearly illustrated by the solid black curves in Figure 8.5e–h. The well-behaved symmetric and antisymmetric superposition of four subenvelope functions has led to the absence of interpair interband excitations. Furthermore, by using the intrapair interband excitations, the well-known parabolic dispersions near the band-edge states induce shoulder/logarithmic peak structures in the $\mathrm{Im}[P_{11}]$ & $\mathrm{Im}[P_{12}]/\mathrm{Re}[P_{11}]$ & $\mathrm{Re}[P_{12}]$ (Figure 8.5f and h/Figure 8.5e and g). The special structures at $\omega \sim 0.01$ & 0.91 eV's, respectively, originate from $\pi_1^v \to \pi_1^c$ & $\pi_2^v \to \pi_2^c$. Specifically, the imaginary parts of bare response functions could survive in most of (momentum, frequency)-phase diagram. The 2D plasmon modes are expected to experience the rather strong Landau dampings; that is, they might be difficult to present the prominent peak structures in the energy loss spectra.

The further large deviation from the AB stacking, for example, $\delta = 11/8$ (the black solid curves in Figure 8.6a–d), also creates complicated and unique Coulomb excitations, as revealed in $\delta = 6/8$ (Figure 8.5a–d). According to the range of frequency, the diverse single-particle Coulomb excitations cover the intrapair interband transitions $[\pi_1^v \to \pi_1^v$ & $\pi_2^c \to \pi_2^c]$, the interpair interband channels $[\pi_2^v \to \pi_1^c$ & $\pi_2^v \to \pi_1^c]$, and the intrapair interband scatterings $[\pi_1^v \to \pi_1^c$ & $\pi_2^v \to \pi_2^c]$. Their special structures, which directly reflect the unusual band structure (Figure 8.2e), respectively, exist at $\omega \sim 0.03$, 0.30–0.32, and 0.74–0.84 eV. The low-lying, distorted, and nonvertical Dirac cones, which are accompanied with the higher-/deeper-energy parabolic conduction/valence band in the first/second one, are responsible for the threshold/first square-root asymmetric peak. Furthermore, this rather asymmetric energy spectrum about $E_F = 0$ induces the second structures arising from $\pi_2^v \to \pi_1^c$ & $\pi_2^v \to \pi_1^c$, but not due to $\pi_1^v \to \pi_2^c$ & $\pi_1^v \to \pi_2^c$. It should be noticed that the excitation channels of the latter cannot generate any special structures under the absence of Fermi-momentum and parabolic band-edge states. Finally, the intrapair interband transitions dominate the higher-frequency polarization functions and present two neighboring special structures (red or black circles) for the specific $\pi_1^v \to \pi_1^c/\pi_2^v \to \pi_2^c$. The splitting excitation

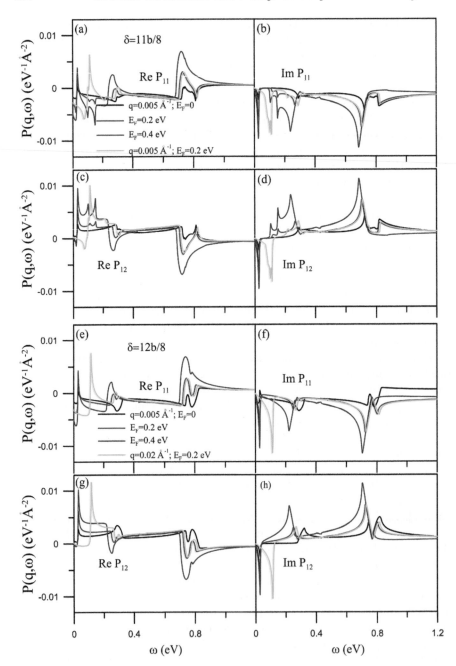

FIGURE 8.6
Similar single-particle excitation functions in Fiure 8.4, but illustrated by the $\delta = 11/8$ and $\delta = 12$ bilayer graphene systems, respectively, in (a–d) and (e–h).

phenomena purely come from the highly hybridized energy bands, since the conduction/valence band of the first/second pair possesses separated Dirac-cone and parabolic dispersions.

After the final transformation of $\delta = 11/8 \to 12/8$, the composite energy dispersions become the nonvertical two Dirac-cone structures, leading to the disappearance of certain excitation channels/phenomena. The bare response functions of the AA$'$ stacking, being clearly illustrated in Figure 8.6e–h, exhibit the unusual square-root asymmetric peaks at $\omega \sim 0.03$ eV, 0.36 eV; 0.78 and 0.86 eV. They arise from the Fermi-momentum states in the linear energy dispersions, as discussed earlier in the AA stacking. Such structures, respectively, correspond to $[\pi_1^v \to \pi_1^v \ \& \ \pi_2^v \to \pi_2^c]$, the interpair interband channels $[\pi_2^v \to \pi_1^c]$, and the intrapair interband scatterings $[\pi_1^v \to \pi_1^c \ \& \ \pi_2^v \to \pi_2^c]$. The interpair interband channel of $\pi_2^c \to \pi_1^c$ is forbidden as a result of the well-behaved functions in the linear superposition of four subenvelope functions. Moreover, there are no excitation splittings in the higher-frequency channels, since the parabolic conduction/valence band disappears in the first/second Dirac cone.

The doping effects create a great diversity in the single- and many-particle electronic excitations. The sliding bilayer graphene is assumed to keep a rigid band structure after the electron/hole doping. The blueshift of E_F is chosen for a model study. From the viewpoints of Coulomb excitations, the energy spectra about the Fermi level become more asymmetric. This might lead to dramatic changes in the bare response functions. According to three categories of excitation frequency ranges, the drastic/negligible variations of the special structures are described as follows. As for the AA stacking, the square-root asymmetric peak at $\omega \sim 3\gamma_0 \, bq$ almost remains the same in the presence of any dopings, e.g., the approximately identical P_{11} and P_{12} under $E_F = 0.4$, 0.2 and 0 eV (the blue, red, black curves in Figure 8.4a–d). Such result directly reflects the fact that the two Dirac-cone structures are highly overlapping and straightly vertical (Figure 8.2a). However, the strength of polarization functions in the $\delta = 1/8$ stacking is enhanced by the increase of E_F, as shown in Figure 8.4e–h. It is sensitive to the number of Fermi-momentum states (Figure 8.2b). With the further increase of shift, the $\delta \, 6/8$ stacking presents unusual response functions in Figure 8.5a–d. The two kinds of excitation channels, both intrapair intraband and interband transitions $[\pi_1^c \to \pi_1^c \ \& \ \pi_1^v \to \pi_1^c]$, replace the original intrapair interband transition associated with the Fermi-momentum and saddle-point states (Figure 8.2c). The enlarged square-root peak (the first structure) is attributed to more Fermi-momentum states. Furthermore, the logarithmically symmetric peak in the imaginary part of polarization function (the second structure) is roughly substituted by the shoulder structure. Specially, the AB stacking (Figure 8.5e and f), with two pairs of parabolic bands (Figure 8.2d), exhibits a dramatic transformation, the variation from the shoulder/symmetric peak structure into the square-root asymmetric peak for Im$[P_{11}]$ & Im$[P_{12}]$/Re$[P_{11}]$ & Re$[P_{12}]$, where the original mechanism of $\pi_1^v \to \pi_1^c$ is replaced by $\pi_1^v \to \pi_1^c$. It should be noticed that the former channel is transferred to the next frequency range. Finally, Figure 8.7a–h shows that the identical doping effects are revealed in the $\delta = 11/8$ & $12/8$ bilayer graphene systems, being related to similar non-vertical Dirac-cone structures (Figure 8.2e–f). That is to say, the square-root asymmetric peak is strengthened by the increment of free carrier density.

Concerning the second category of special structures, the $\delta = 0$ and $\delta = 1/8$ stackings remain featureless in the range of 0.2 eV $< \omega <$ 0.6 eV under any dopings/different E_F's, as clearly shown in Figure 8.4a–h. The unchanged wave functions under the rigid-band approximation are responsible for the vanishing interpair interband excitations. Such transitions in the $\delta = 6/8$ stacking appear at the higher-frequency range of 0.4 eV$< \omega <$ 0.6 eV. Three kinds of channels, $[\pi_1^v \to \pi_2^c, \ \pi_2^v \to \pi_1^c]$ & $\pi_1^c \to \pi_2^c]$, present more special structures, in which the third one is purely induced by the doping effect. Specially, the AB stacking can display strong responses within the range of 0.36 eV $< \omega <$ 0.48 eV, strongly depending on free carrier densities. The new channels are associated with intrapair interband and interpair excitations, namely, $[\pi_1^v \to \pi_1^c, \ \& \ \pi_1^c \to \pi_2^c]$. They only survive at a finite E_F; furthermore, the latter is not suppressed by the well-behaved wave functions, being thoroughly different from the vanishing $\pi_1^v \to \pi_2^c$ & $\pi_2^v \to \pi_1^c$. The interpair transitions of $\pi_2^c \to \pi_1^c$ & $\pi_2^v \to \pi_1^c$, corresponding to 0.10 eV $< \omega <$ 0.36 eV, keep the same for the $\delta = 11/8$ stacking. However, there are more special structures and wider distributions, further illustrating the close relation between energy spectrum and E_F. Also, these two channels occur in the AA$'$ system, while the former one is dominated by carrier dopings. For any sliding bilayer systems in Figures 8.4–8.6 the critical excitation channels at 0.75 eV $< \omega <$ 0.95 eV (the third category) are not affected by a finite E_F, mainly owing to the higher/deeper-energy bands with the specific energy dispersions. Apparently, the intensity, number, form, and frequency of the special structures are sensitive to the change of free carrier densities. In addition, the main features of the bare response functions also depend on the magnitude of transferred momenta, e.g., those under $q = 0.02$ Å^{-1} and $E_F = 0.2$ eV (the green curves in Figures 8.4–8.6).

The energy loss spectra of the sliding bilayer systems could provide diverse screened phenomena, in which the 2D plasmon modes might be damped by the various electron–hole pair excitations. As to the pristine AA stacking, Figure 8.7a clearly illustrates that there are two significant peaks at $\omega_P \sim 0.22$ eV and 1.095 eV under $q = 0.005$ Å^{-1}, especially for the very prominent former. The first plasmon mode, which is due to the intrapair intraband excitations (Figure 8.4a–d), has the excitation frequency much higher than the single-particle one (0.055 eV). This plasmon is the collective charge oscillations of free electrons and holes purely induced by the rather strong interlayer hopping integrals. The collective excitations do not experience any Landau dampings, since both intrapair interband and interpair transitions are absent at small transferred momenta. Another higher-frequency plasmon mode arises from the intrapair interband channels; furthermore, their strength is weakened

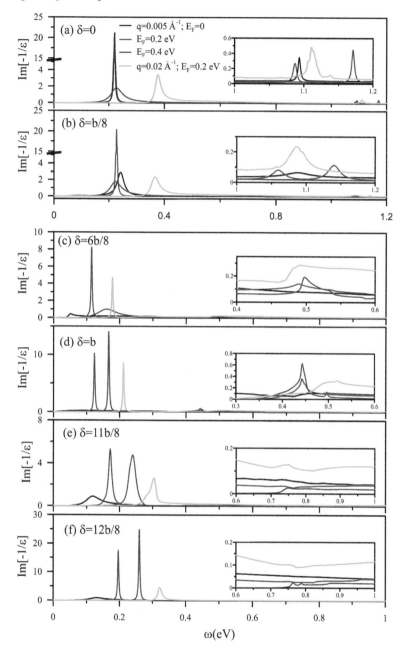

FIGURE 8.7

The energy loss spectra under $[q = 0.005\ \text{Å}^{-1}, \phi = 0°, E_F = 0, 0.2\ \&\ 0.4\ \text{eV's}]$ and $[q = 0.02\ \text{Å}^{-1}; E_F = 0, 2\ \text{eV}]$ for the sliding bilayer systems: (a) $\delta = 0$, (b) $\delta = 1/8$, (c) $\delta = 6/8$, (d) $\delta = 1$, (e) $\delta = 11/8$, and (f) $\delta = 12/8$. Also shown in the insets are the enlarged higher-frequency results.

by the single-particle excitations (inset in Figure 8.8a). However, the intensities of plasmon peaks are obviously reduced in the $\delta = 1/8$ stacking, as indicated in Figure 8.7b. This result directly reflects the low-lying distorted Dirac-cone structures (Figure 8.2b) with the lower free carrier density and the stronger Landau dampings. As to the plasmon frequencies, only a small enhancement/decrease is revealed in the first/second mode. With the further increase in shift, such plasmon modes present dramatic changes in their intensities. They might be very weak in the $\delta = 6/8$ and $\delta = 1$ stackings (Figure 8.7c and d), or they would be rather difficult to be observed in the energy loss spectra. The main reason is too low DOS at E_F/free carrier density. Finally, the second plasmon peak almost disappears for the $\delta = 11/8$ and $\delta = 12/8$ systems (Figure 8.7e and f). The intensity of the first plasmon mode is much lower/higher than that of AA/AB. Such results should be attributed to the nonvertical Dirac-cone band structures.

The electron or hole doping can create a rich and unique phenomena in the plasmon peaks of the energy loss spectra, mainly owing to the enhanced asymmetry of energy spectra. For the AA stacking (Figure 8.8a), the frequency and strength of two plasmon modes do not present a simple and monotonous relation in the increase of Fermi energy, especially for those of the former. For example, the first plasmon mode at $E_F = 0.4$ eV (the blue curve) only presents a weak peak. The main mechanism is that the Fermi level, being above and close to the Dirac point (Figure 8.2a), has led to nonuniform distributions of free electrons and holes. Similar results are revealed in the $\delta = 1/8$ stacking, as indicated in Figure 8.8b. However, the first/second plasmon of the $\delta = 6/8$ stacking is largely/somewhat strengthened; furthermore, it exists for the AB stacking. In these two systems, the increase of free conduction electrons is responsible for the enhanced/emergent plasmon mode. At last, the doping effect on the $\delta = 11/8$ and $\delta = 12/8$ stackings is to increase the intensity and frequency of the first plasmon, while it cannot generate the second plasmon mode. Such effect is much stronger than that in the AA stacking (Figure 8.7a), being related to the smaller Dirac-point energies (Figure 8.2b).

The sliding bilayer systems obviously exhibit the geometry- and doping-diversified Coulomb excitation phenomena by illustrating the rich (momentum, frequency)-phase diagrams, as indicated in Figures 8.8–8.10. For pristine graphenes, Figure 8.8 clearly displays the shift-induced dramatic variations in electronic excitations, where two plasmon modes and electron–hole excitation boundaries are very sensitive to the stacking configurations/the low-lying energy bands. The regions of single-particle excitations, as indicated by the dashed and solid lines, are mainly determined by the Fermi momenta and band-edge states. Two kinds of collective excitations, corresponding to the red/yellow/green curves, could only survive in the $\delta = 0$ and $\delta = 1/8$ systems (Figure 8.8a and b). From the momentum dispersion relation, the first and second plasmons, respectively, belong to the acoustic and optical modes. The frequency of the former has the $\sqrt{q}$ relation at small momenta, being similar to that of 2D electron gas. That is to say, free electrons and holes,

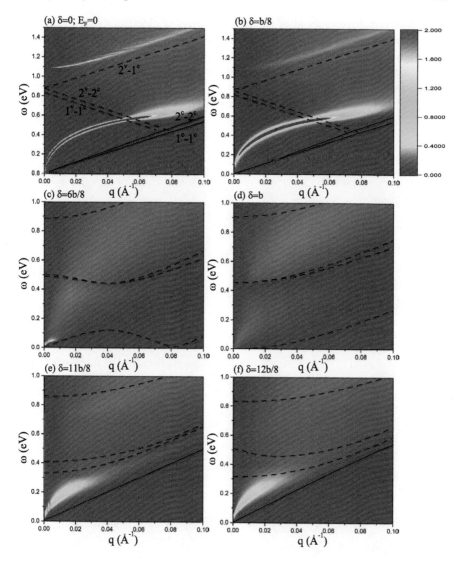

FIGURE 8.8

The (momentum, frequency)-phase diagrams of pristine sliding bilayer graphenes: (a) $\delta = 0$, (b) $\delta = 1/8$, (c) $\delta = 6/8$, (d) $\delta = 1$, (e) $\delta = 11/8$, and (f) $\delta = 12/8$.

respectively, in two distinct Dirac cones simultaneously contribute to the collective charge oscillations at long wavelengths. This plasmon is completely undamped within $q < 0.05$ Å^{-1} and experiences rather strong intrapair interband dampings after a critical momentum of $q_c > 0.1$ AA^{-1} Apparently, it is relatively difficult to observe the latter from the weaker energy loss spectra. Similar results in these two systems suggest that a small shift of two graphene layers does not drastically alter the main features of electronic excitations. As a result, two plasmon modes almost disappear for the $\delta = 6/8$ and $\delta = 1$ stackings (Figure 8.8c and d), except that the acoustic plasmon exists in the former at very small q's. This result reflects the fact that the free carrier density due to the interlayer atomic interactions is very low. Concerning the $\delta = 11/8$ and $\delta = 12/8$ bilayer graphenes (Figure 8.8e and f), only the first plasmon modes, with the lower intensity, are revealed at $q_c \sim 0.05$ Å^{-1}. Apparently, the AA′

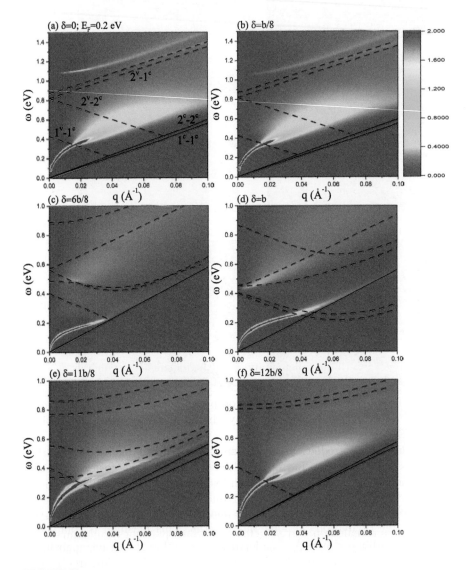

FIGURE 8.9

Similar plots as Figure 8.8a–f, but shown for $E_F = 0.2$.

stacking is quite different from the AA system in the regions of electron–hole excitations and the existence of plasmon modes. The great differences are attributed to the tilted/nontilted axis of two distinct Dirac-cone structures and the energy spacing between the Dirac points/ free carrier density.

Apparently, the blueshift of the Fermi level results in dramatic changes and different effects on the Coulomb excitation spectra, as clearly shown in Figures 8.9a–f and 8.10a–f. Of course, the electron–hole boundaries largely vary with the increase of E_F (the dashed and solid curves). As to the AA and $\delta = 1/8$ bilayer stackings, the same carrier density of free electrons and holes becomes nonhomogenous during the increment of E_F (Figure 8.2a and b), in which the former and the latter, respectively, increases and decreases (even disappears at $E_F = 0.4$). This hinders the efficiency of the collective charge oscillations and thus reduces the plasmon strength in the screened response spectrum. The acoustic plasmon mode directly presents the decrease of the undamped momentum range and the critical transferred momentum, e.g., the most significant loss spectra in Figures 8.8a and b, 8.9a and b, 8.10a and b), respectively, under $E_F = 0$, 0.2 eV and 0.4 eV. On the other hand, the adjustment of E_F has greatly enhanced the carrier density of free conduction electrons in the $\delta = 6/8$ and AB systems (Figure 8.2c and d). The first plasmon mode exists at any doping levels (Figures 8.9c,d and 8.10c,d), while the second one appears only at sufficiently high Fermi level (e.g., $E_F = 0.4$ eV in Figure 8.10c and d). A simple relation between the acoustic plasmon strength and the Fermi energy is absent for these two kinds of stackings. The E_F-enhanced asymmetric effects on the energy bands of the $\delta = 11/8$ and $\delta = 12/8$ bilayer stackings (Figure 8.2e and f) are to largely strengthen the collective oscillation frequency of the acoustic plasmon, but it cannot create the higher-frequency optical one. Also noticed ω_P of the first mode increases with an increase in Fermi level, and its strength does not present a concise E_F-dependence.

Obviously, for the sliding pristine bilayer systems, the Coulomb and electromagnetic wave perturbations present rather different bare response functions, mainly due to the nonvertical and vertical scatterings. The electronic excitations frequently appear at a finite transferred momentum, while the optical absorption spectra correspond to the $q \to 0$ transitions. The intrapair intraband excitations are forbidden in the latter, so that the optical gaps exist in the semimetallic AA and AA′ stackings with the higher free carrier density. There are no square-root asymmetric peaks in the optical spectral functions even at the higher frequency. Only simple shoulders and few logarithmically symmetric peaks are revealed as optical absorption structures. That is to say, the splitting special structures are absent under vertical transitions. In short, the Coulomb excitations could provide more information about the shift-enriched energy bands and wave functions, compared with the optical absorptions.

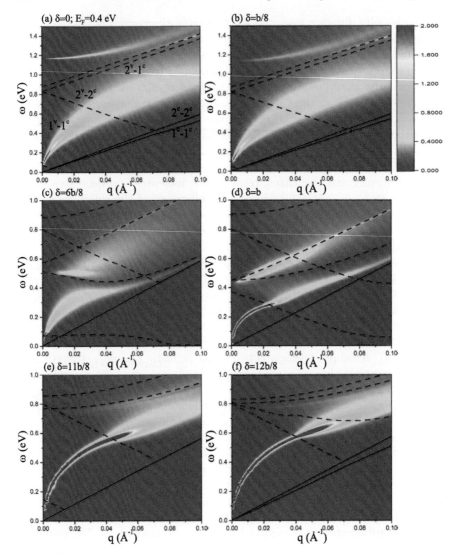

FIGURE 8.10
Similar plots as Figure 8.9a–f, indicating the distinct doping effects at $E_F = 0.4$.

9

Diversified Effects due to a Perpendicular Electric Field

An external electric field perpendicular to the graphene plane $(F\hat{z})$ is an efficient factor in creating diverse fundamental properties. There are a lot of theoretical and experimental studies on the electronic properties [106, 320, 322, 349], optical absorption spectra [320, 350–355], and quantum transports [356–360]. For few-layer graphene systems, this field plays a critical role in drastic/dramatic changes of energy dispersions and energy gaps [361–363], and the unusual quantum Hall effects [356]. On the other hand, only few calculated results on the Coulomb excitations [21, 23, 26, 189, 364], and the experimental measurements from the inelastic light scatterings (details in Section 2.5) are required to examine them. The previous calculations are fully conducted on the AA bilayer stacking [20], the AA-stacked N-layer graphenes [21], and the AB bilayer stacking [26]. In addition, the partial results are associated with the ABA and ABC trilayer graphenes [21]. In this work, the electronic excitations in the AB- and ABC-stacked graphenes deserve thorough investigations, including those in the AA-stacked ones.

Specifically, for the bilayer AA stacking [20], the analytic results are obtained for single-particle excitation regions and momentum-dependent plasmon modes in the presence of $F\hat{z}$ using the generalized tight-binding model and the modified random-phase approximation (RPA). This study shows that such a field destroys the uniform probability distribution of the four sublattices, so this drives a symmetry breaking in the intralayer and interlayer polarization intensities from intrapair band excitations. A $F\hat{z}$-induced acoustic plasmon thus emerges in addition to the strong field-tunable intrinsic acoustic and optical plasmons. At long wavelengths, the three modes present different dispersions and field dependence. In this work, the excitation properties have been discussed. Concerning an N-layer AA stacking, the numerical calculations indicate that the low-lying band structure is like a combination of N pairs of linear bands. The electric field shifts the Dirac points of linear subbands, enhances the density of free electrons and holes, alters the charge distributions on the distinct layers, and then, as a result, drastically changes the layer-dependent bare response functions and energy loss functions. In the absence of $F\hat{z}$, it displays one acoustic plasmon and $N-1$ optical plasmons at low frequency. The electric field increases the plasmon frequencies, alters their energy spacings, and rearranges their spectral weights. This field also creates extra plasmon modes, including few optical plasmon (OP) modes and an acoustic one in certain momentum ranges; the latter lowers the threshold excitation frequency. The similar $F\hat{z}$-created effects are observed in the bilayer and trilayer AB stackings [21, 26] and the trilayer ABC stacking [21], such as the creation, modification, and/or replacement of plasmon modes. These significant changes imply that plasmon effects in layered graphene systems could be electrically tunable, a feature that could be useful in electronic applications and might attract more theoretical and experimental researches to this area.

Apparently, the perpendicular electric field will greatly diversify the Coulomb excitation phenomena in layered graphene systems, especially for the well-behaved few-layer AAA, ABA, and ABC stackings. The complex calculation results mainly originate from the generalized tight-binding model and the layer-dependent RPA, in which these two modes are directly linked together, as done for the absence of $F\hat{z}$. The low-lying energy bands, density of state (DOS), and wave functions are fully investigated for their electric-field dependence, such as the obvious variations in the band overlap, the energy dispersions, the large DOSs due to the unusual energy bands, and the enhanced nonequivalence of layer-dependent sublattices. And then, the effects of $F\hat{z}$ on the bare response functions and the energy loss spectra are explored in detail, in which the stacking configurations, the layer number, and the transferred momenta are included in the numerical calculations. The changes of electron–hole (e–h) excitation boundaries, the creation/destruction/modification of acoustic & optical plasmons, the relations between the number of layers and plasmon modes, and the enhanced/reduced critical momenta are the main focuses of study. Moreover, the analytic essential properties are illustrated for the AA bilayer stacking, covering the electric-field-dependent electronic energy spectra, wave functions, intraband & interband polarization functions/various Landau dampings, and dispersion relations of the plasmon frequency with the transferred momentum. This study could provide more concise physical pictures in thoroughly understanding the elementary excitations. For example, the determinant of the real part of the dielectric tensor can determine the momentum-dependent plasmon frequencies and thus distinguish the acoustic and optical modes. Finally, how to examine the predicted results from the inelastic light scatterings is also discussed.

9.1   AAA Stacking

The trilayer AAA stacking is very suitable for a model study, since it has three pairs of π-electronic energy bands, in which the middle one corresponds to that of monolayer graphene (detailed discussions in Section 4.1). The low-lying linear and isotropic energy dispersions keep similar in the presence of an external electric field, as clearly displayed in Figure 9.1. That is, three pairs of

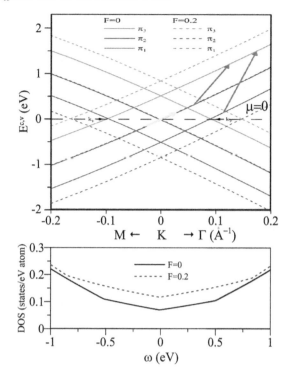

FIGURE 9.1
(a) Geometric structure of trilayer AAA stacking under a perpendicular electric field and (b) band structures in the presence/absence of $F\hat{z}$ by the dashed/solid curves. The blue, red, and green, respectively, indicate the first, second, and third groups of excitation channels.

Dirac-cone structures $[\pi_1^{c,v}, \pi_2^{c,v}, \pi_3^{c,v}]$, being initiated from the K/K′ points, reveal a similar energy spectrum under $F\hat{z}$. The Fermi level also crosses the Dirac points of the second pair/the middle one, in which the charge distributions are spatially localized on the two outmost layers with only very weak interlayer interaction of α_2. The first and third (deeper- and higher-energy) pairs present significant changes on the Fermi momenta, the free carrier density, and the Dirac-point energies, but with slight modification of the Fermi velocity ($k_F \sim 3\alpha_0/2$). With the increase in the strength of electric field, there are more free electrons and holes in the π_1 and π_3 energy bands, respectively. With respect to the π_2 band, the π (π_3) band has an energy downshift (upshift), with its Fermi momentum moving away from the K point. By the detailed calculations and analysis, the electric-field-dependent Fermi momentum is roughly given by $F \sim 2\sqrt{2\alpha_1^2 + (eFI_c)^2}/(3\alpha_0 b)$. The Dirac-point energies of the first and third pairs are, respectively, $-v_F k_F$ and $+v_F k_F$. The Fermi velocities are slightly affected by the interlayer nonvertical interaction of α_3 [20–22], leading to the weak e–h asymmetry. It should be noted that the linearly symmetric and antisymmetric superposition of the six sublattices are gradually destroyed by the increment of F-field strength [20, 21], leading to the significant changes of bare and screened response functions.

A perpendicular electric field shifts the Dirac points of the linear π_1 and π_3 bands, enhances the density of free carriers, alters the charge distributions on three layers, and thus alters the frequency, height, and number of asymmetric peaks in the polarization functions, as shown by the brown/orange curves in Figure 9.2 at $F = 0.2$ V/Å. The intralayer polarization functions $[P_{11} = P_{33}, P_{22}]$ and the interlayer ones $[P_{12} = P_{21} = P_{23} = P_{32}, P_{13} = P_{31}]$, in which they, respectively, characterize the charge correlations on the same and distinct layers through Coulomb interactions. In the absence of F, there are three groups of peak structures in bare response functions, corresponding to the specific excitation frequencies: $\omega_{ex}^1 \sim 3\alpha_0 bq/2$, $\omega_{ex}^2 \sim \sqrt{2\alpha_1^2} \pm 3\alpha_0 bq/2$, and $\omega_{ex}^2 \sim 2\sqrt{2\alpha_1^2} \pm 3\alpha_0 bq/2$. Such single-particle excitations are mainly due to the special Dirac points even in the presence of electric field. The first group, which comes from the intrapair intraband/interband excitations $[\pi_1 \rightarrow \pi_1, \pi_2 \rightarrow \pi_2, \pi_3 \rightarrow \pi_3]$ (the yellow arrows in Figure 9.1), exhibit the threshold excitations. Its specific ω_{ex}^1 is almost independent of F, but the peak height clearly displays the F-induced drastic changes by the variation of wave functions. The second group, being dominated by $[\pi_1 \rightarrow \pi_2 \ \& \ \pi_2 \rightarrow \pi_3$ (the brown arrows in Figure 9.1), present the double-peak structures centered at $\sqrt{2\alpha_1^2} - 3\alpha_0 bq/2$ and $\sqrt{2\alpha_1^2} + 3\alpha_0 bq/2$. The neighboring two peaks are induced by the disparity of the energy spacing between π_1/π_2 and π_2/π_3 (the distinct slopes in the left- and right-hand linear energy bands). The Dirac points of the first and third linear Dirac-cone structures are greatly enhanced by the increasing strength of electric field, and so do the specific excitation frequencies. Furthermore, asymmetric peaks exist in P_{22} and P_{12}; that is, they are, respectively, absent and present under $F = 0$ and $F \neq 0$ (the brown and orange curves in Figure 9.2c). As to the $\pi_1 \rightarrow \pi_3$ transition channels (the purple arrow in Figure 9.1), they belong to the highest-frequency excitations. Apparently, the frequency of the double asymmetric peaks is increased by F, and the field effect leads to a drastic change in peak intensities.

The screened response functions are greatly modified by an electric field, such as the frequency, strength and number of the prominent plasmon peaks. Without F (the brown curve in Figure 9.3), the first plasmon peak, with the lowest frequency, is strongest among three peak structures. It is purely due to the intrapair and intraband excitations of the π_1 and π_3 energy bands; that is, the strongest collective excitations originate from all free electrons and holes. Apparently, the frequency and intensity of this plasmon peak rapidly increases as the electric field increases from zero, e.g., the energy loss spectrum at $F = 0.2$ V/Å by the orange curve. Such 2D acoustic plasmon, which corresponds to the free carriers oscillating in-phase and the energy vanishing at zero momentum, could survive under any electric-field strength. The second plasmon peak is closely related to the second group of asymmetric peaks in P_{11} and P_{13} (Figure 9.2a and d), clearly indicating the collective charge

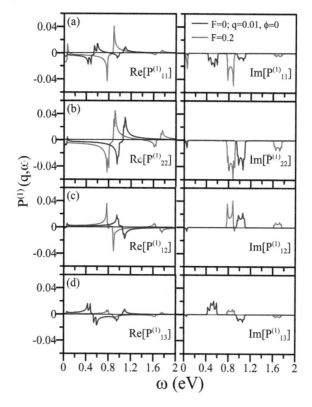

FIGURE 9.2
The independent matrix elements of layer-dependent polarization functions for the trilayer AAA stacking under $q = 0.01$ Å^{-1}, $\phi = 0°$, and $F = 0.2$ V/Å/& $F = 0$ by the orange/brown curves: (a) Re[P_{11}] & Im[P_{11}], (b) Re[P_{22}] & Im[P_{22}], (c) Re[P_{12}] & Im[P_{12}], and (d) Re[P_{13}] & Im[P_{13}].

oscillations on the first and third graphene layers. It is classified into the optical mode because of the finite frequency at $q \to 0$. Its frequency and intensity are largely enhanced by the electric field, since the occurrence of the second group of asymmetric peaks P_{22} and P_{12} (Figure 9.2b and c) brings about extra charge oscillations on the middle layer. The external field also creates an extra peak (marked by the orange star) to the left of the second plasmon peak. The generation of this new plasmon peak is the counterpart of the peak splitting of the second group of asymmetric peaks in polarization functions as discussed earlier, and the field enhances the splitting effects on plasmon excitations. As for the third plasmon peak, it has the highest oscillation frequency, but suffers quite a Landau damping. Such mode is associated with the third group of asymmetric peaks existing in all types of polarization functions (Figure 9.2a–d), which means that the collective oscillations occur on all layers. Oppositely, this optical plasmon strength is weakened by F (the orange

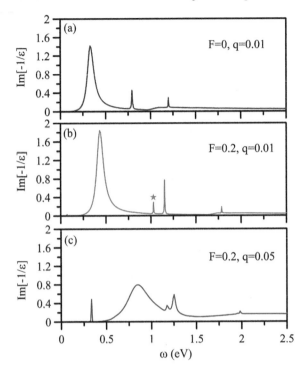

FIGURE 9.3
The energy loss spectra of the trilayer AAA stacking under various cases: (a) $F = 0$ & $q = 0.01$ Å^{-1}, (b) $F = 0.2$ V/Å & $q = 0.01$ Å^{-1}, and (c) $F = 0.2$ V/Å & $q = 0.05$ Å^{-1} by the brown, orange, and red curves, respectively.

curve), mainly owing to the descent of bare response function. By changing the transferred momentum to a proper range of 0.04 Å^{-1} $< q < 0.07$ Å^{-1}, another peak could become visible below the first major peak, as shown by the green line below a green star. Its appearance might lower the threshold excitation frequency and could be potentially important in real transport applications.

The momentum dependence of the single-particle and collective excitations is very important in fully understanding the diverse Coulomb excitation phenomena, especially for strong effects due to an electric field (Figure 9.4). The e–h excitation boundaries are mainly determined by Dirac points and Fermi-momentum states, as observed in most of the layered graphene systems. According to excitation frequencies in polarization function (Figure 9.2a–d), they, respectively, correspond to the $[\pi_1 \to \pi_1, \pi_3 \to \pi_3]$ $[\pi_1 \to \pi_2, \pi_2 \to \pi_3]$, and $[\pi_1 \to \pi_3]$ (the solid and dashed lines). Apparently, the second and third kinds of e–h boundaries present drastic changes during the variation of field strength. There exist three plasmon modes in the absence of F, in which the lowest one and the other two, respectively, belong to the acoustic and optical modes. Each plasmon mode experiences serious Landau damping due to the

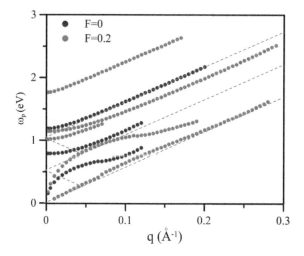

FIGURE 9.4

The momentum-dependent plasmon frequencies and intensities under $F = 0$ and $F = 0.2$ V/Å, respectively. Also shown are single-particle excitation boundaries of the dominating transition channels under the former case (blue dashed lines).

higher-frequency interband e–h excitations. The critical momentum is smallest (largest) for the acoustic plasmon (the higher-frequency optical plasmon). All the plasmons exhibit the F-enhanced oscillation frequencies, On the other hand, the critical momenta increase quickly for the acoustic plasmon and the lower-frequency optical plasmon, but the opposite is true for another optical mode. Specifically, an extra acoustic plasmon mode and a new optical plasmon mode are created by the external field simultaneously, in which the former is easily observed at a large momentum ($q > 0.02$ Å^{-1}).

The electric-field dependence of the plasmon modes in the N-layer AAA stacking deserves a closer examination, as clearly indicated in Figures 9.5a,b and 9.6a–c. The pristine AA-stacked graphene has one major acoustic plasmon (Figure 9.5a) and $N - 1$ optical plasmons (Figure 9.6a–c without the green arrows). The former is due to the free electrons/holes in the lower/higher Dirac-cone structures. Its frequency steadily increases with the increase in field strength (the orange, green, and purple dots). For bilayer and trilayer systems, it is well fitted by the relation of $\omega_p \propto N_d^{1/4}$ (the dashed orange and green curves), where N_d is the electron/hole density of the π_1^c/π_3^v [π_1^c/π_2^v] energy band for $N = 3$ ($N = 2$). N_d is proportional to the square of the Fermi momentum, being sensitive to the strength of the electric field. For layer number $N = 4$, the dependence of ω_p on f cannot be well characterized by a simple equation because the low-energy bands are obviously distorted. In other words, it is always necessary to consider the exact π-electronic structure.

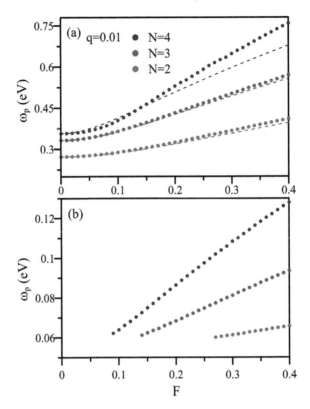

FIGURE 9.5
The F-dependent plasmon frequencies in the $N = 2, 3; 4$ few-layer graphene systems from the low to high curves for (a) the major acoustic plasmon and (b) the electric-field-induced one. Also shown by the dashed curves in (a) are those from the approximate relations.

Another lower-frequency acoustic plasmon, being shown in Figure 9.5b, is purely created by the splitting of the intrapair intraband and interband transitions under a sufficient high electric field. Its frequency depends on F in a roughly linear relation, directly reflecting the F-enhanced disparity of the linear Dirac cones. Furthermore, the critical field strength is mainly determined by the number of layer and transferred momentum.

As for the optical plasmons, the significant dependence on the electric-field strength in terms of their frequency and intensity is shown in Figure 9.6a–c. The frequencies of optical modes, being closely related to the energy differences between various Dirac-point states in linear bands (Figure 9.1), increase monotonously with the increasing F. Whether the intensity of energy loss spectrum is enhanced or reduced depends on the transition channel. The lowest optical plasmon dominated by the nearest-interpair interband transitions

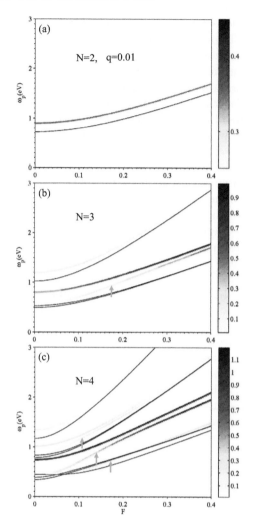

FIGURE 9.6
Similar plots as Figure 9.5a and b, but displayed for the optical plasmon frequencies and intensities in (a) $N = 2$, (b) 3, and (c) 4 graphene systems. The single-particle excitation frequencies are plotted by black solid lines. The green arrows indicate the field-induced optical modes.

is largely enhanced with its frequency gradually turning away from the single-particle boundary. Other higher-frequency optical plasmons, in contrast, are quickly weakened with their frequencies approaching the e–h boundary. Furthermore, they will disappear at larger Fs; that is, they are absent under serious Landau dampings beyond the critical electric fields. The extra branches indicated by the green arrows exist only for the cases of $N > 2$ and a sufficiently strong electric field. They carry strong spectral weights during the variation of F. Their appearance mainly originate from the nonuniform subband slopes and energy spacings associated with the different K states, which are the important result of considering all the significant interlayer atomic interactions and the exact π-band structures. With the increase of the layer number, the dispersions of the field-induced branches become more complex since more pairs of linear subbands are involved. In general, it could be observed in the energy loss spectra that most of the extra branches approach the major ones (those at $F = 0$) as F keeps on increasing, because the uniformity of the subband slopes and the energy spacings initiated from the K states is improved. In addition, there is no simple relation between the number of extra optical plasmon modes and layers.

To fully explore the concise pictures of the diverse Coulomb excitation phenomena, the AA-stacked bilayer graphene, which possesses the most simple stacking configuration, is suitable for analytic researches. The analytical derivations have been roughly done by the tight-binding model and the RPA (details in Refs. [20, 22]). These two methods are, respectively, adopted to evaluate the energy bands and wave functions, and the dynamic Coulomb screenings. An electric field, the intralayer & interlayer atomic interactions, and the intralayer & interlayer Coulomb interactions are taken simultaneously into account. According to the detailed calculations, a uniform perpendicular electric field could induce site energies in the diagonal Hamiltonian matrix elements, so that the $|\mathbf{k}|$ and F-dependent energy bands and wave functions are directly obtained by solving the 4×4 Hermitian matrix in the analytic forms. Apparently, there are two pairs of vertical valence and conduction Dirac-cone structures, in which the Dirac-point energies and the Fermi velocities are greatly modified by an electric field. Furthermore, the symmetric and antisymmetric linear superpositions of the four-sublattice tight-binding functions gradually disappear in the increment of F; that is, the equivalence of four sublattices would be lost during the variation F. And then, the layer-dependent polarization functions are expressed as analytic functions of (q, ω) under the long wavelength limit, in which the electronic energy spectrum is assumed to be isotropic, and so do the electronic excitations (independent of ϕ). They are classified into two kinds of transition channels: the intra-pair and interpair ones. The complicated single-particle excitation boundaries, being largely enriched by F, could be presented smoothly. They are useful in providing the full information about the spectral weight of the Landau damping. By setting the zero points of the determinant of the dielectric-function tensor matrix, three/two are revealed in the presence/absence of an electric field, and the q-, ω-, and F-dependent plasmon frequencies are given by the analytic formulas. The original acoustic and optical plasmons, respectively, present the frequency dispersions of $\sim c_1 q^{1/2} + c_2 q^{3/2}$ and $\sim c_3 + c_4 q + c_5 q^2$ Also, the extra acoustic mode exhibits $\sim c_{6_q} + c_7 q^2$. c_is are constants related to Fermi velocity, the Fermi momentum, the vertical interlayer hopping, and the electric-field strength. These collective excitations include an electrically inducible and tunable lowest-frequency mode, which could open possibilities for electronic nanodevice applications. The clear physical mechanism of the electrically inducible and tunable mode can be expected to exist in other AA-stacked few-layer graphenes.

9.2 ABA Stacking

The bilayer Bernal stacking is suitable for illustrating the electric-field-diversified Coulomb excitation phenomena in AB-stacked layered graphenes; furthermore, the trilayer ABA stacking is investigated in detail. It is difficult to observe the low-frequency acoustic and optical plasmon modes in the pristine semimetallic graphene systems, since only few electrons and holes are induced by the interlayer atomic interactions, or the DOSs of the band-edge states are not sufficiently high (details in Chapter 5). Although the electric field would induce the oscillatory energy bands and even open energy gap in bilayer systems, their very large DOSs are capable of creating very strong single-particle and collective excitations. The critical mechanisms, field strength, and transferred momenta will be thoroughly discussed in this section. Specifically, the momentum- and field-strength-dependent dispersion relations of the low plasmon frequencies are also evaluated by the zero determinant of the dielectric function matrix.

The AB-stacked bilayer graphene owns two pairs of parabolic conduction and valence bands, as clearly shown in Figure 9.7 by the brown curves $[\pi_1^{c,v}, \pi_2^{c,v}]$. The higher and deeper $\pi_2^{c,v}$ energy bands do not make significant contributions to the lower-frequency electronic excitations and thus is not a focus of the current discussion. The first pair, being nearest to the Fermi level, exhibits a very small band overlap. The minimum of the π_1^c subband and the maximum of the π_1^v subband are located at almost the same wave vector ($\sim$ the K point). These two subbands are greatly modified during the variation of field strength, in which energy dispersions change from monotonic parabolic dispersions into oscillating ones. This shape could also be regarded as a Mexican hat, being similar to those of AAB stacking in Chapter 7. In addition to the K point, two extra band-edge states exist along any direction. Concerning the π_1^c (π_1^v), there exists a band-edge state with the minimum (maximum) energy along the KΓ direction. Another band-edge state along the KM direction belongs to the saddle point. Such specific states are the critical points in the energy-wave vector space and thus have a rather high DOS. The energy difference between the lowest band-edge state of π_1^c and the highest one of π_1^v determines the size of bandgap, which is indicated by the blue dashed double arrows. E_g at first increases quickly and then slowly decreases with the increase of F. At the same time, the band curvatures of the saddle points along the specified direction appear larger and those of the

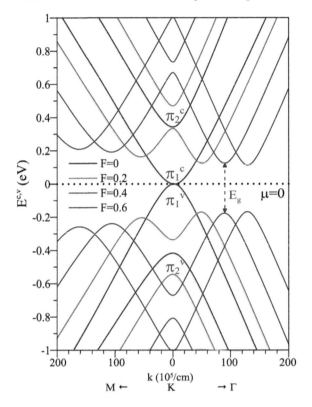

FIGURE 9.7
The low-lying valence and conduction bands of the bilayer AB stacking under various strengths of the electric field, and energy gap indicated by the blue dashed double arrows.

local extreme become smaller. Also, the asymmetry between the conduction and valence bands about E_F is enhanced under any F's.

The bare polarization functions are drastically modified by the electric field and transferred momentum. In a zero field, the layer-related independent response functions, P_{11} and P_{12}, are very weak within $\omega < 0.8$ eV (the brown curves in Figure 9.8a–d), mainly owing to the almost vanishing Coulomb matrix elements. However, the significant low-frequency responses are created by the sufficiently higher electric fields, such as those at $F = 0.2, 0.4$; 0.6 V/Å (the orange, blue, and purple curves). As to the imaginary part of the polarization functions, a pair of discontinuous step structures (a shoulder structure) and a pair of logarithmically divergent peaks (a symmetric peak) clearly exist at $F \geq 0.4$ V/Å ($F \leq 0.2$ V/Å), e.g. those indicated by the dotted and dashed lines in Figure 9.8a. The step structures mainly come from the local extrema (KΓ), while the logarithmically divergent peaks are due to the saddle points (KM). The main reason for the formation of twin structures

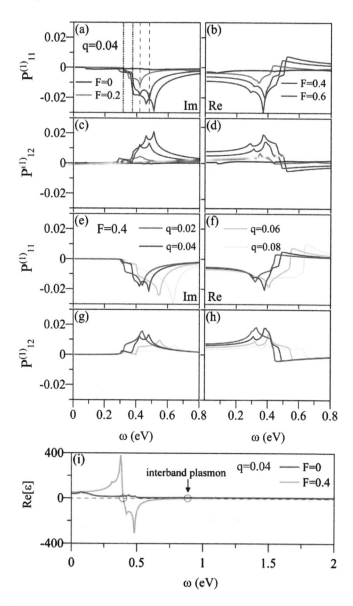

FIGURE 9.8
Polarization functions of bilayer AB stacking for the (a) imaginary and (b) real parts of P_{11} at $q = 0.01$ Å^{-1} and different F's. (c) and (d), respectively, correspond to those of P_{12}. (e)−(h) are related plots at $F = 0.4$ and distinct q's. The real part determinant of the dielectric-function 2×2 matrix is shown in (i). The temperature is zero and the energy width due to deexcitation mechanisms is 2 meV. The vertical lines in (a) indicate the positions of special structures at $F = 0.4$.

is the asymmetry of the band structures about the Fermi level and the non-identical band slopes at both sides of the critical points. By the well-known Kramers–Kronig relations, the logarithmic peaks in $Im[P_{11}]$ & $Im[P_{12}]$ correspond to the step discontinuities in $Re[P_{11}]$ & $Re[P_{12}]$, and vice versa. The intensities of the special structures increases with the increment of F, being associated with the rise in DOS or the number of excitation channels around the critical points. The polarization functions are also sensitive to the transferred momentum, as clearly indicated in Figure 9.8e–h. The intensity of P_{11} rises with an escalation of q; however, the opposite is true for that of P_{12}. Their opposite behaviors might be ascribed to the F-induced charge transfer. The first pair of twin structures, with the lower frequency, could survive at large q's, while the second pair is reduced to a single main structure. It should be noticed that as long as the excitation frequency is higher than the energy gap, the imaginary parts of the polarization functions are nonzero. This implies that both plasmons and single-particle excitations will always coexist in the momentum-frequency phase diagram. Most of the interband plasmon modes are expected to exhibit an unusual behavior.

The screened response function $Im[-1/\epsilon]$ or the dielectric function tensor is useful in determining the low-frequency collective excitations due to the partial π electrons in the first valence band (not the whole valence π electrons). Figure 9.8i clearly illustrates the real-part determinants of dielectric function matrix under the specific electric fields. From the comparison of $F = 0.4$ and 0, it can be seen that this field obviously creates divergent peaks, which represent the major excitation channels of $\pi_1^v \rightarrow \pi_1^c$, further resulting in two zero points in $Re[\epsilon]$ (marked with black circles). That zero point away from the peak structures (indicated by the black arrow) suffers a weaker Landau damping (judged from the imaginary-part value) and could induce plasmon mode, i.e., a prominent peak in the loss function (Figure 9.10). Assuming an equal charge distribution on the two layers (under weak field strengths and small transferred momenta), the intralayer and interlayer polarization functions have almost same weights, i.e., $P_{11} = -P_{12}$. The determinantal equation of the 2×2 dielectric-function matrix is thus expressed as

$$\epsilon = \epsilon_0[\epsilon_0 - V_{12}P_{12} - V_{22}P_{22} - V_{11}P_{11} + \text{(square terms)}]$$
$$\rightarrow \epsilon = \epsilon_0[\epsilon_0 - 2(V_{11} - V_{12})P_{11}]. \tag{9.1}$$

The dispersion relation of plasmon frequency can be obtained by setting the earlier equation to zero. The subtraction of the intralayer and interlayer Coulomb interactions ($V_{11} - V_{12}$) means that the field-induced interband plasmon should belong to an antibonding (out-of-phase) mode [26].

The energy loss spectrum, defined in Eq. (2.8), could provide the full information on the measured excitation spectra, especially for many-particle excitations. In a zero field, the screened response function has no prominent structures as a result of the too low DOSs (the brown curve in Figure 9.9a). The electric field creates one prominent peak and certain weak peaks or shoulders. The former corresponds to the higher-frequency zero point in $Re[\epsilon]$ with the weaker Landau damping (the arrow-indicated circle in Figure 9.8i). They are easily enhanced by the increasing electric-field strength. The prominent loss peaks (diamonds in Figure 9.9a) could be considered as collective excitations according to their apparent polarization shifts (peak frequency is much higher than the excitation energies of saddle points). Such interband plasmons resemble those in carbon nanotubes (discussed in Chapter 12), the π-plasmons in graphite (investigated in Chapter 11), and the inter-Landau-level plasmons in monolayer graphene and silicene (explored in Chapter 13; [32,34]). Whether they could be observed in the experimental measurements is directly inferred from the peak intensities of energy loss spectra. The weak peaks or shoulders, marked by triangles and crosses, possess frequencies close to those of $\pi_1^v \to \pi_1^c$ & $\pi_1^v \to \pi_2^c$ ($\pi_2^v \to \pi_1^c$) single-particle excitations, respectively. They are always obscure because of the rather strong Landau damping. On the other hand, the dependence of the plasmon strength on q is not monotonous. The most prominent plasmon peak becomes stronger as q increases from zero. However, when q goes through a specific critical momentum, it induces more single-particle transitions arising from $\pi_1^v \to \pi_2^c/\pi_2^v \to \pi_1^c$. The enhanced Landau

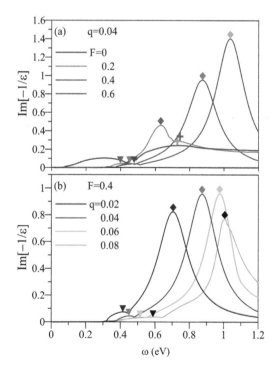

FIGURE 9.9
The energy loss spectra for bilayer AB stacking (a) at $q = 0.04$ Å^{-1} and various F's and (b) under $F = 0.4$ V/Å and distinct q's.

damping reduces the plasmon peak. This specific critical momentum could be raised by increasing the electric-field strength (discussed in Figure 9.10a).

It is worthy of a detailed comparison between this study and other works (done by other groups) on AB-stacked graphenes. The energy loss function discussed earlier is similar to the imaginary part of the trace of the full response function in the previously published papers [68]. However, there are some significant differences in Coulomb excitations. This work is focused on the occupied electronic states in undoped graphenes, while the others are on the conduction carriers in the doped cases. Moreover, their results have neglected certain important band-structure effects when evaluating the screened/bare excitation spectra. In this study, all the energy bands with the important inter-layer atomic interactions make contributions to the dynamic charge screening. As a result, more reliable plasmon frequencies and intensities are obtained from complicated calculations. Since the complete energy-band effects are considered, the external electric and magnetic fields can be added simultaneously in the energy loss spectra. That is, the field-diversified excitation

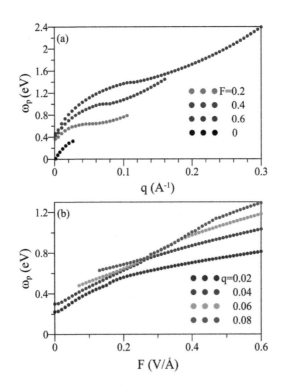

FIGURE 9.10
For the bilayer AB stacking, the dispersion relations of plasmon frequency with (a) the transferred momentum at various electric fields and (b) the electric field under distinct transferred momenta.

phenomena are directly revealed from the calculation results, e.g., the diverse magnetoexcitation spectra in Chapter 10.

There is a possibility of creating collective charge fluctuations on the two graphene layers by the applied electric field, being similar to that in the gated double-layer graphene with the opposite but nominally same potential. Under an extreme value of field strength, the large potential difference between layers might suppress the interlayer atomic interactions. In such conditions, the system might be considered as two independent graphene layers (without energy-band hybridizations), and the two layers interact with each other only by the interlayer Coulomb potentials. The other possible condition is that when the two graphene layers are separated far apart, the interlayer atomic interactions become negligible, as seen in Refs. [273, 274, 365]. The transformation of the excitation properties from the coupled layered graphene to the uncoupled system is another interesting topic.

The strong dispersions of the plasmon frequency and intensity with the transferred momentum and the electric-field strength are clearly indicated in Figure 9.10a and b. The frequency of the prominent interband plasmon induced by an electric field starts from a finite value under the long wavelength limit $q \to 0$. Apparently, such collective excitations belong to optical plasmon modes. According to the relation between the intralayer and interlayer polarization functions, $Im[P_{11}]$ and $Im[P_{12}]$ are, respectively, negative and positive at $F \neq 0$, meaning the density difference of two individual layers and the out-of-phase charge density fluctuations on the two graphene layers (Eq. 9.1). Apparently, the plasmon frequency displays a strong dependence on the transferred momentum, which indicates a propagating plasma wave. Increasing F could enhance the group velocity $[d\omega_p(q)/dq]$ at long wavelengths and prolong the propagation lifetime. The intensity of each plasmon peak at first increases with an increase in momentum. However, when q goes through a critical value, the frequency enters into the $\pi_1^v \to \pi_2^c$ & $\pi_1^v \to \pi_2^c$ excitation region, and the plasmon begins to decline and finally damps out. To create the prominent interband plasmon, the field strength needs to exceed a minimum value (Figure 9.10b). This value is lowered when q gets smaller. The main reason for this reduction is a weakening of the Landau damping from the higher-frequency interpair e–h excitations. It is also noted that this field-induced plasmon mode arises from $\pi_1^v \to \pi_1^c$ interband excitations, and its peak strength is relatively weak at $F \to 0$. The intensity of plasmon peak gradually increases with an increment of field strength, clearly indicating that the prominent plasmon peak hardly depends on other transition channels. The significant dependence of the plasmon mode on F and q could be further examined by inelastic X-ray scattering [165, 222, 225, 300].

The trilayer ABA stacking is easily modulated by a uniform perpendicular electric field, when compared with the bilayer one. The electronic properties and Coulomb excitations are more sensitive to F. The effects due to the layer-dependent Coulomb potentials are enhanced by the number of graphene layers. That the essential properties of a pristine ABA system could be regarded as the superposition of those in monolayer and bilayer subsystems [28, 31] is dramatically changed under electric fields but not under magnetic fields (discussed in Chapter 10). In general, for an odd-N multilayer ABA stacking, it is impossible to find electronic states, without contributions from the even-layer graphene, in the presence of F. That is to say, all electronic states are composed of tight-binding functions on the various graphene layers through very strong layer-dominated site energies. This will be clearly revealed in single-particle polarization functions.

Band structures of trilayer ABA stacking are thoroughly changed by the external electric fields, as obviously indicated in Figure 9.11. Apparently, two pairs of valence and conduction bands near the Fermi level are replaced by one pair of oscillator energy bands, with a very narrow energy gap (the first pair under F's). Furthermore, the second and third pairs have deeper/higher state energies, corresponding to sombrero-shaped/oscillatory and parabolic energy dispersions, respectively. In addition to the K/K' points, some extra band-edge states exist. Such critical points belong to the extreme points, and the constant-energy loops, respectively, leading to the shoulder or V-shaped structures (the parabolic or linear dispersions), the divergent peaks, and the square-root divergent peaks as the van Hove singularities in DOSs [28]. The low-energy DOS, which are clearly shown in Figure 9.12 under the various electric fields, exhibit the composite structures. At zero field, there exists the V-shape and shoulder structure near the Fermi level simultaneously

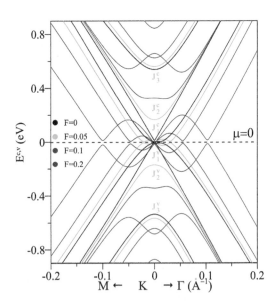

FIGURE 9.11
The low-lying valence and conduction bands of the trilayer ABA stacking under various fields: (a) $F = 0$, (b) 0.05, (c) 0.1, and (d) 0.2 (V/Å).

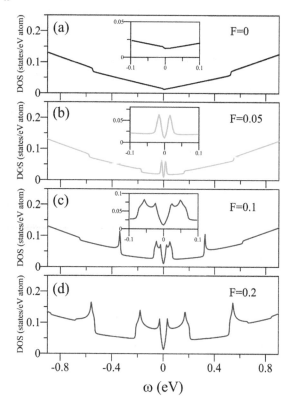

FIGURE 9.12
The low-energy density of states for trilayer ABA stacking at different electric fields: (a) $F = 0$, (b) 0.05, (c) 0.1, and (d) 0.2 (V/Å).

(Figure 9.12a). Furthermore, the composite of the V-shaped structure and asymmetric peaks are centered about E_F for a finite electric field (insets in Figure 9.12b–d). Such structures are directly reflected in the imaginary part of bare polarization functions. Most importantly, the slightly distorted Dirac cones (the black curves in Figure 9.12), mainly arising from the contributions of the first and third layers, quickly disappears by the action of an electric field. This clearly indicates that electronic states similar to those in monolayer graphene could not survive as a result of the strong combination of different graphene layers through various Coulomb on-site energies.

The electric field in trilayer ABA stacking can induce a very strong dynamic screening response, as obviously revealed in the bare polarization functions (Figure 9.13a–d). Under zero field, only an obvious shoulder structure in $\text{Im}[P_{lm}]$ appears at the higher frequency ($\omega \sim 1.10$ eV (the black solid curves), mainly owing to the interband transitions of the third pair of energy bands. With the increasing electric-field strength, the extra special

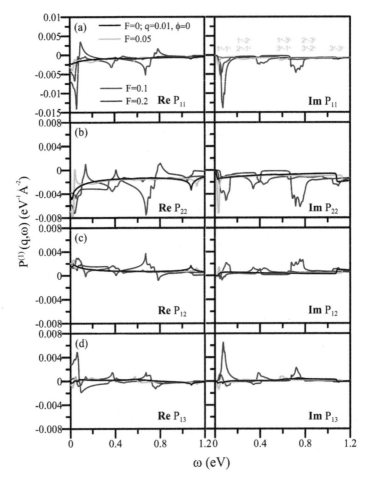

FIGURE 9.13
The independent matrix elements of the layer-dependent polarization functions for the trilayer ABA stacking under $q = 0.01$ Å^{-1}, $\phi = 0°$, and various F's: (a) Re$[P_{11}]$ & Im$[P_{11}]$, (b) Re$[P_{22}]$ & Im$[P_{22}]$, (c) Re$[P_{12}]$ & Im$[P_{12}]$, and (d) Re$[P_{13}]$ & Im$[P_{13}]$.

structures exist, or the weak structures are greatly enhanced. The observable structures cover shoulders and asymmetric peaks, being determined by the band-edge states in three pairs of energy bands. For example, at small F's (the green curves), there exist five excitation categories: (I) $\pi_1^v \to \pi_1^c$ at $\omega < 0.12$ eV, (II) $\pi_1^v \to \pi_2^c$ & $\pi_2^v \to \pi_1^c$ at 0.2 V $< \omega < 0.4$ eV, (III) $\pi_1^v \to \pi_3^c$ & $\pi_3^v \to \pi_1^c$ at 0.55 eV $< \omega < 0.7$ eV, (IV) $\pi_2^v \to \pi_3^c$ & $\pi_3^v \to \pi_2^c$ at 0.7 eV $< \omega < 0.85$ eV, and (V) $\pi_3^v \to \pi_3^c$ at $\omega > 1.0$ eV. It is also noticed that the $\pi_2^v \to \pi_2^c$ interband transitions disappear because of the vanishing Coulomb matrix elements. The first special structure, which belongs to an asymmetric peak with the specific twin form, is strongest among all van Hove singularities. The twin-peak structure is largely strengthened in the increment of F, since the first pair of energy bands becomes more anisotropic. The excitation regions of (III) and (IV) are merged together at stronger electric fields, e.g., those at $F \geq 0.2$ V/Å. In addition, the real-part polarization functions only exhibit corresponding responses according to the Kramers–Kronig relations.

Apparently, the energy loss spectra of trilayer ABA stacking are very sensitive to the electric-field strength and transferred momentum, as clearly shown in Figure 9.14a and b. A pronounced plasmon peak appears only under the sufficiently high F's (Figure 9.14a), in which it is closely related to the $\pi_1^v \to \pi_1^c$ interband transitions. The collective excitations experience the Landau damping of $\pi_1^v \to \pi_1^c$ at small transferred momenta, e.g., the F-dependent loss peaks at $q = 0.001$ Å^{-1}. The high-F plasmon modes obviously reveal the twin-peak structures, directly reflecting the two neighboring asymmetric peaks in the bare polarization functions (Figure 9.13) of the anisotropic energy spectra (Figure 9.11). Furthermore, there are weak but observable plasmon peaks at higher frequencies, such as the twin peak at

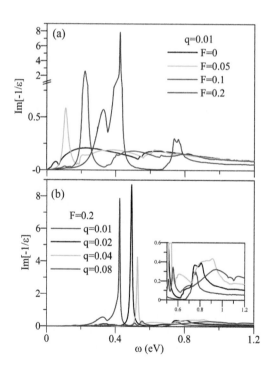

FIGURE 9.14

The energy loss spectra of the trilayer ABA stacking (a) at $q = 0.01$ Å^{-1} and various F's, and (b) at $F = 0.2$ V/Å and distinct q's.

$\omega_p \sim 0.75$ eV under $F = 0.2$ V/Å, mainly owing to the $\pi_1^v \to \pi_2^c$ and $\pi_2^v \to \pi_1^c$ interband excitation channels. Concerning the momentum dependence, the frequency of plasmon mode increases as q increases. However, there exist a nonmonotonous relation between intensity of plasmon peak and momentum. The main reason is that different e–h excitation channels exist during the variation of q.

The (q, ω)- and (F, ω)-phase diagrams, which cover the energy loss functions, could provide the full information about the existence of the single-particle and collective excitations, the great modification of plasmon modes, the boundaries of the available e–h transition channels, the critical momentum, and the critical electric-field strength. For the former in Figure 9.15a, the momentum-dependent boundaries of the distinct interband excitations clearly show the corresponding channels related to the most prominent plasmon mode, i.e., the lowest $\pi_1^v \to \pi_1^c$ transitions, with the largest joint DOSs (or the largest $\text{Im}[P_{lm}]$ in Figure 9.13), are responsible for the lower-frequency and strong plasmon mode. The plasmon under a finite F should belong to an acoustic mode because of $\omega_P \to 0$ at long wavelength limit. When the transferred momentum increases from zero, the plasmon intensity, being suppressed by the higher-ω $\pi_1^v \to \pi_1^c$ e–h excitations, increases quickly and even displays the twin-peak structure in loss function, e.g., the splitting, wide and high peak intensity at $q \sim 10$, as a result of the anisotropic energy spectrum (Figure 9.11). With an increase in q, the plasmon peak starts to experience the Landau damping of $\pi_1^v \to \pi_2^c$ & $\pi_2^v \to \pi_1^c$ at $q = 0.0013$ Å^{-1}. It reaches a maximum value in the further increment of q, and then the quick decay of plasmon peak leads to the vanishing at $q \sim 0.009$ Å^{-1}. Moreover, another weaker plasmon mode, which mainly comes from the $\pi_1^v \to \pi_2^c$ & $\pi_2^v \to \pi_1^c$ interband excitations, occurs at $\omega_p \sim 0.71$ eV at small q's. This optical mode is strongly damped by the $\pi_1^v \to \pi_2^c$ & $\pi_2^v \to \pi_1^c$ transitions, but not the almost vanishing $\pi_2^v \to \pi_2^c$ channels.

The splitting of the strong plasmon modes deserves a closer examination, being clearly illustrated by the electric-field-dominated phase diagram at small transferred momenta (Figure 9.15b). When the electric field gradually increases from zero, the only plasmon peak is quickly enhanced until $F \sim 0.15$ V/Å, in which the critical one in observing the significant response is about $F = 0.025$ V/Å. And then, this prominent peak starts to become the twin-peak structure, in which the upper and lower branches, respectively, present the decaying and amplifying behaviors. At sufficiently high electric fields ($F \geq 0.4$ V/Å), the lower-frequency mode are thoroughly changed into the dominating plasmon. That is, the splitting plasmon peaks could survive in the range of 0.15 V/Å $\leq F \leq 0.4$ V/Å. The F-induced dramatic transformation, the drastic change from the isotropic energy spectrum into the anisotropic one (Figure 9.11), is responsible for the unusual variation of plasmon mode. The first pair of valence and conduction bands is almost isotropic for $F < 0.15$ V/Å, so that all the constant-energy loops have the same contributions. But for the twin-peak region, such loops are splitting into those close

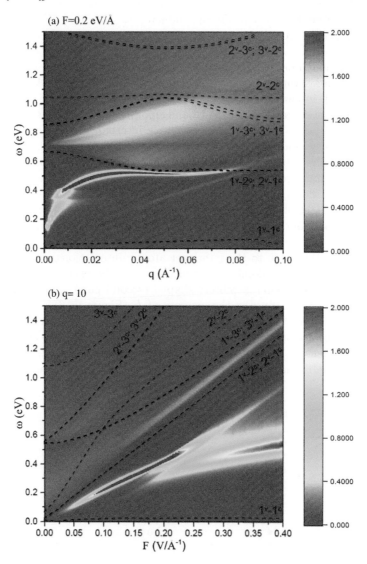

FIGURE 9.15
Concerning the trilayer ABA stacking, the dispersion relations of plasmon frequency with (a) q at various F's and (b) F under distinct q's.

to the KM and KΓ directions, respectively, leading to the lower and upper branches. The higher-frequency one is seriously suppressed by the interband $\pi_1^v \rightarrow \pi_1^c$ e–h excitations with the further increase of F. In addition, another weaker plasmon mode is observable only beyond the critical field of $F = 0.15$ V/Å.

9.3   ABC Stacking

To fully explore the electric-field-diversified Coulomb excitation phenomena, the significant differences among the trilayer ABC, ABA, and AAA stackings are thoroughly investigated, especially for the (q, ω)- and (F, ω) phase diagrams. They cover the F-dependent bandgap, energy dispersions, band-edge states (critical points), categories of available interband transitions, and frequency, intensity, and the number of plasmon modes. The predicted results could provide the delicate information for the experimental examinations.

The electronic properties of trilayer ABC stacking are greatly modified by the electric fields, as clearly illustrated in Figure 9.16. The semimetal–semiconductor transition exists in the presence of F, revealing the oscillatory field dependence, but not the monotonous gap variation. The F-induced gap opening could not survive in AAA and ABA stackings. This suggests the strong competitions due to the interlayer atomic interactions, stacking symmetries, and on-site Coulomb potential energies. The energy dispersions of the partial flat (first pair), sombrero-shaped (second pair), and linear bands present the drastic changes in the increase of F (Figure 9.16). They, respectively, become the oscillatory, more-deep sombrero-shaped, and parabolic bands. The surface states at $F = 0$ (details in Chapter 6), corresponding to very weak dispersions of the first pair (the black curves in Figure 9.16), mainly comes from the carriers on the first and third layers. They disappear under

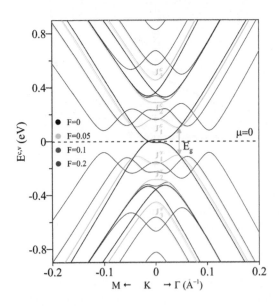

FIGURE 9.16
The low-lying three pairs of energy band for the trilayer ABC stacking at different electric fields.

a finite F, indicating the field-induced charge transfer among different layers. Such states are replaced by the extreme points and the constant-energy loops nearest to the Fermi level. The similar critical points are created in the second pair, while only the local maxima or minima appear for the third pair.

The electric field can greatly enrich the special structures in DOSs, as obviously displayed in Figure 9.17a–d. DOS at $F = 0$ presents two pairs of asymmetric peaks near the Fermi level and $\omega + 0.34$ eV and -0.34 eV (Figure 9.17a), respectively, corresponding to the partially and sombrero-shaped energy dispersions. A finite DOS at $E_F = 0$ clearly indicates a semimetallic behavior. F creates the vanishing DOS at E_F and two forms of van Hove singularities, in which the latter includes asymmetric peaks and shoulder structures (Figure 9.17b–d). It should be noticed that an asymmetric peak could also be regarded as a composite structure of the logarithmically divergent peak and shoulder. That is to say, a constant-energy loop, with the slight anisotropy, is the superposition of the extreme and saddle points. The special structures become more complicated with the increase of field strength. The important differences among the trilayer ABC, ABA, and AAA stackings

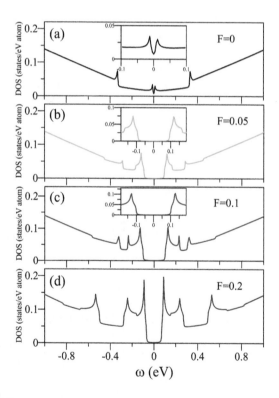

FIGURE 9.17
The low-energy density of states under various electric fields: (a) 0, (b) 0.05 (c) 0.1, and (d) 0.2 V/Å.

cover the number, form, and height of van Hove singularities, especially for the low-lying ones near the Fermi level. They could be verified by the scanning tunneling spectroscopy (STS) experimental measurements [80, 121, 281, 306].

The electric-field-dependent polarization functions exhibit the stacking-enriched bare response spectra, as clearly illustrated in Figures 9.2, 9.13, and 9.18. For the ABC trilayer stacking (Figure 9.18), $\text{Im}[P_{lm}]$ presents the shoulder structures and asymmetric peaks (or the composite structures of the shoulders and logarithmically divergent peaks), being similar to those in the trilayer AB-stacked system (Figure 9.13). However, the available excitation categories quite differ from each other. Under small F's (the green curves), the significant interband transitions are classified into four categories: (I) $\pi_1^v \to \pi_1^c$ at

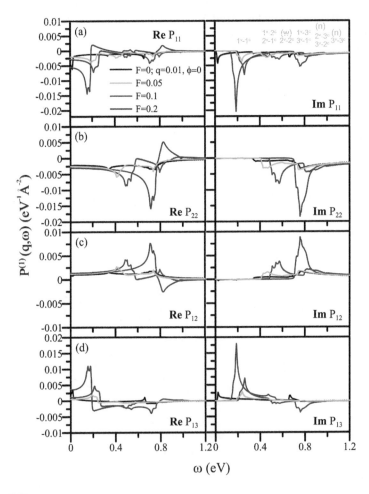

FIGURE 9.18
Same plot as Figure 9.14, but shown for the trilayer ABC stacking.

$\omega < 0.30$ eV, (II) $\pi_1^v \to \pi_2^c$ & $\pi_2^v \to \pi_1^c$ at 0.3 V $< \omega < 0.5$ eV, (III) $\pi_2^v \to \pi_2^c$ at 0.55 eV $< \omega < 0.65$ eV, and (IV) $\pi_1^v \to \pi_3^c$ & $\pi_3^v \to \pi_1^c$ at 0.7 eV $< \omega < 0.85$ eV. The interband excitation channels of $\pi_2^v \to \pi_3^c$, $\pi_3^v \to \pi_2^c$ & $\pi_3^v \to \pi_3^c$ only show the almost vanishing screening response. Among three kinds of stacking configurations, the important differences of polarizations lie in the intensities, categories, and special structures. At $F = 0$, the AAA trilayer stacking displays the strongest response (Figure 9.2), mainly owing to the highest free carrier density. The ABC, ABA, and AAA stackings, respectively, have four, five, and three excitation categories. The $\pi_2^v \to pi_2^c$ transition channel only in the ABC stacking can create special structures, and similar behaviors are revealed in the $\pi_2^v \to \pi_3^c$ & $\pi_3^v \to \pi_3^c$ ones for the AAA and ABA stackings. Moreover, the shoulder structures are absent in the AAA stacking as a result of linear energy dispersions under various F's. The predicted results on the diverse polarization functions could be verified by the optical spectroscopies [310–313].

There are certain important differences between the trilayer ABC and ABA stackings in the energy loss spectra. At $F = 0$, the screened response function is stronger for the former (the black curves in Figures 9.14 and 9.19). Furthermore, there are three F-induced observable plasmon peaks that exist simultaneously. They are closely related to the first pair of oscillatory energy bands

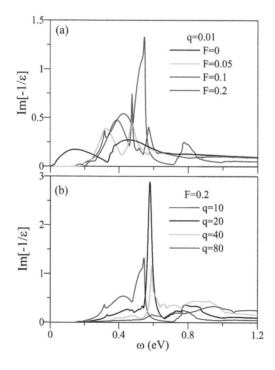

FIGURE 9.19
Same plot as Figure 9.15, but displayed for the ABC-stacked trilayer graphene.

(green, red, and blue curves in Figure 9.16). Such dispersions can induce the shoulder, symmetric peak and asymmetric peak/shoulder in the conduction (valence) DOS (Figure 9.17b–d); that is, three (four) van Hove singularities appear in the right hand of the Fermi level under $F \leq 0.1$ V/Å ($F > 0.1$ V/Å). The strong $\pi_1^v \rightarrow \pi_1^c$ interband transitions due to these large DOSs are responsible for the neighboring three plasmon peaks. However, only two splitting plasmon peaks are revealed in the ABA stacking (Figure 9.14), since the first pair of energy bands has less obvious oscillation (Figures 9.11 and 9.16). Moreover, another weaker plasmon mode, which mainly originate from the $\pi_1^v \rightarrow \pi_2^c$ & $\pi_2^v \rightarrow \pi_1^c$ excitations, occurs at a higher frequency (>0.7 eV). It is relatively easy to observe in the ABA stacking.

Apparently, the (q, ω)- and (F, ω)-dependent phase diagrams, as displayed in Figure 9.20a and b, are more complex in the ABC-stacked trilayer graphene. All the plasmon modes belong to the optical ones, since the available interband channels have excitation frequencies higher than the bandgap. The lower-frequency plasmons exhibit three splitting modes at $q \sim 0.001$ Å^{-1} (Figure 9.20a), directly reflecting the large and different joint DOSs in the $\pi_1^v \rightarrow \pi_1^c$ transition. Furthermore, only the highest-frequency submode could survive until the critical momentum of $q \sim 0.007$ Å^{-1}. The dispersion relation of plasmon frequency with momentum presents a monotonous increase at small q's, while it shows a slight decrease after $q > 0.003$ Å^{-1} and is even merged with the boundary of the $\pi_1^v \rightarrow \pi_2^c$ & $\pi_2^v \rightarrow \pi_1^c$ excitation channels. That is to say, such e–h excitations create serious Landau dampings and thus lead to the vanishing of plasmon mode. As to the electric-field dependence, the observable plasmons appear at the critical strength of $F \sim 0.05$ V/Å (Figure 9.20b). The splitting of plasmon modes could be observed only under sufficient high and suitable field strengths, e.g., three plasmon peaks at $F \sim 0.2$ V/Å and a specific $q = 0.001$ Å^{-1}. It should be noticed that the plasmon modes are damped by the $\pi_1^v \rightarrow \pi_2^c$ single-particle excitations at small q's. Another weak optical plasmon has $\omega_p > 0.4$ eV at $F = 0$ (Figure 9.20b), and its frequency shows a strong dependence on F, but not q (Figure 9.20a). This mode mainly comes from the $\pi_1^v \rightarrow \pi_2^c$ & $\pi_2^v \rightarrow \pi_1^c$ transitions and is also damped by themselves.

There are certain important differences among the trilayer AAA, ABA, and ABC trilayer graphene systems in the essential properties, clearly illustrating the stacking-diversified physical phenomena. Their interlayer hopping integrals are quite different from one another, and so do the low-lying energy bands (Figures 9.1, 9.11, and 9.16). For the electric-field-dependent AAA system, three pairs of vertical Dirac-cone structures remain similar, while the main changes are revealed in the first and third Dirac-point energies and the splitting of the Fermi velocities in each pair. Obviously, the free electrons and holes due to the interlayer atomic interactions can create one acoustic plasmon even at $F = 0$. The F-enhanced free carrier density and anisotropic energy spectrum even induce another weaker acoustic plasmon, and furthermore, two optical plasmons become three modes in the presence of F. The prominent

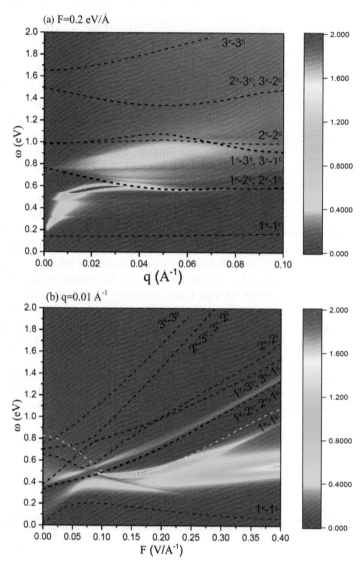

FIGURE 9.20
Same plot as Figure 9.16, but corresponding to the trilayer ABC stacking.

lower-frequency plasmon is absent in the pristine ABA and ABC stackings, except for the latter under very small transferred momenta, mainly owing to the quite low free carrier densities. The observable plasmon peaks are generated only under sufficiently high electric field, in which the critical field strength is 0.025 V/Å (0.05 V/Å) for the ABA (ABC) stacking. Specifically, this plasmon, respectively, belongs to the acoustic and optical modes in the

ABA and ABC stackings, since the semimetal–semiconductor transition only appears in the latter. Within a certain range of (F, q), such mode is split into two/three neighboring plasmons in the trilayer ABA/ABC stacking. As to another higher-frequency optical plasmon, it is strongly affected by F, but only a weak dispersion relation with q. Finally, the special structures of the bare polarization functions, being associated with the van Hove singularities in joint DOSs, cover shoulder structures and square-root divergent peaks (or composite structures of shoulder and logarithmically divergent peak) for the ABA and ABC systems, while only asymmetric peaks are revealed in the AAA one. All the significant single-particle responses of the ABA-, ABC-, and AAA-stacked systems are, respectively, classified into five interband excitation categories, four ones, and three interband and two intraband excitation categories. The theoretical predictions on the F-enriched Coulomb excitation phenomena are worthy of thorough examinations on the energy loss spectra [188, 189, 191–193, 195–198].

10

Magnetoelectronic Excitations: Monolayer and Bilayer Graphenes

A uniform perpendicular magnetic field plays a critical role on the essential physical properties. The magnetic quantization arising from $B_z\hat{z}$ belongs to a diverse but not monotonous phenomena [28–31, 82, 91, 339, 366–372]. This means that the Landau-level (LL) energy spectra greatly exhibit various B_z-dependences and noncrossing/crossing/anticrossing behaviors during the variation of field strength [29,30,339], and the magnetowave functions apparently show regular or extremely irregular probability distribution with a specific zero-point number [30], the coexistent major and minor mode [82,91], or the B_z-dependent zero points [30], and the magneto-optical absorption spectra clearly present well-behaved and extra absorption selection rules [28,31], the transport properties directly reveal the normal and unusual quantum Hall effects [366,368–372], and the magneto-Coulomb electron–electron (e–e) interactions obviously create rich and unique single-particle and collective excitations [32, 373–383]. The fundamental magnetoelectronic properties might be solved by the generalized model [28, 31] and the effective-mass model [12, 318, 327, 369, 384–386]. It should be noticed that the latter is very difficult to deal with the magnetically quantized states induced by the complex energy dispersions (e.g., the oscillatory, partially flat, and sombrero-shaped ones) [82, 339], the crossing/mixed energy bands [91], and the multipairs of valence and conduction bands near the Fermi level [82]. Unfortunately, most of the layered condensed-matter systems display such electronic structures. On the other hand, the former is successful in thoroughly exploring the diversified electronic, optical, transport, and Coulomb-excitation properties associated with the emergent 2D materials. For example, this model has conducted a full and systematic study on the few-layer graphene [28, 30, 31], silicene [29,387], germanene [29], phosphorene [29,388], and bismuthene [29,389]. Most important, the lattice symmetries, layer numbers, stacking configurations, planar/buckled structures, spin–orbital couplings, and single- and multiorbital hybridizations (hopping integrals) are taken into account simultaneously, being automatically consistent with the requirements of quantum statistics [29]. When the generalized tight-binding model is reformulated to link the linear static/dynamic Kubo formula, quantum transport behaviors/selection rules of absorption spectra could be obtained and analyzed. It is also combined with the modified random-phase approximation (RPA) to completely investigate novel magnetoelectronic excitations in layered graphene [23, 25]. Moreover, the concise physical pictures are proposed to comprehend the unusual essential properties.

Up to now, certain features in magnetoelectronic energy spectra and wave functions of few-layer graphene are identified from experimental measurements [299, 316, 390–395]. The powerful scanning tunneling spectroscopy (STS) has confirmed the square-root B_z-dependent LL energies in monolayer graphene [305, 316, 391], the linear B_z-dependence in bilayer AB stacking, the coexistent square-root and linear B_z-dependences in trilayer ABA-stacked graphene [316], and the 3D and 2D characteristics of Landau subbands in Bernal graphite [305, 306]. The aforementioned diverse phenomena, which are attributed to the massless and massive fermions, are directly reflected in the inter-LL vertical optical excitations. They have been verified by the magneto-Raman [299, 394, 396] and magneto-optical [390, 395] spectroscopies. The measured magneto-optical spectra show that there are a plenty of delta-function-like prominent absorption peaks corresponding to a specific selection rule ($\Delta n = |n^v - n^c| = \pm 1$). This provides useful information about the well-behaved LL wave functions. Moreover, magnetic transport measurements are conducted on monolayer [366, 368], bilayer, & trilayer AB-stacked [369–371] and ABC-stacked graphene systems [372]. The unconventional integer/half-integer Hall effects, with the step height of $4e^2/h$ or $2e^2/h$, are clearly shown to strongly depend on the layer number and stacking configuration. It is also noticed that the static scatterings arising from initial and final LLs dominate the Hall conductivities, so that they obey the same selection rule, as observed in the magneto-optical excitations. On the other side, the measured magneto-electronic excitation spectra are absent. This indicates the huge difficulties in measuring the inelastic electron/photon scatterings with a $B_z \hat{z}$-enriched sample.

The main focus of Chapter 10 covers the magneto single-particle and collective Coulomb excitations of monolayer graphene, and the layer- and configuration-enriched phenomena in bilayer AA and AB stackings. The temperature effects are included in the calculations of pristine systems without electron/hole dopings. This study clearly shows a lot of inter-LL electronic excitations from the occupied valence states to the unoccupied ones, as observed in the momentum-dependent polarization functions. The magnetoplasmon modes are precisely identified based on the resonant peaks in the dimensionless loss functions. The critical transferred momentum at which plasmon damping occurs is clearly established. It strongly depends on the magnetic field strength as well as LLs between which the transition takes place. At finite temperature, there exist plasma resonances purely induced by the Fermi distribution function. Whether such plasmons exist is mainly determined by the magnetic-field strength, temperature, and momentum. The magneto-Coulomb excitations are greatly diversified in bilayer AB and AA stackings. The energy-loss spectrum of the former is mainly dominated by the discrete inter-LL transitions, while in the latter a 2D-like plasmon involving the entire low-frequency Landau states is found. The plasmon, being similar to that of 2D electron-gas systems, is a result of highly symmetric stacking order and the dense LL distribution around E_F. The inter-LL plasmon and the 2D-like plasmon have a different dependence on q and B_z. Their behaviors are well understood from the fundamental single-particle excitations and screening effects in polarization functions. The calculated results presented in this chapter suggest that tunable collective-excitation properties in few-layer graphene could be achieved by the combined effects of the stacking order and the magnetic field.

10.1 Magneto Single-Particle and Plasma Excitations in Monolayer Graphene

The magnetodielectric function is thoroughly different from that in the absence of B_z. At zero temperature, a monolayer graphene without B_z exhibits the interband transitions due to the linear valence and conduction bands. The frequency-dependent $\text{Im}[\epsilon]$ and $\text{Re}[\epsilon]$ under a specific q are, respectively, divergent in the right-hand and left-hand square-root forms ($1/\sqrt{\omega - v_f q}$ and $1/\sqrt{v_f q - \omega}$, as clearly indicated by the red and blue curves in Figure 10.1a. This is purely due to the linear valence and conduction energy dispersions. On the other hand, the inter-LL excitation structures in the magneto-$\text{Im}[\epsilon]$ present a lot of delta-function-like symmetric peaks in Figure 10.1b–d. The main reason is that each LL is dispersionless and highly degenerate. Such structures further lead to many pairs of asymmetric peaks in the magneto-$\text{Re}[\epsilon]$ through the well-known Kramers–Kronig relations. All the special structures appear at specific excitation frequencies $v_f[\sqrt{n^c} + \sqrt{n^v}]\sqrt{B_z}$, regardless of the transferred momentum. However, the magnitudes of $\text{Im}[\epsilon]$ and $\text{Re}[\epsilon]$ are nonuniform functions of frequencies and transferred momenta, in which they are mainly determined by the LL-wave function-dependent Coulomb matrix elements in Eq. (2.13). In the single-particle excitation spectra, all the special structures are similar and appear at the same frequencies, while they might be strong or weak. If the zero point of $\text{Re}[\epsilon]$ occurs under a very small $\text{Im}[\epsilon]$, the almost undamped plasma resonance will represent in the screened response function.

A uniform perpendicular magnetic field in monolayer graphene creates a plenty of dispersionless LLs at low energy. For the specific magnetic states at $(k_x = 0, k_y = 0)$, such LLs are sufficient in exploring all the physical properties, since the contributions due to the distinct $\mathbf{k}$'s are identical. Each LL is eight-fold degenerate, corresponding to the $(1/6, 2/6, 4/6, 5/6)$ four localization centers accompanied with the degree of freedom. On the basis of the node structure of the well-behaved Landau wave functions (Figure 10.2b), the

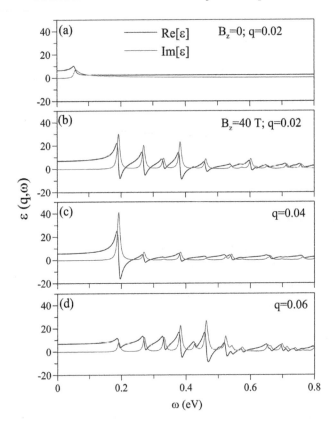

FIGURE 10.1
The magnetic real and imaginary parts of the dielectric function as a function
of frequency for various field strengths and transferred momenta: (a) $B_z = 0$
and $q = 0.04$, (b) $B_z = 40$ T and $q = 0.02$ Å^{-1}, (c) $B_z = 40$ T and $q = 0.04$
Å^{-1}, and (d) $B_z = 40$ T and $q = 0.06$ Å^{-1}.

critical quantum number n^v (n^c) for each valence (conductance) LL is characterized by counting the number of zero points in the spatial distribution;
this is the same as the number labeling the n-th occupied (unoccupied) LL
below (above). The low-lying LLs have the energy spectra approximately proportional to the square root of $n^{c,v}B_z$, being attributed to the linear Dirac-
cone structure [28, 30, 31]. In pristine monolayer graphene, fermion electrons
will be excited from valence LLs to conduction LLs through absorption of
an electromagnetic field, for example. However, the e–e Coulomb interactions could affect the single-particle transition mode with excitation frequency
$\omega = E(n^c, k + q) - E(n^v, k)$ and a momentum transfer of q. Each inter-LL
excitation channel is labeled (n^v, n^c) and the transition order $\Delta n = n^v - n^c$,
as depicted in Figure 10.2a. For example, (0,1) represents the transition from

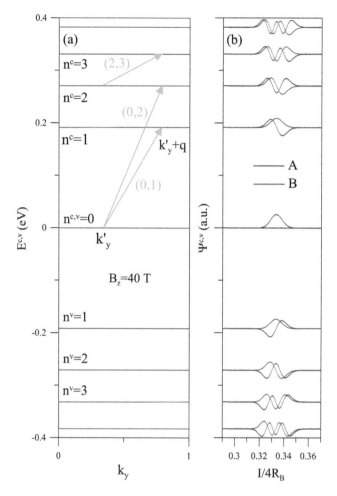

FIGURE 10.2
(a) The LL energy spectrum and (b) wave functions of monolayer graphene
at $B_z = 40$ T. Two subenvelope functions on the A and B sublattices show
well-behaved spatial distributions centered at 2/6.

the highest occupied LL to the second low unoccupied one and possesses the
same excitation energy as (1,0) because of the inversion symmetry between
the conduction and valence LLs. In collective mode excitations, the pair number denotes the channel with the largest contribution; this channel dominates
the excitation at a large or small limit of q. Specially, the $n^{c,v} = 0$ LLs at the
Fermi level are half filled in undoped graphene. The spin-up and spin-down
states are, respectively, situated at the conduction and valence bands if the
Zeeman effects were considered. Under charge density fluctuations, the (0,0)
excitation branch is absent because of the spin conservation.

The magnetic energy loss function, $\mathrm{Im}[-1/\epsilon\,(q,\omega,B_z)]$, is very useful for fully exploring the magnetoplasmon modes and the measured excitation spectra. A pristine monolayer system, as displayed in Figure 10.3a by the green curve, does not present any prominent peak at low frequency. However, an external magnetic field gives rise to certain prominent/observable peaks, e.g., the loss spectra under various q's and B_z's in Figure 10.3a–c. These peaks could be further classified into weak or strong collective magnetoplasma excitations based on the strength of charge resonance, or might be ascertained by their magnetoexcitation frequencies. The former modes exhibit frequencies close to the single-particle excitation ones (red dashed lines) subject to rather strong Landau dampings with a finite value in $\mathrm{Im}[\epsilon]$, or without an obvious zero point for $\mathrm{Re}[\epsilon]$. On the other side, the latter corresponds to a zero point in $\mathrm{Re}[\epsilon]$ and an almost vanishing value in $\mathrm{Im}[\epsilon]$ for the gap region between two single-particle inter-LL energies. The smaller the first derivative of $\mathrm{Re}[\epsilon]$ versus frequency at the plasmon frequency (ω_p), the higher will be the plasmon peak. Similar results have been revealed in electron energy loss calculations for plasma excitations of the 2D electron gas as well as 1D carbon nanotubes. The magnetopeak distribution, covering position, intensity, and number, is very sensitive to the transferred momenta (Figure 10.3a). For example, a simple

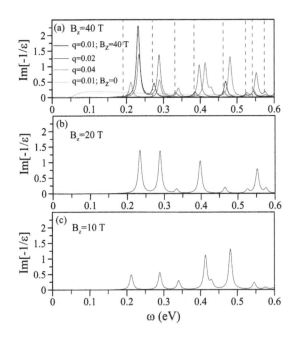

FIGURE 10.3
The energy loss functions at $B_z = 40$ T (a) for different transferred momenta in (a), and under various field strengths in (b) and (c). The green curve in (a) shows that of pristine graphene.

relation between the peak intensity/frequency and q is absent, being very different from that in the doped case (detailed in Section 3.1). This clearly demonstrates the strong competition of the longitudinal Coulomb interactions and transverse magnetic forces. Also, the energy loss functions are easily modulated by the magnetic-field strength, as indicated in Figure 10.3b and c. The threshold magnetopeak frequency declines and the number of significant peaks increases by lowering B_z. The main mechanism is that the LL energy spacing becomes narrow, or there are more dense LLs.

The magnetoplasmon frequencies, which arise from the significant peak positions in the energy loss function, obviously exhibit an abnormal dispersion relation with the transferred momentum. Figure 10.4 clearly shows the nonmonotonous dependence on $q's$ for the initial five collective excitation modes under $B_z = 40$ T. The inter-LL excitation frequency, being proportional to $\sqrt{n^c} + \sqrt{n^v}$, is higher, and so does the plasmon frequency. At long wavelength limit, the magnetoplasmon frequency is very close that of the inter-LL excitation energy (the red dashed line). With the increasing $q's$, it gradually increases and reaches a maximum value at the magnetic-field-dependent critical momentum (q_B). And then, ω_p declines and recovers to the original single-particle excitation frequency. When the transferred momentum is lower/higher than q_B, the magnetoplasmon wave propagates in the positive/negative group velocity. This directly reflects the dominating longitudinal forces due to the e–e Coulomb interactions under $q < q_B$, and the opposite is true for the transverse magnetic forces. The latter is fixed for a specific magnetic field strength. Furthermore, the former depends on the

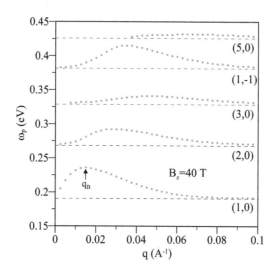

FIGURE 10.4
The transferred momentum-dependent magnetoplasmon frequencies for the initial five modes under $B_z = 40$ T.

bare Coulomb potential inversely proportional to q, and the Coulomb matrix elements associated with the initial and final LL wave functions.

The main features of magnetoplasmon modes are also sensitive to the magnetic-field strength, as clearly indicated for the critical momenta and the collective excitation frequencies, respectively, in Figure 10.5a and b, due to the initial four modes. The critical transferred momentum q_B's monotonously increases with the increase of B_z, directly reflecting the B_z-generated LL degeneracy (Figure 10.5a). The more carrier density in each LL will greatly enhance the Coulomb interactions, so that they could effective suppress the linearly increased magnetic forces. This is also the main mechanism for the enlarged magnetoplasmon frequencies (Figure 10.5b). ω_p is roughly characterized by the square-root B_z-dependence, as identified from the single-particle excitation energies of the inter-LL peaks in $\text{Im}[\epsilon]$ (details in Figure 10.1).

It is worthy of examining the important differences of magnetoelectronic excitations arising from the e–e Coulomb interactions and the

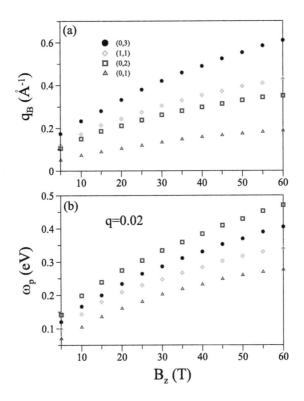

FIGURE 10.5
For the initial four collective excitation modes, the magnetic-field-dependent (a) critical momenta with zero group velocities and (b) magnetoplasmon frequencies under $q = 0.02\ \text{Å}^{-1}$.

electromagnetic wave. These two kinds of perturbations need to satisfy the conservation of energy and momentum and the Pauli principle (the Fermion statistics). The joint density of states, being proportional to the number of excitation channels, is due to the initial and final LLs, so it exhibits similar special structures (the delta-function-like prominent peaks) in Coulomb and optical excitations. As a result, the critical factor lies in the Coulomb matrix elements or the dipole/velocity ones. The former cannot prohibit any available channels, while the latter can induce the specific selection rule associated with the spatial distribution symmetries of the initial and final LLs, e.g., $\Delta n = n^c - n^v = \pm 1$ for the well-behaved LLs [30]. There are more strong peaks with a nonuniform intensity in the Coulomb response function. Moreover, the Coulomb interactions induce the transferred-momentum-dependent electronic excitations, in which the fluctuations of charge density have a wavelength $2\pi/q$. However, the optical processes belong to the vertical transitions under a long wavelength limit.

Free electrons and holes, which are, respectively, generated in the conduction and valence LLs, are increased with temperature. In addition to the very strong interband inter-LL transitions at higher frequencies, the intraband excitations from the $n^c \rightarrow n^{c'}$ LLs (the $n^v \rightarrow n^{v'}$ LLs) gradually increases with T, as clearly identified from the magnetodielectric function at lower frequency (<0.05 eV) in Figure 10.6a and b at $B_z = 5$ T. There are certain symmetric peaks in $\text{Im}[\epsilon]$ arising from the intraband inter-LL excitations (Figure 10.6b), since the conduction/valence LLs have sufficient electron/hole occupation probability at higher temperatures. The intraband magnetoplasmons are revealed as significant peaks in the energy loss functions of Figure 10.6c, in which the intensity/frequency of the most prominent one is obviously enhanced by an increase in temperature. The opposite behavior is true for the interband channels, e.g., the lowered intensity and the redshift frequency in the initial inter-LL of $n^v = 0 \rightarrow n^c = 1$ at $\omega \sim 0.07$ eV (Figure 10.6c). Moreover, the strength of the magnetic field has a strong effect on the intraband magnetoplasmons, since the free carrier density is reduced by the increased B_z/the enlarged LL energy spacing. It needs to require a sufficiently high temperature to observe the prominent plasmon peak. A critical temperature T_c monotonously increases with the increment of B_z, as shown in Figure 10.6d.

10.2 Layer- and Stacking-Enriched Magneto-Coulomb Excitations

The electronic excitations of bilayer graphene under a magnetic field are explored thoroughly using the generalized tight-binding model in conjunction with the modified RPA. The interlayer atomic interactions, interlayer Coulomb interactions, and magnetic field effects are simultaneously included

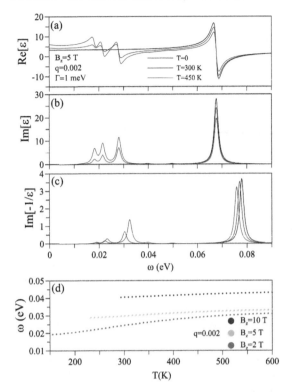

FIGURE 10.6

The temperature-dependent (a) Re[ϵ], (b) Im[ϵ]. (c) energy loss function at $B_z = 5$ T, and (d) intraband magnetoplasmon frequency, with the most strong intensity, at $q = 0.002$ Å^{-1} under the various B_z's.

in the dielectric-function matrix. That enables us to derive the magneto-Coulomb-excitation spectrum of different stacking structures. The two typical arrangements of bilayer graphenes, AB and AA, are considered in this chapter.

The two pairs of valence and conduction energy bands in bilayer graphene are magnetically quantized into two groups of dispersionless valence and conduction LLs, as clearly indicated in Figures 10.7 and 10.8. AB and AA stackings are very different from each other in energy bands and highly degenerate LLs. The former includes the following atomic interactions in the generalized tight-binding model calculations [301]: $\alpha_0 = -3.12$ eV, $\alpha_1 = 0.38$ eV, $\alpha_2 = -0.021$ eV, $\alpha_3 = 0.28$ eV, $\alpha_4 = 0.12$ eV, $\alpha_5 = -0.003$ eV, and $\alpha_6 = -0.0366$ eV (the definitions in Section 5.1). The method in solving the bilayer magnetic quantization is the same, when compared with monolayer graphene. Also, similar properties cover the hexagonal symmetry, the LL degeneracy, and the well-behaved LL wave functions. Specially, Figure 10.7a shows that the first/second group of valence and conduction LLs ($Sn_1^{c,v}S/Sn_2^{c,v}S$) are initiated from the electronic states near the Fermi level (roughly $-\alpha_1$ and $+\alpha_1$). The two LLs

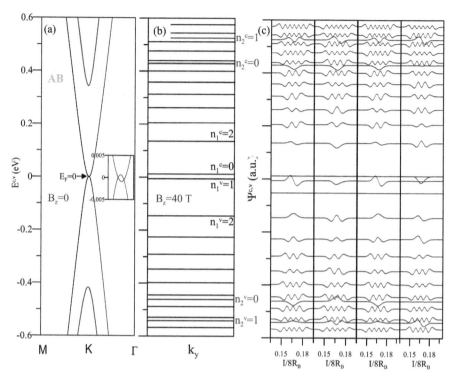

FIGURE 10.7

The pristine bilayer AB stacking: (a) the low-lying energy bands, (b) the two groups of LLs, and (c) the magnetic wave functions under $B_z = 40$ T.

nearest to $E_F = 0$ belong to the occupied $n^v = 1$ and unoccupied $n^c = 0$ ones. However, the higher/deeper-energy LLs of the second group starts from $n_2^v = 0$ and $n_2^c = 0$. Apparently, the LL energies directly reflect the low-lying parabolic dispersions (Figure 10.7a). As to the magnetic wave functions, they are characterized by the subenvelope functions based on the four (A^1, B^1, A^2, B^2) sublattices in the enlarged unit cell. Since A_1/B_1 and A_2 sublattices have the same/different (x, y)-projections, their spatial distributions exhibit similar modes/the distinct modes with the specific difference of zero-point number (± 2), as shown in Figure 10.7c. The quantum numbers are determined by the dominating sublattice. That is to say, $n_1^{c,v}$ and $n_2^{c,v}$, respectively, correspond to the $(B_1, B_2)/(A_1, A_2)/(B_1, B_2)/(A_1, A_2)$ and $(A_1, A_2)/(B_1, B_2)/(A_1, A_2)/(B_1, B_2)$ sublattices for the 1/6/2/6/4/6/5/6 localization center.

On the other hand, the two groups of LLs in the AA stacking strongly overlap near the Fermi level (Figure 10.8b), clearly arising from the magnetic quantization of two vertical Dirac-cone structures (Figure 10.8a). The C–C hopping integrals used in the generalized tight-binding model cover $\alpha_0 = 2.569$ eV, $\alpha_1 = 0.361$ eV, and $\alpha_3 = 0.032$ eV (details in Section 4.1). The electronic

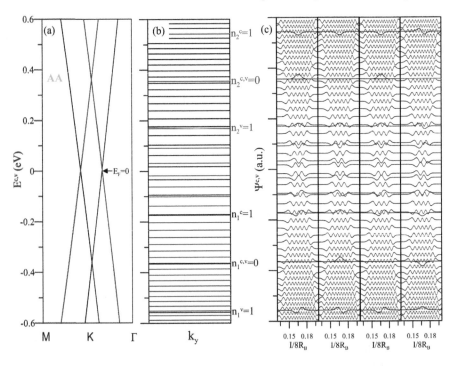

FIGURE 10.8
Similar plot Figure 10.7, but displayed bilayer for the bilayer AA stacking.

states, which are close to the two Dirac points with the energies $+\alpha_1$ and $-\alpha_1$, will become the initial $n_1^{c,v} = 0/n_2^{c,v} = 0$ LLs in the first/second groups. The LL energies, as measured from those of the initial ones, exhibit the square-root dependences on the quantum numbers and the magnetic-field strength, as observed in the monolayer graphene [30,367]. The Fermi level is approximately situated at the middle of two vertical Dirac points. After the magnetic quantization, E_F, which is determined by the first/second group, lies in between the highest occupied and lowest unoccupied valence/conduction LLs. It should be noticed that the unoccupied (occupied) valence LLs of the first (second) group above (below) E_F have highly degenerate free holes (electrons). Such free carriers are expected to induce unusual magnetoplasmon modes (discussed later in Figures 10.10 and 10.11). Moreover, there are simple relations between the A^1 and B^1 sublattices/the A^1 and A^2 sublattices, mainly owing to the identical chemical environment for any carbon atoms. The linear superpositions of their subenvelope functions are symmetric and antisymmetric (antisymmetric and antisymmetric) for the first groups of valence (conduction) LLs, and such combinations become symmetric and symmetric (antisymmetric and symmetric) for the second group of valence (conduction) LLs. The similar properties could be found in the zero-field wave functions.

The normal bilayer graphenes have four layer-dependent magnetopolarization functions, $P_{11} = P_{22}$ and $P_{12} = P_{21}$. P_{11} is sufficient in fully understanding the low-energy single-particle excitation properties. AB and AA stackings, as shown in Figure 10.9a–d, are very different in the inter-LL excitation channels. The imaginary/real part of P_{11} presents a lot of symmetric peaks/pairs of asymmetric peaks due to the available inter-LL transitions, being similar to the monolayer case. Such peak structures in the AB-stacked bilayer graphene mostly arise from the valence to conduction LLs of the first group. That is to say, the dominating intraband inter-LL excitations correspond to $n_1^v \rightarrow n_1^c$, such as the threshold channel of $(1,0)$. They might show twin-peak structures, e.g., the $(2,3)$ and $(3,2)$ excitations under the slightly antisymmetric LL spectrum about E_F (Figure 10.9a and b). The intergroup inter-LL excitations, corresponding to the first and second groups, are almost vanishing because of the negligible Coulomb matrix elements. As for the available channels due to the second group, they frequently appear at $\omega > 0.8$ eV ($\sim 2\gamma_1$). In addition, the peak structures in the interlayer polarization function (P_{12}), as shown in Figure 10.9e, are similar to those in the intralayer one. The differences between them only lie in their strengths because of the Coulomb matrix elements. Generally speaking, it cannot understand the direct relations of the layer-dependent polarization functions and the energy loss spectra.

The peak distributions in the layer-dependent polarization functions are drastically changed as the stacking configuration is transformed from AB to AA. As to the latter, a rather strong overlap of the first and second groups (Figure 10.8b) will induce unique low-lying inter-LL transitions (Figure 10.9c and d). The low-energy electronic excitations prohibit the intergroup channels, the $n_2^c \rightarrow n_1^v$ and $n_1^v \rightarrow n_2^c$ transitions, since the linearly symmetric or antisymmetric superpositions of the initial and final LLs just lead to the zero Coulomb matrix elements. They only come from the intragroup excitations of $n_1^v \rightarrow (n_1^v)'$ and $n_2^c \rightarrow (n_2^c)'$. The threshold channel at $B_z = 40$ T, as clearly shown in Figure 10.9c, corresponds to $(4,3)$ and $(3,4)$ with an almost identical excitation energy. It is very sensitive to the magnetic-field strength, depending on the conservation of electronic density. That is to say, the highest occupied LLs dramatically vary with B_z. The neighboring inter-LL excitation channels are relatively easy to observe under the larger transferred momenta. For example, certain significant peaks exist at $q = 0.03$ Å^{-1} in Figure 10.9d. On the other hand, the higher-energy Coulomb excitations present the intergroup inter-LL transitions originating from $n_2^c \rightarrow n_1^c$ and $n_2^v \rightarrow n_1^v$, e.g., the $(0,0)$, $(1,1)$, $(2,2)$ and $(3,3)$ peak structures near $\omega \sim 0.7$ eV. They hardly make important contributions to the magnetoplasmon modes (Figure 10.11b). Specifically, all the layer-dependent polarization functions of the bilayer AA stacking are featureless in the range of 0.1 eV $\leq \omega \leq 0.5$ eV at $q = 0.01$ Å^{-1} (0.2 eV $\leq \omega \leq 0.4$ eV at $q = 0.03$ Å^{-1}) (Figure 10.9c and d), while those of the bilayer AB stacking possess prominent peak structures within the whole frequency range. The former is thus expected to exhibit the quasi-2D plasmon

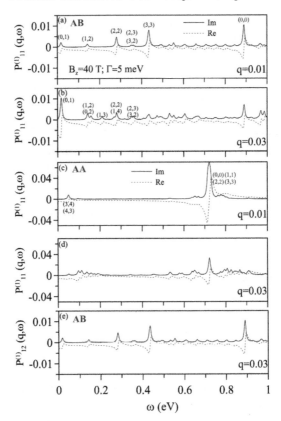

FIGURE 10.9
The first-layer-dependent polarization functions of $B_z = 40$ T for the AB stacking at (a) $q = 0.01$ Å^{-1} & (b) $q = 0.03$ Å^{-1}; for the AA stacking under (c) $q = 0.01$ Å^{-1} & (d) $q = 0.03$ Å^{-1}. Also shown in (e) is the interlayer-dependent ones of the AB stacking under $q = 0.03$ Å^{-1}.

modes even under the magnetic quantization, being in sharp contrast to the monolayer-like behavior in the latter.

The bilayer magnetoenergy loss function, defined by the layer-dependent polarization functions (details in Sections 2.1 and 2.4), could provide the dynamically screened response from the highly degenerate LLs involved by the charge carriers on the first and second layers. However, it is difficult/might be meaningless to identify whether there exist the in-phase or out-of-phase magnetoplasma oscillations on two layers. The bilayer AB stacking, as obviously displayed in Figure 10.10a, exhibits a lot of significant peaks in EELS. They could be identified as discrete magnetoplasmon modes due to the interband inter-LL transitions ($n_1^v \rightarrow n_1^c$) from the first group of LLs. Such peaks are

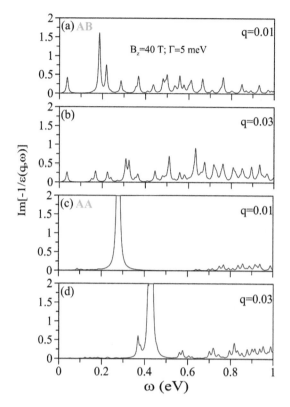

FIGURE 10.10
The magnetoenergy loss functions of $B_z = 40$ T for the AB stacking at (a) $q = 0.01$ Å^{-1} & (b) $q = 0.03$ Å^{-1}; for the AA stacking under (c) $q = 0.01$ Å^{-1} & (d) $q = 0.03$ Å^{-1}.

deduced to be similar to those in monolayer graphene (Figure 10.3). Apparently, both their intensities and frequencies are very sensitive to the transferred momenta, e.g., the $q = 0.01$ Å^{-1} and $q = 0.03$ Å^{-1} energy loss spectra in Figure 10.10a and b, respectively. For example, the threshold mode related to the $(0, 1)$ channel is observable within a large q-range. On the other side, the bilayer AA stacking displays fully different features. The magnetoenergy loss functions, which are indicated in Figure 10.10c/Figure 10.10d, present a rather unique and prominent peak at low frequency, being surrounded by the featureless spectrum/few lower-intensity peaks. This special plasmon peak is attributed to the direct superposition of all intraband and intragroup excitations due to the valence/conduction LLs of the first/second group; that is, it comes from all the free holes/electrons covered in the highly degenerate LLs. The discrete/quantized LLs can exhibit a plasmon behavior of 2D electron gas. The long-wave oscillations (small q's) might be the critical factor, since the longitudinal Coulomb forces are much stronger than the transverse magnetic ones.

The transferred-moment-dependent magnetoplasmon frequencies, which are directly obtained from the intensities of energy loss functions, clearly illustrate the spatial collective charge oscillations. In general, the hump-like abnormal q-dependences appear in the AB bilayer stacking (Figure 10.11a), as observed in monolayer graphene (Figure 10.4). There are a lot of discrete magnetoplasmon modes, indicating the strong competition of the Coulomb and Lorentz forces. The collective excitations are dominated by magnetic quantization; furthermore, they only originate from charge carriers in each LL, but not those of the neighboring LLs. The critical transferred momenta will become very complicated, except that the initial mode of $(0,1)$ hardly depends on the other modes because of the sufficiently large energy spacing. Two magnetoplasmon modes might have close energies, so that they are seriously affected by each other under the variations of the Coulomb matrix elements. For example, the entangled dispersion relations might exist in pairs of $[(0,2)$ & $(1,2)]$ and $[(0,4)$ & $(2,2)]$.

On the other hand, the magnetoplasmon frequencies of the AA bilayer stacking exhibits a specific dispersion relation for the insufficiently large momenta (e.g., $q < 0.01$ Å^{-1} at $B_z = 40$), as clearly displayed in Figure 10.11b. The $\sqrt{q}$ dependence of ω_p is relatively easy to be observed under a weaker magnetic field. Apparently, the magnetoplasmon modes at long wavelengths extremely resemble those of 2D electron gas, in which their oscillation frequencies are almost identical in the presence and absence of a magnetic field. This unusual phenomenon directly reflects the characteristics of free holes/electrons in the unoccupied/occupied valence/conduction LLs; that is, the magnetic quantization does not affect the full cooperation of all the free carriers. Also noticed that ω_p at $q \to 0$ is finite as a result of a small energy gap/a specific inter-LL excitation frequency. However, the discrete magnetoplasmons gradually appear with the increase in momentum, e.g., the initial four modes due to $(3,4)$, $(4,5)$, and $(5,6)$.

The magnetoplasmon modes might dramatically alter during the variation of the magnetic field strength, further illustrating the diverse quantization phenomena due to the stacking configurations. With the increase in field strength, the enhanced LL energies and degeneracies in the bilayer AB stacking create a monotonous dependence of ω_p and B_z, as clearly shown in Figure 10.11c. The magnetoplasmon frequencies gradually increase with the increment of B_z, while a simple relation between their intensities and B_z is absent. Moreover, under a specific transferred momentum (e.g., $q = 0.02$ Å^{-1}), certain magnetoplasmons could survive at lower field strengths, and then they are replaced by the neighboring modes at higher ones. The discontinuous B_z-dependences frequently appear, reflecting the strong effects of the magnetic field on the critical momentum and Landau dampings. For example, the first magnetoplasmon modes present a very weak loss spectra at $B_z < 20$ T because of the rather small q_B's. As AB stacking transforms into AA one (Figure 10.11d),

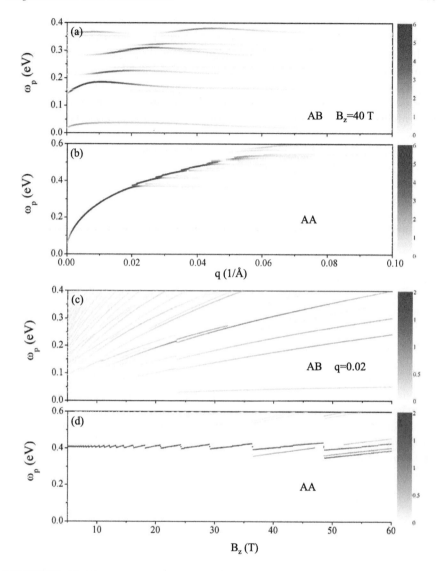

FIGURE 10.11

The magnetoplasmon frequency dispersion with the transferred momenta for the (a) AB and (b) AA bilayer stackings at $B_z = 40$ T. Similar plots in (c) and (d), but associated with the B_z-dependences at $q = 0.02$ Å^{-1}.

many discrete magnetoplasmons become a prominent one (a 2D-like plasmon), accompanied with certain quantized modes at higher B_z's. The discontinuous and oscillatory B_z-dependence apparently indicate the dramatic variation of the highest/lowest occupied/unoccupied LLs, as discussed earlier.

Furthermore, the magnetic dominance is revealed/cannot be deleted at strong fields.

The magnetoelectronic excitations under the Coulomb and electromagnetic (EM) wave perturbations are greatly diversified by the stacking configurations; that is, their differences are enriched by the bilayer AB and AA stackings. The similar available transition channels lie in the low-frequency spectral functions associated with the interband/intraband & intragroup inter-LL excitations in the AB/AA stacking, only reflecting the symmetric or antisymmetric linear superpositions of the subenvelope functions on four sublattices. However, most of excitation channels are further forbidden by the dipole matrix elements [33], but not by the Coulomb ones. The Coulomb excitation spectra present more prominent peaks, in which their magnetoplasmon modes of the AB/AA stacking exhibit a hump-like/square-root q-dependence. The 2D-like plasmon of an electron gas cannot survive in the optical absorption spectra. Moreover, the critical transferred momenta for the AB stacking are absent during the optical excitation process, so does the discontinuous B_z-dependence of magnetoabsorption frequencies. The aforementioned diverse excitation phenomena are closely related to the rather strong competition/cooperation between the longitudinal Coulomb field/the transverse EM wave and the transverse Lorentz force.

11

3D Coulomb Excitations of Simple Hexagonal, Bernal, and Rhombohedral Graphites

For three kinds of graphites, there are a lot of theoretical and experimental researches on the electronic properties and Coulomb excitations, clearly illustrating the diversified phenomena by distinct stacking configurations. According to the first-principles calculations [51, 338, 397], the AA-, AB-, and ABC-stacked graphites, respectively, possess the free carrier densities of $\sim 3.5 \times 10^{20}$, $\sim 10^{19}$, and 3.0×10^{18} e/cm^3, being in sharp contrast with one another. As a result of configuration-induced free electrons and holes, such semimetallic systems are predicted/expected to display the unusual low-frequency single-particle excitations and plasmon modes [113, 168, 234, 398–402]. The simple hexagonal graphite exhibits parallel and perpendicular collective oscillations relative to the graphitic planes, with frequency higher than 0.5 eV [31, 397, 403]. Furthermore, certain important differences between two distinct oscillation modes lie in the frequency, intensity, and critical momentum, as revealed in Bernal graphites [113, 401, 402]. Such plasmon modes are strongly modified by doping effects. The interlayer bondings become weaker in the natural graphite, so that few free carriers only show the lower-frequency plasmons of $\omega_p < 0.2$ eV. In addition to the transferred momentum, the collective excitations are very sensitive to the changes in temperature (T) [404, 405]. They could survive at larger momenta in the increase of temperature, and the frequencies are enhanced by T. The T-dependent plasmons are the prominent peaks in the energy loss spectra as well as abrupt edge structures of the optical reflectance spectra. Up to now, there is absence of theoretical predictions on the low-frequency Coulomb excitations of the ABC-stacked graphite. This study will provide the full information from random-phase approximation (RPA) calculations, e.g., the lowest plasmon frequency among three systems and the most difficult observations using electron energy loss spectroscopy (EELS) and optical spectroscopies. After the intercalation of atoms or molecules, the donor-type (acceptor-type) graphite intercalation compounds possess much conduction electrons (valence holes), so that their electrical conductivity might be high as copper [44–47]. The Coulomb excitations of free carriers have been investigated by the 2D superlattice model [234, 261], being responsible for the threshold edge in the measured optical spectra [231–233] and the $\omega_p \sim$ 1-eV optical plasmon in the momentum-dependent EELS spectra [37, 38, 167, 168, 215, 231–233].

The high-, middle-, and low-frequency plasmon modes in Bernal graphite have been measured and identified by optical spectroscopies [231–233, 406] (including reflectance, absorption, and transmission spectroscopies) and EELS [37, 38, 167, 168, 215, 231–233]. The π and σ bands in graphite are, respectively, formed by $2p_z$ and $(2s, 2p_x, 2p_y)$. The single-particle excitations and plasmons are examined by the optical absorption spectra [231, 232] and the energy loss spectra [168], respectively. Zeppenfeld and Buchner utilize the q-related transmission EELS, in which the transferred momentum is along the graphitic layers [37, 166]. These measurements clearly show that the π plasmon frequency is $\sim$ 7–12 eV and the $\pi + \sigma$ plasmon frequency is $\sim$ 27–32 eV. Two kinds of spectroscopies can identify the low-frequency plasmons of $\omega_p < 0.2$ eV. From the measured loss spectra (reflectance spectra) at room temperature, the AB-stacked graphite exhibits low-frequency plasmons at 45–50 meV (50 meV) and 128 meV for the electric polarizations parallel and perpendicular to the z-axis, respectively. Furthermore, the latter is strongly affected by temperature, as verified by the detailed T-dependent energy loss spectra [404]. Compared with graphite, graphite intercalation compounds possess rich excitation phenomena. Ritsko and Rice [169] thoroughly investigate the q-dependent energy loss spectra of stage-1 graphite $FeCl_3$ (acceptor-type system) and verify the intraband plasmon of $\omega_p \sim 1$ eV and the interband π plasmon. Also, there are certain EELS measurements of the stage-1 C_8M [170] (M for K, Rb, and Cs; donor-type systems), but under specific momenta. For example, the EELS [171] and reflectance measurements [172] on the stage-1 LiC_6 present $\omega_p(q_{\parallel} = 0.1 \text{ Å}^{-1}, q_z = 0) = 2.85$ eV, $\omega_p(q_{\parallel} \to 0, q_z = 0) = 2.65$ eV, $\omega_p(q_{\parallel} = 0, q_z \to 0) = 1.50$ eV [231].

The intralayer and interlayer atomic interactions due to the $2p_z$ orbitals are sufficient in thoroughly exploring the diverse excitation phenomena for three kinds of graphites. The 3D bare Coulomb interactions and the band-structure effects are fully included in the detailed RPA calculations. The electronic properties are evaluated by the tight-binding model, in which the significant parameters are fitted from the first-principles method. The 3D transferred momentum, $\mathbf{q} = [\mathbf{q}_{\parallel}, \mathbf{q_z}]$, is decomposed into the parallel and perpendicular components relative to $\hat{z}$, according to experimental measurements. The dependence of single-particle and collective excitations on the stacking configurations, the transferred momenta, temperatures, and doping levels are investigated in detail, especially for the magnitude and direction of $\mathbf{q}$. That is to say, the focuses cover the geometry-diversified fundamental properties, the rather strong anisotropy, the temperature-created/temperature-reduced intraband/interband excitations, and the doping-created intraband optical plasmons. To completely comprehend the mechanisms of the optical absorption arising from the low-frequency plasmon modes, temperature-dependent reflectance spectra are studied for AB-stacked graphite. For the stage-1 LiC_6 with the AA stacking, a lot of free conduction electrons are shown to drastically change the frequency and strength of the low-frequency intraband plasmons. Moreover, a 2D superlattice model is introduced to explain the features of optical intraband plasmons in stage-1 graphite $FeCl_3$. A detailed comparison with the previous experimental measurements using EELS [169] and reflectance spectroscopy [44, 231, 232] is also made. As for the low-frequency plasmons and π plasmons, the significant differences among the AA-, AB-, and ABC-stacked graphites (between graphene and graphite) are worthy of a thorough discussion.

11.1 Simple Hexagonal Graphite

The $2p_z$-orbital tight-binding model is suitable for studying the low- and middle-energy electronic properties of three kinds of graphites. For simple hexagonal graphite, all carbon atoms, as clearly shown in Figure 11.1a, possess the same chemical environment; therefore, there are two carbon atoms in a primitive hexagonal unit cell. The tight-binding π-band calculations are similar to those done for monolayer graphene. The Bloch wave function has been expressed in Eq. (3.3). The 2×2 Hamiltonian matrix in the subspace built by the two tight-binding functions is given by

$$H = \begin{pmatrix} \alpha_1\Gamma + \alpha_2(\Gamma^2 - 2) & (\alpha_0 + \alpha_3\Gamma)h(k_x, k_y) \\ (\alpha_0 + \alpha_3\Gamma)h^*(k_x, k_y) & \alpha_1\Gamma + \alpha_2(\Gamma^2 - 2) \end{pmatrix}, \qquad (11.1)$$

where $h(k_x, k_y) = \sum_{j=1}^{3} \exp(i\mathbf{k} \cdot \mathbf{r}_j) = \exp(ibk_x) + \exp(ibk_x/2)\cos(\sqrt{3}bk_y/2)$ represents the phase summation arising from the three nearest neighbors, $\Gamma = 2cos(k_zI_c)$ and $I_c = 3.35$ Å the periodical distance along the z-axis. From a detailed comparison between the tight-binding model and the first-principles method, the significant interaction parameters [397] cover $\alpha_0 = 2.569$ eV, $\alpha_1 = 0.361$ eV, $\alpha_3 = -0.032$ eV, and $\alpha_2 = 0.013$ eV (Figure 11.1a). They, respectively, originate from the intralayer nearest-neighbor hopping integral, the interlayer vertical atomic interactions of the neighboring layers, the interlayer nonvertical nearest-neighbor hopping integral, and the next-neighboring-layer vertical atomic interactions. By diagonalizing the 2×2 Hamiltonian matrix, the energy dispersions are characterized by

$$E_{\pm}^{c,v}(k_x, k_y, k_z) = \alpha_1\Gamma + 2\alpha_2(\Gamma^2/2 - 1) \pm (\alpha_0 + \alpha_3\Gamma)|h(k_x, k_y)|. \qquad (11.2)$$

Also note that both 3D AA-stacked graphite and 2D graphene exhibit identical wave functions. Their essential properties are different from each other, since the former shows a strong k_z-dependence.

There exist unusual low- and middle-energy electronic properties. One can do the expansion of energy dispersion about the corners, e. g., the K [$(k_x = 2\pi/3b, k_y = 2\pi/3\sqrt{3}b, k_z = 0)$] and H [$(k_x = 2\pi/3b,$

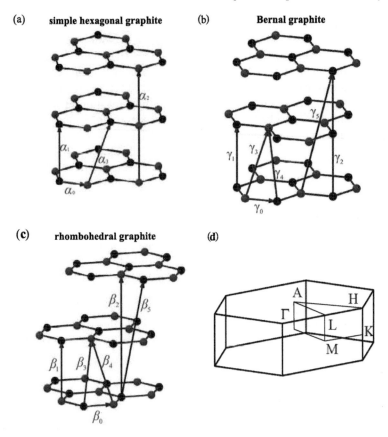

FIGURE 11.1
Geometric structures and various hopping integrals for the (a) the simple hexagonal graphite, (b) Bernal graphite, and (c) rhombohedral graphite, and (d) their first Brillouin zone.

$k_y = 2\pi/3\sqrt{3}b$, $k_z = \pi/I_c)$] points in the first Brillouin zone (Figure 11.2a). The linear Dirac-cone structures remain on the $k_x - k_y$ plane; furthermore, their Dirac-point energies strongly depend on the $\hat{k}_z$-direction wave vector. For the low-lying electronic states near the Fermi level (the dashed red curve), the K and H points are, respectively, situated at $2(\alpha_1 + \alpha_2)$ & $-2(\alpha_1 - \alpha_2)$; that is, the k_z-dependent Dirac-point energy width is $4(\alpha_1 - \alpha_2)$. There are a lot of free carriers due to the strong interlayer atomic interactions, in which valence holes and conduction electrons are, respectively, distributed near the K and H points. They create a U-shape van Hove singularity centered at E_F in density of states (DOSs) (Figure 11.3a; [31]). This structure could be directly verified by scanning tunneling spectroscopy (STS) measurements. A lot of conduction electrons/valence holes correspond to the neighboring states of K/H points, mainly owing to significant interlayer hopping integrals. They are expected to

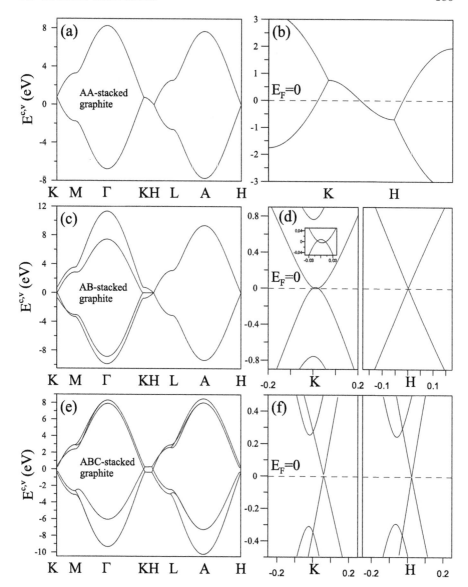

FIGURE 11.2
The whole-energy/low-lying band structures of the (a)/(b) AA-, (c)/(d) AB-, and (e)/(f) ABC-stacked graphites. The dashed red curves indicate the Fermi level, and the inset in (d) shows the band overlap.

create rich low-frequency electronic excitations. Concerning the middle-energy states, they are centered about the M and L points (the middle points between two hexagonal corners), in which their valence state energies (conduction ones) are, respectively, $-(\alpha_0 + 2\alpha_3) + 2(\alpha_1 + \alpha_2)$ and $-(\alpha_0 - 2\alpha_3) - 2(\alpha_1 + \alpha_2)$

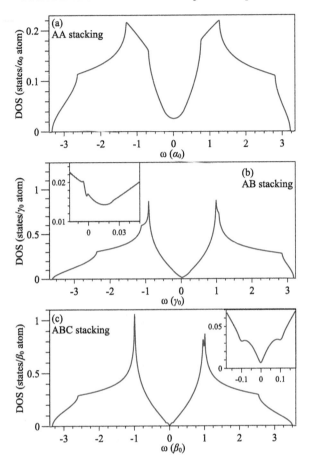

FIGURE 11.3
The density of states for the (a) AA- (b) AB-, and (c) ABC-stacked graphites.
The insets show the van Hove singularities near the Fermi level.

$[(\alpha_0 + 2\alpha_3) + 2(\alpha_1 + \alpha_2)$ and $(\alpha_0 - 2\alpha_3) - 2(\alpha_1 + \alpha_2)]$. Such k_z-dependent
points belong to the saddle points on the (k_x, k_y)-plane, This leads to a pair
of tilted temple structures in valence/conduction DOS (Figure 11.3a; [31]).
They will create the rather strong π-electronic excitations and π plasmon
modes, as revealed in few-layer graphene systems [20–23].

The low-frequency single-particle and collective excitations of the simple
hexagonal graphite are mainly determined by the low-lying electronic states
along the $K \to H$ line (Figure 11.2a). The isotropic energy spectrum clearly
indicates that Coulomb excitations are almost independent of the direction of
2D transferred momentum $\mathbf{q}_\parallel$; that is, $\phi = 0°$ is sufficient in understanding
the rich excitation phenomena. Furthermore, the temperature dependence is
negligible, mainly owing to the sufficiently high carrier density of free electrons

and holes. The zero-temperature dielectric functions, with the transferred momentum parallel and perpendicular to the graphitic plane, are chosen to clearly illustrate the fundamental electronic excitations. For the electronic excitations of $q_z = 0$, they are associated with the π-electronic states along the $M \to K/H \to L$ (Figure 11.2a). There exist two kinds of transition channels: intra-π-band $[\Psi_- \to \Psi_-/\Psi_+ \to \Psi_+]$ and inter-π-band ones $[\Psi_- \to \Psi_+]$. The Fermi-momentum states are the critical points in the energy-wave-vector space, leading to the special structure in the bare response function, as clearly shown in Figure 11.4a. The imaginary part of dielectric function (the dashed curve) exhibits a positively singular structure in the right-hand form arising from the dominating intra-π-band excitations. The specific excitation frequency is estimated to be $\omega_{ex} \sim 3\alpha_0 bq_\parallel/2$, almost identical to that of monolayer graphene (Section 3.1). On the other hand, the real part of dielectric function is negatively divergent near ω_{ex} and then slowly approaches to zero

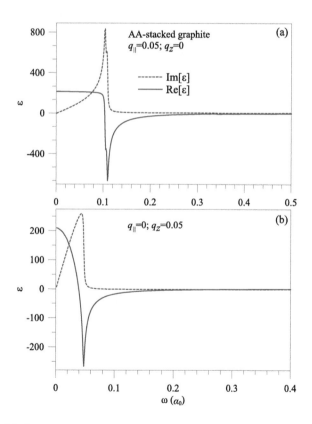

FIGURE 11.4
The low-frequency real and imaginary parts of the dielectric function for a simple hexagonal graphite under a broadening factor of $0.004\ \alpha_0$ and (a) $q_\parallel = 0.05$ Å^{-1} & $q_z = 0$; (b) $q_\parallel = 0$ & $q_z = 0.05$ Å^{-1}.

with an increase in frequency. $\text{Re}[\epsilon]$ is only vanishing at large $\text{Im}[\epsilon]$; that is, the plasmon mode experiences serious Landau damping.

When the transferred momenta are transformed in the perpendicular direction, the excitation phenomena are drastically changed. The dielectric function under the $q_\parallel = 0$ and q_z cases (Figure 11.4b and a) are quite different from each other. The former only presents one excitation channel, the intra-π-band excitations mainly due to the k_z-dependent Dirac points along KH (Figure 11.2a). The Fermi-surface states, corresponding to $E_\pm^{c,v}(k_x, k_y, k_z = 0.49\pi/I_c) = 0$, make the bare response function display a singular structure at $w_{ex} = E_\pm^{c,v}(k_x, k_y, k_z - q_z)$. The specific excitation frequency appears at lower frequency, when compared with the $q_z = 0$ case. As a result, the zero of $\text{Re}[\epsilon]$ behaves so. This result clearly indicates that collective charge oscillations along the z-axis have lower plasmon frequencies. $\text{Re}[\epsilon]$ might vanish at a small $\text{Im}[\epsilon]$; that is, the Landau dampings due to the intra-π-band transitions is very weak at w_p. Apparently, the aforementioned results in Figure 11.4a and b are responsible for the strong anisotropic Coulomb excitations in 3D graphite.

Obviously, the energy loss spectra in Figure 11.5a and b, being a direct measurement of the screened electronic excitations, exhibit a strong dependence on the direction and magnitude of the transferred momentum. The low and broad plasmon peaks are revealed in those under the planar momenta (Figure 11.5a), mainly owing to very small $\text{Re}[\epsilon]$ and a large $\text{Im}[\epsilon]$ at $\omega \sim \omega_p$. The collective charge oscillations on the graphitic planes are heavily damped by the inter-π-band electron–hole (e–h) excitations. The plasmon frequency increases with an increase in momentum, while the opposite is true for its intensity. The single-particle excitations rapidly increase with momentum, so that the plasmon peak would almost disappear beyond the critical momentum, e.g., the screened response spectrum at $q_\parallel = 0.15 \text{ Å}^{-1}$. Moreover, the 3D layered graphite presents a significant anisotropy. The plasmon modes, which correspond to the collective carrier oscillations along the z-axis, display a prominent peak (Figure 11.5b), and they can survive at large momenta. These are attributed to the absence of inter-πband single-particle excitations. In short, the strength of the low-frequency plasmon is mainly determined by the existence of higher-energy inter-π-band transitions.

The low-frequency plasmons, which are due to the collective excitations of free electron and holes, exhibit a rich momentum dependence. The collective oscillation modes parallel and perpendicular to the graphitic layers are, respectively, displayed in Figure 11.6a and b. The significant dispersion relations directly reflect the π-band characteristic, the strong wave-vector dependence. The plasmon frequency approaches to a finite value under $q \to 0$, being similar to an optical mode in a 3D electron gas. The important differences between parallel and perpendicular modes cover the critical momentum, the plasmon frequency, and the strength. The former and the latter could survive at $q_c \sim 0.12$ and $q_c \sim 0.8 \text{ Å}^{-1}$, in which the $q_\parallel = 0$ case presents the prominent plasmon peak and a very strong q-dependence. In short, the main features of

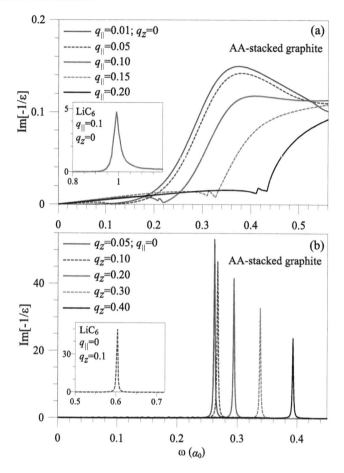

FIGURE 11.5
The energy loss spectra of the simple hexagonal graphite under (a) various $q_{\parallel}$'s and $q_z = 0$ and (b) various q_z's and $q_{\parallel} = 0$. For a detailed comparison, the insets shows those of LiC_6.

3D energy bands are responsible for the unusual behaviors of the dielectric function, the energy loss spectrum, and the momentum-dependent plasmon frequency.

A semimetallic hexagonal graphite sharply contrasts with a semiconducting monolayer graphene in Coulomb excitation phenomena. The significant differences lie in the existence of plasmon, the feature of oscillation mode, the plasmon frequency, and the critical momentum. Free carriers in graphite, being induced by interlayer atomic interactions, could create a low frequency even at zero temperature. However, the low-frequency plasmon in graphene is purely due to a finite temperature. Plasmons in these two systems, respectively, belong to the optical and acoustic modes, since the collective carrier

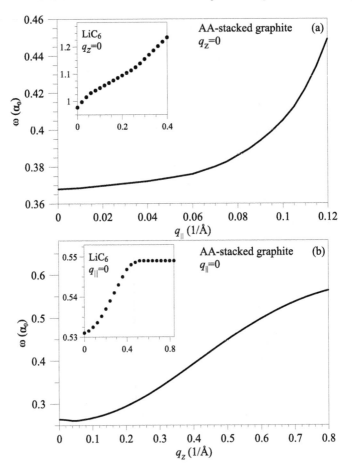

FIGURE 11.6

The momentum-dependent plasmon frequencies for simple hexagonal graphite at (a) $q_z = 0$ & (b) $q_\| = 0$, and those of LiC_6 in the insets.

oscillations of graphite mainly come from the superposition of all hexagonal layers. As a result, optical plasmons have higher frequency [$\omega_p \sim 0.5 - 1.0$ eV & $\omega_p \sim 0.05 - 0.10$ eV] and higher critical momentum [$q_c \sim 0.1 - 0.8$ Å^{-1} & $q_c \sim 0.02 - 0.04$ Å^{-1}].

Lithium graphite intercalation compound, LiC_6, has an AA stacking configuration in which the periodical distance between two neighboring graphitic layers is about 3.37 Å, being close to 3.35 Å of Bernal graphite. Apparently, I_c is slightly increased after the intercalation of Li atoms, but the free electrons are largely transferred from Li to carbon atoms. According to the modified I_c, the interlayer hopping integrals are reduced by the Harrison's rule [44]; that is, their strengths are inversely proportional to the distance between two carbon atoms. This compound belongs to the donor-type system. The charge transfer per Li atom estimated from various experimental measurements covers a wide range of $\sim 0.43 - 1.0$ [44], and the average value of ~ 0.7 is taken for a model study, corresponding to the Fermi energy of about 1.98 eV. The 3D conduction electron density is evaluated to be $\sim 3.6 \times 10^{22}$ e/cm^3. Electrons excited from the occupied states are excited to the unoccupied ones with energies higher than E_F, so the range and intensity of inter-π-band excitations largely decline. The plasmon peaks are greatly enhanced in the energy loss spectra (the insets in Figure 11.5a and b), since the Landau dampings due to higher-energy inter-π-band excitations exhibit an opposite behavior. More high free carrier density also leads to a great enhancement of plasmon frequency and critical momentum, as indicated in the insets of Figure 11.6a and b. The plasmon frequencies have been measured by EELS [171] and reflectance spectra [172] under the specific momenta. The experimental results $[\omega_p (q_\| = 0.1)$ Å^{-1}, $q_z = 0)$=2.85 eV [171], $\omega_p (q_\| \to 0, q_z = 0)$=2.65 eV [171,172] and $\omega_p (q_\| = 0$, $q_z \to 0)$=1.50 eV [171] are roughly consistent with the theoretical predictions [2.6 eV, 2.45 eV; 1.33 eV].

The π-electronic excitations, which are due to all the valence carriers or are associated with the middle-energy saddle points with the high DOSs (Figure 11.3a–c), are worthy of detailed examination. The large broadening factor of Γ=0.1 α_0 (Eq. 2.9), being responsible for various deexcitations of high-energy excited states, is used in the calculations. The wide-frequency-range dielectric functions strongly depend on the direction and magnitude of transferred momentum, as clearly displayed in Figure 11.7a–d. There are two or three special structures in ϵ. For example, at $[q_\| = 0.2$ Å^{-1}, $\phi = 0°$ & $q_z = 0]$ and $[q_\| = 0.6$ Å^{-1}, $\phi = 0°$ & $q_z = 0]$ (Figure 11.7a and b), Im[ϵ]/Re[ϵ], respectively, presents the asymmetric peaks/the dip-like structures at [0.5 α_0 & 2.0 α_0] [1.2 α_0, 1.95 α_0 & 2.65 α_0]. For $q_z = 0$ and $\phi = 0°$, the lower-frequency special structures mainly originate from the excitation channels along K→K and H→H. Furthermore, the higher-frequency ones are due to the M→M' and M→ Γ channels, in which the former forms a constant-energy loop with a specific energy difference of $2\alpha_0$, and both channels might exhibit the merged special structure at small transferred momenta (Figure 11.7a). It should be noticed that at large $q_\|$'s, the M→M' and K→K channels would make comparable contributions to the lower-frequency special structure [115]. The bare response function is sensitive to the direction of the planar momentum, such as two special structures in ϵ under $q_\| = 0.2$ Å^{-1}, $\phi = 30°$ & $q_z = 0$ (Figure 11.7c). Specifically, the lower- and higher-frequency ones are, respectively, created by K→Γ and M→Γ. When the transferred momenta become perpendicular to the (k_x, k_y)-planes, the electronic excitations dramatically change. At small q_z's and $q_\| = 0$ (the blue curves in Figure 11.7c), there is only one low-frequency special structure in ϵ; however, the π-electronic excitations arising from the M and L saddle points almost disappear, owing to the vanishing Coulomb matrix elements (the first term in the integration in Eq. (2.9)).

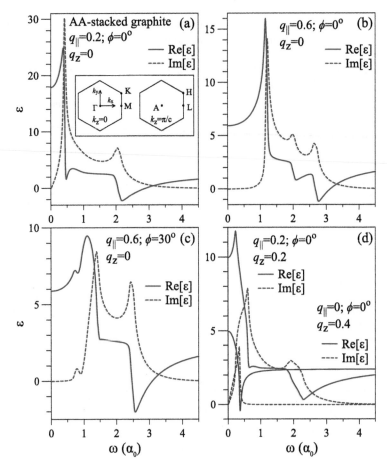

FIGURE 11.7
The wide-frequency-range dielectric function of the simple hexagonal graphite under (a) $q_{\parallel} = 0.2$ Å^{-1}, $\phi = 0°$ & $q_z = 0$, (b) $q_{\parallel} = 0.6$ Å^{-1}, $\phi = 0°$ & $q_z = 0$, (c) $q_{\parallel} = 0.6$ Å^{-1}, $\phi = 30°$E & $q_z = 0$, and (d) $q_{\parallel} = 0.2$ Å^{-1}, $\phi = 0°$ & $q_z = 0.2$ Å^{-1}, corresponding to the real and imaginary parts by the solid and dashed red curves. $q_{\parallel} = 0$ Å^{-1}, $\phi = 0°$ & $q_z = 0.4$ Å^{-1} is also shown for comparison (the blue curves). The insets indicate the first Brillouin zones on the $k_z = 0$ and π planes.

They exist only under a sufficiently large $q_{\parallel}$, such as $q_{\parallel} = 0.2$ Å^{-1}, $\phi = 0°$ & $q_z = 0.2$ Å^{-1} in Figure 11.7d by the blue curves. The middle-frequency special structure is closely related to the nonvertical saddle-point transitions initiated from electronic states near the M/L points.

The π plasmons of the AA-stacked graphite, being due to the collective excitations of all valence electrons, exhibit unusual peaks in the energy loss spectra at ω_p higher than $2\alpha_0$ (Figure 11.8a–f). For the planar momentum transfer (Figure 11.8a and b), the frequencies of the π-plasmon peaks rapidly increase with an increase of $q_\|$. However, their intensities at $\phi = 0°/\phi = 30°$ present a quick/gradual decrease, mainly owing to very strong (weak) Landau damping arising from the higher-frequency inter-π e–h excitations. Similar collective excitations are revealed in monolayer graphene, while the π-plasmon strengths might have an opposite $q_\|$-dependence (Figure 11.8c). Such plasmon modes are expected to vanish under the long wavelength limit $q_\| \to 0$ as a result of very low plasmon peaks. On the other hand, as to the perpendicular momentum transfer, the π plasmon peaks obviously appear in the energy loss spectra only when the parallel component is sufficient large, such as $q_\| = 0.2$ Å^{-1} in Figure 11.8d, but not $q_\| \to 0$ (the black circles). The normal relation between the frequencies of the π-plasmon peaks and q_z under a specific $q_\|$ is absent in Figure 11.8d–f. That is to say, the decrease or enhancement of ω_p with q_z is closely related to $q_\|$.

The theoretical predictions on the momentum-dependent π plasmon frequencies, as clearly shown in Figure 11.9a–c, could provide for experimental examinations. A simple hexagonal graphite exhibits a positive dispersion relation in the variation of the π-plasmon frequency with the parallel momentum transfer ($q_z = 0$ in Figure 11.9a). The anisotropic property, the dependence of ω_p on ϕ, is revealed only at $q_\| > 0.5$ Å^{-1}. Such plasmons could survive at $q_\| \to 0$ and have $\omega \sim 2.90 \, \alpha_0$ much higher than $2\alpha_0$. Their critical momentum ($q_{\| \, c}$ is longer than 1.5 Å^{-1}. However, the π plasmons in monolayer graphene are absent for $q_\| < 0.03$ Å^{-1} (Figure 11.9b). The plasmon frequencies increase quickly with the increment of $q_\|$, when they increase from $\omega_p \sim 2\alpha_0$. This is attributed to the prominent energy dispersions. The difference of plasmon frequency at long wavelength limit between 3D graphite and 2D graphene mainly lies in the fact that the former has stronger Coulomb interactions. As for the π plasmons with the component of perpendicular momentum transfer, their existence needs to have high enough $q_\|$ (> 0.1 Å^{-1}; not shown). The plasmon frequencies are greatly enhanced by an increase in $q_\|$ (Figure 11.9c), while their dependence on q_z might be gradual increase, slow decrease, or rapid decrease. Such result directly reflects the complicated nonperpendicular and nonparallel Coulomb excitations.

11.2 Bernal Graphite

Bernal graphite has two layers and four carbon atoms in a primitive unit cell, in which the A and B sublattices on two layers will experience different chemical environments. The 4×4 Hamiltonian matrix associated with the four tight-binding functions of $2p_z$ orbitals is expressed as

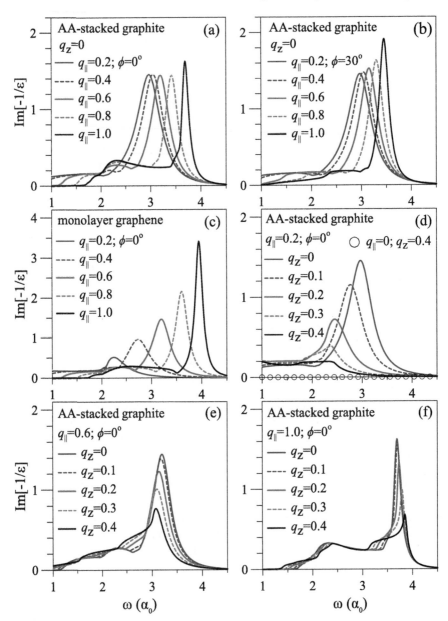

FIGURE 11.8

The higher-frequency energy loss spectra of the simple hexagonal graphite at various transferred momenta under specific conditions: (a) $\phi = 0°$ & $q_z = 0$, (b) $\phi = 30°$ & $q_z = 0$, (d) $\phi = 0°$ & $q_{\parallel} = 0.2$ Å^{-1}, (e) $\phi = 0°$ & $q_{\parallel} = 0.6$ Å^{-1}, and (f) $\phi = 0°$ & $q_{\parallel} = 1.0$ Å^{-1}. Also indicated in (d) under $q_{\parallel} = 0$. Specifically, (c) $\phi = 0°$ is for monolayer graphene.

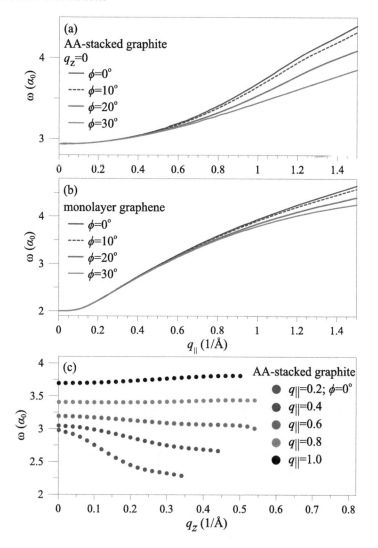

FIGURE 11.9

The momentum-dependent π plasmon frequencies under the specific cases: (a) $q_z = 0$ & various ϕ's for the AA-stacked graphite, (b) various ϕ's for monolayer graphene, (c) various $q_\parallel$'s and $\phi = 0°$ for the first one.

$$H = \begin{pmatrix} E_A & \gamma_0 h(k_x, k_y) & \gamma_1\Gamma & \gamma_4\Gamma h^*(k_x, k_y) \\ \gamma_0 h^*(k_x, k_y) & E_B & \gamma_4\Gamma h^*(k_x, k_y) & \gamma_3\Gamma h(k_x, k_y) \\ \gamma_1\Gamma & \gamma_4\Gamma h(k_x, k_y) & E_A & \gamma_0 h^*(k_x, k_y) \\ \gamma_4\Gamma h(k_x, k_y) & \gamma_3\Gamma h^*(k_x, k_y) & \gamma_0 h(k_x, k_y) & E_B \end{pmatrix},$$

$$(11.3)$$

where $E_A = \gamma_6 + \gamma_5 h^2/2$ and $E_B = \gamma_2 h^2/2$ indicate the sum of the on-site energy. The hopping integrals, as indicated in Figure 11.1b, related to the A and B sublattices on three layers are, respectively, γ_0, ($\gamma_1 = 2.598$ eV, $\gamma_3 = 0.319$ eV, $\gamma_4 = 0.177$ eV), ($\gamma_2 = -0.014$ eV, $\gamma_5 = 0.036$ eV) and $\gamma_6 = -0.026$ eV [51], corresponding to the intralayer nearest-neighbor interaction, (the vertical A-A, the nonvertical B-B & A-B interactions of the neighboring layers), (the vertical and nonvertical interactions from the next-neighboring layers), and the different chemical environments experienced by two sublattices.

Band structure of Bernal graphite, as clearly shown in Figure 11.2c and d, exhibits a significant wave vector dependence and the strong anisotropy. There are two pairs (one pair) of valence and conduction band on the $k_z = 0$ plane (the $k_z = \pi$ plane) due to the four $2p_z$ orbitals. They are highly asymmetric about the Fermi level, as observed in those of simple hexagonal graphite. The electronic states, which are centered at the K point on the $k_z = 0$ plane, have parabolic dispersions and are nondegenerate. However, those initiated from the H point possess a Dirac-cone structure and are doubly degenerate. That is to say, the K and H points, respectively, correspond to the bilayer- and monolayer-like graphene systems. It should be noticed that the second pair of conduction and valence bands near the K point have the initial state energies of $\sim \pm 2\gamma_1$ (Figure 11.2d). The overall k_z-dependent energy width is about 0.1 eV ($\sim 0.05\ \gamma_0$), while the interlayer-interaction-induced free electrons and holes are approximately situated in a small overlay range of ~ 10 meV. The free carrier density is much smaller than that in the AA-stacked graphite (Figure 11.2b). A similar nondegeneracy/double degeneracy is revealed in the middle-energy saddle point [M/L], in which the conduction and valence state energies are $(\gamma_0 \pm \gamma_1)$ and $-(\gamma_0 \pm \gamma_1)/\pm\gamma_0$. This will induce prominent structures (strong shoulders and asymmetric peaks) in middle-frequency electronic and optical excitations [31].

The whole DOSs due to the $2p_z$ orbitals will dominate the essential physical properties. The Bernal graphite exhibits three kinds of van Hove singularities, as clearly shown in Figure 11.3b. There exists a V-shape structure crossing the Fermi level (the inset), directly reflecting the k_z-dependent first pair of valence and conduction bands with the dominating linear dispersions and the minor parabolic ones (Figure 11.2d). DOS is a finite value at E_F, so that this system is a low-carrier-density semimetal. The low-energy DOS is much smaller than that of the simple hexagonal graphite (Figure 11.3a). At about $E = \pm 2\gamma_1$ (the inset), DOS displays a pair of shoulder structures to represent the emergence of the second pair of parabolic bands (Figure 11.2d). More-over, the composite structures, the shoulder and asymmetric peak, are obviously revealed in the middle-frequency ranges of $-1.12\gamma_0 \leq E \leq -0.91\gamma_0$ and $0.99\gamma_0 \leq E \leq 1.07\gamma_0$. The interlayer atomic interactions are expected to have significant effects on the π-electronic single-particle and collective excitations.

According to the theoretical predictions on the low-frequency electronic excitations in the AA-stacked graphite (Figures 11.4–11.6), similar behaviors are expected to be revealed by Bernal graphite. The dominating Coulomb screenings might belong to the perpendicular momentum transfer; furthermore, there is low 2D free carrier density. The direction and magnitude of transferred momenta and temperature will play important roles in the low-frequency Coulomb excitation phenomena. The very delicate calculations on the 3D first Brillouin zone are required to fully explore the $\mathbf{q}$- and T-dependent single-particle and collective excitations. The AB- and ABC-stacked graphites (Section 11.3) sharply contrast with the AA-stacked system in the low-frequency electronic excitations, especially for the temperature dependence. In addition, the T factor can be neglected for the latter, as discussed in Section 11.1.

It is almost impossible to observe the free-carrier- and temperature-induced the low-frequency plasmon mode, when the transferred momentum is parallel to the graphene plane [$\mathbf{q}_\parallel$, $q_z = 0$]. The imaginary part of the dielectric function, as shown Figure 11.10a, exhibits a very strong peak, with an order of ~ 1000, at small $q_\parallel$ and various temperatures. It mainly comes from the interband transition of the first pair of valence and conduction bands (Figure 11.2d). Its height increases with an increase in temperature; that is, the single-particle excitations are enhanced by T. Moreover, the real part of the dielectric function presents a drastic change near the peak position (Figure 11.10b), in which $\mathrm{Re}[\epsilon]$ is transformed from a giant positive value to a large negative one. This function approaches to zero in the further increase of frequency. However, its zero point is accompanied with a non-negligible $\mathrm{Im}[\epsilon]$ of ~ 100. The low-frequency energy loss spectra only reveal very low intensities, without the prominent plasmon peaks. The thermal excitations cannot create the acoustic plasmons in the Bernal graphite, being thoroughly different from the purely T-induced ones in monolayer graphene (details in Section 3.1).

When the direction of the transferred momentum is changed from the graphene plane to the periodical one, the low-frequency plasmon mode exists (Figure 11.11a–c). At $q_\parallel = 0$ and small q_z's, $\mathrm{Im}[\epsilon]$ exhibits a very strong peak under the assistance of the room-temperature thermal excitations (Figure 11.11a), in which the specific excitation frequency (the intensity) increases (decreases) with an increase in q_z. This single-particle excitation peak is due to the intrapair intraband and interband transition channels of the energy bands along the KH line (Figure 11.2d); that is, the strong e–h excitations are closely related to the free electrons and holes. Also, the simple relations with q_z are also revealed in $\mathrm{Re}[\epsilon]$ (Figure 11.11b). The second zero point of $\mathrm{Re}[\epsilon]$ appears at a very small $\mathrm{Im}[\epsilon]$ (Figure 11.11a), so that there exists a significant peak in the energy loss spectrum (Figure 11.11c) if q_z is not too large (e.g., $q_z < 0.035$ Å^{-1}). This plasmon mode could be regarded as the collective charge oscillations along the z-direction by the free carriers due to the interlayer atomic interactions and the thermal excitations. As to the frequency and intensity of the plasmon peak (Figure 11.11c), their q_z dependences are roughly similar to those of the single-particle excitations (Figure 11.11a).

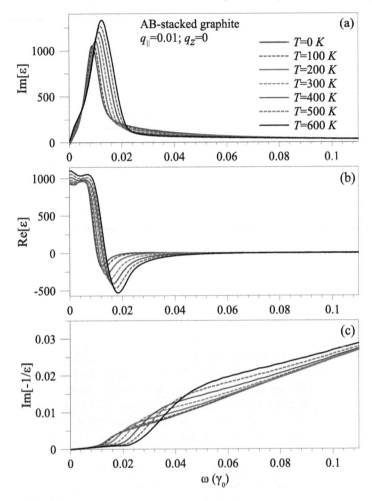

FIGURE 11.10
The (a) imaginary and real (b) parts of the dielectric function and the energy loss functions for the Bernal graphite at $[q_\parallel = 0.01, q_z = 0]$ and various temperatures.

Also, both e–h and collective excitations are very sensitive to the change in T. In general, temperature leads to the enhancement in the bare and screened response functions under various temperatures in Figure 11.11d–f. However, at enough low temperature ($T \leq 100$ K), the energy loss spectra might reveal the abnormal T-dependence on the plasmon intensity and frequency, relying on the significant competition between the band-structure and thermal effects.

The dispersion relations of the plasmon frequencies with the transferred momentum and temperature could provide more information about the collective excitations due to the free electrons and holes, as clearly shown

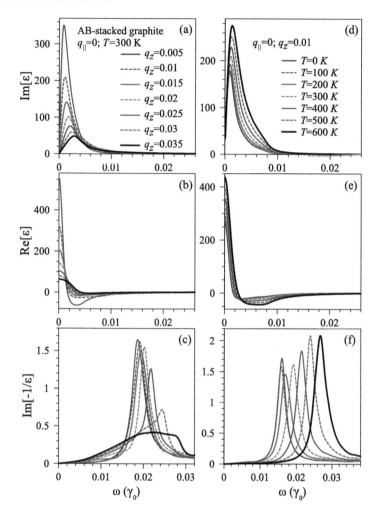

FIGURE 11.11
Same plot as Figure 11.10, but shown for the perpendicular transferred momentum: [(a), (b), (c)]/[(d), (e), (f)] for [Im(ϵ), Re(ϵ) & Im($-1/\epsilon$)] at $q_{\parallel} = 0$ under $T = 300$ K and different q_z's/under $q_z = 0.01$ and different T's.

in Figure 11.12a–c. Generally speaking, the momentum dependence of ω_p (Figure 11.12a) directly reflects the cooperative relation of the weak energy dispersion along KH and the thermal excitations. It exhibits a slow variation at $q < 0.01$; furthermore, ω_p approaches to a finite value at $q \to 0$. The low-frequency plasmon in Bernal graphite belongs to an optical mode, with a well-fitted formula $\omega_0 + c_0 q^2$. Such plasmon mode, which is similar to that in a 3D electron gas [269], could survive at larger critical momenta as temperature increases. Also, the plasmon frequency strongly depends on

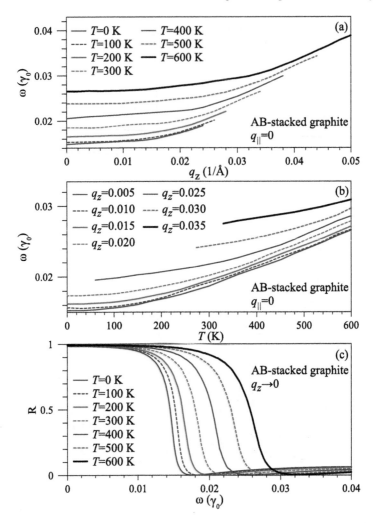

FIGURE 11.12
The momentum- and temperature-dependent plasmon frequencies and the temperature-related reflectance spectra, respectively, shown at (a) different T's, (b) various q's, and (c) distinct T's.

temperature (Figure 11.12b), mainly owing to the rapid enhancement in the free carrier density with temperature. To support the low-frequency plasmons at higher transferred momenta, the temperature needs to be high enough. For example, at $q=0.03$ Å^{-1}, this plasmon exists in Bernal graphite only at $T \geq 250$ K. The aforementioned plasmon frequencies could provide a reasonable explanation for the experimental measurements [404, 407]. Jensen et al. [404] and Geiger et al. [407] examine the plasmon frequencies at different temperatures (room temperature) using the EELS technique. ω_P's are (36, 48, 53, 58 meV) under (152, 263, 310, 384 K) [45 meV at 300 K]. They agree with the calculated results for $q_z < 0.01$ Å^{-1}. In addition, EELS measurements are required for q_z-dependent plasmon frequencies.

The dielectric function under the long wavelength limit [$\mathbf{q} \to \mathbf{0}$] is very useful in fully comprehending the optical reflectance spectra. When an electromagnetic field is normally incident on the graphite surface [262], it induces the dynamic charge screening due to the free electrons and holes. The screening ability is characterized by the transverse dielectric function, being identical to the longitudinal one at $\mathbf{q} \to \mathbf{0}$. The aforementioned Im[$\epsilon$] and Re[$\epsilon$] could be directly utilized to evaluate the frequency-dependent reflectance: $R(\omega) = [1 - \sqrt{\epsilon(\omega)}]^2/[1 + \sqrt{\epsilon(\omega)}]^2$. The well-known Drude edges [113, 232], which is purely due to the collective excitations of charge carriers in the condensed-matter systems, are clearly revealed in the reflectance spectra, as clearly shown in Figure 11.12c at different temperatures. $R(\omega)$ decreases quickly as ω increases from zero; furthermore, it drops to a minimum at $\omega \sim \omega_0$ The abrupt edge structure in the reflectance spectrum created a low-frequency plasmon mode; that is, most of the incident electromagnetic (EM) waves under $\omega \sim \omega_0$ are absorbed by the free carriers; therefore, it leads to the propagation of plasmon waves. The Drude edges become less sharp with an increase in temperature as a result of thermal broadening effect. At $T = 300$, the measured reflectance spectrum displays an abrupt edge structure with a minimum value at $\omega \sim 50$ meV [232]. Apparently, the theoretical predictions are consistent with the experimental measurements. The significant T-dependence of the reflectance spectra could be verified by the detailed optical experiments.

The π plasmon modes in the AA-, AB-, and ABC-stacked graphites might be similar to one another, since they possess similar electronic states near the saddle points (e.g., M and L points). The stacking configurations somewhat affect the plasmon frequencies and modify the peak intensity, as clearly indicated in Figures 11.8, 11.9, 11.13, 11.14–11.16. Under the parallel momentum transfer, the plasmon frequency increases with an increase in $q_\parallel$ (Figures 11.8a,b, 11.9a,b, 11.13a,b, 11.14a, 11.15a,b and 11.16a). However, the opposite is true for the perpendicular momentum transfer. $q_\parallel$ needs to be sufficiently large (> 0.1 Å^{-1}) for the existence of the perpendicular collective carrier oscillations (Figures 11.8d–f, 11.9c, 11.13c,d, 11.14b, 11.15c,d and 11.16b). The planar π plasmons depend on the azimuthal angle only for very large $q_\parallel$ (> 0.8 Å^{-1}) (Figures 11.9a, 11.14a and 11.16a). As to the π-plasmon peak intensities, their relations with the transferred momenta are enriched by the distinct stacking configurations and the number of layers. For example, both AB- and AB-stacked graphites display a rapid decrease of π plasmon strength during the increment of $q_\parallel$ (Figures 11.13a and 11.15a), while the comparable/even higher plasmon peaks appear in simple hexagonal graphite and monolayer graphene (Figure 11.8a–c). Moreover, the 3D π plasmon frequency is much higher than the double of

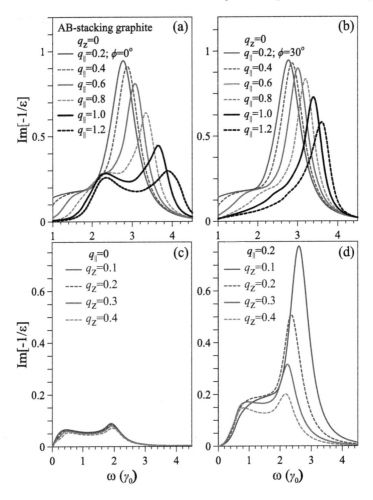

FIGURE 11.13
The wide-frequency-range energy loss spectra for the Bernal graphite under (a) various $q_{\parallel}$'s, $\phi = 0°$ & $q_z = 0$, (b) various $q_{\parallel}$'s, $\phi = 30°$ & $q_z = 0$, (c) various q_z's & $q_{\parallel} = 0$, and (d) distinct q_z's & $q_{\parallel} = 0.2$ Å^{-1}.

the intralayer nearest-neighbor hopping integral. Up to now, EELS measurements are conducted on $q_{\parallel}$-dependent π plasmon frequencies of the Bernal graphite (the open circles in Figure 11.14) [113], and such results are roughly consistent with the theoretical calculations. The 2D superlattice model [234, 261], which covers the 2D band structure and the intralayer & interlayer Coulomb interactions, is also proposed to account for the experimental measurements. In addition, this model is useful in understanding the obvious π plasmons in graphite intercalation compounds, e.g., the stage-1 $FeCl_3$ [169] and C_8M (the alkali atom) [170]. That two different methods

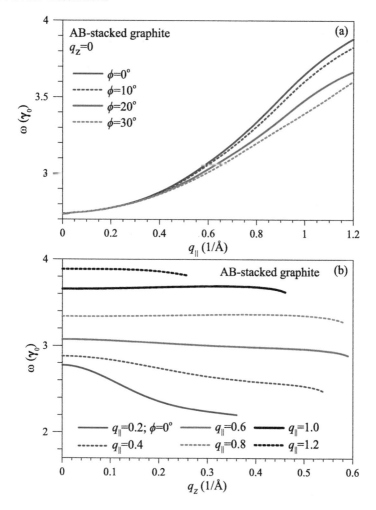

FIGURE 11.14
The momentum-dependent π-plasmon frequencies of the Bernal graphite at
(a) $q_z = 0$ & various ϕ's, and (b) various $q_\parallel$'s and $\phi = 0°$.

could explain the measured results further illustrates the critical mechanism in determining the π plasmons, the saddle points with very large DOS's.

The 3D π-band structure in Bernal graphite exhibits low-frequency plasmons as well as π plasmons. The important differences between these two kinds of plasmon modes lie in the cause, excitation frequency, temperature dependence, and momentum relation. The former and the latter, respectively, arise from the free electrons and holes near the Fermi level and the high-DOS electronic states initiated from the saddle points ($E^{c,v} \sim \pm\gamma_0$). Their frequencies

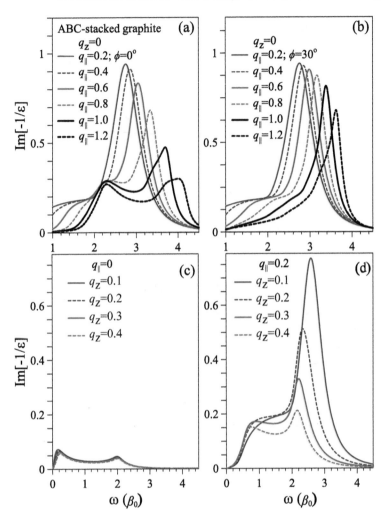

FIGURE 11.15

The wide-frequency-range energy loss spectra of the rhombohedral graphite at (a) different $q_\parallel$'s, $\phi = 0°$ & $q_z = 0$, (b) different $q_\parallel$'s, $\phi = 30°$ & $q_z = 0$, (c) different q_z's & $q_\parallel = 0$, and (d) different q_z's & $q_\parallel = 0.2$ Å^{-1}.

appear at $\omega_p < 0.2$ eV and $\omega_p > 5$ eV. The temperature effect is significant for the former, but negligible for the latter. Specifically, the low-frequency plasmon could survive only under the perpendicular transferred momentum and $q_z < 0.05$ Å^{-1}. However, the π plasmon exists for any momentum direction, and its critical momentum reaches 1.0 Å^{-1}. Obviously, the π plasmon, being related to all the valence $2p_z$ orbitals, are relatively easy to be observed.

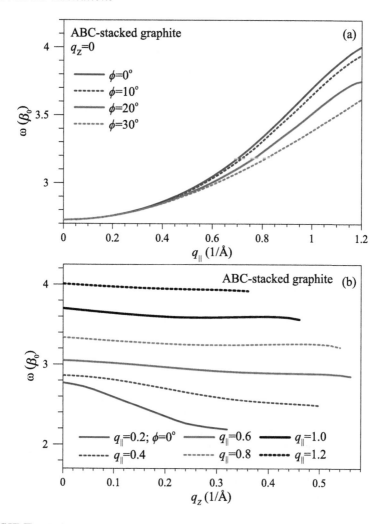

FIGURE 11.16
The momentum-dependent π-plasmon frequencies of the rhombohedral graphite under (a) $q_z = 0$ & distinct ϕ's, and (b) distinct $q_{\parallel}$'s and $\phi = 0°$.

11.3   Rhombohedral Graphite

It is relatively easy to utilize a conventional primitive unit cell to fully explore the essential physical properties, while it is the opposite for the primitive rhombohedral unit cell, except for a few cases [31]. As a result, the Bloch wave functions are composed of six tight-binding functions on three layers. Similar to the calculations of trilayer graphene systems (Chapters 4–7), the

6×6 Hamiltonian matrix could be reduced to consist of two independent block matrices:

$$H = \begin{pmatrix} H_1 & H_2 & H_2^{\star} \\ H_2^{\star} & H_1 & H_2 \\ H_2 & H_2^{\star} & H_1 \end{pmatrix}, \tag{11.4}$$

where 2×2 H_1 and H_2 matrix take the analytic forms in their element forms

$$H_1 = \begin{pmatrix} 0 & \beta_0 h(k_x, k_y) \\ \beta_0 h^*(k_x, k_y) & 0 \end{pmatrix}; \tag{11.5}$$

$$H_2 = \begin{pmatrix} (\beta_4 \exp(ik_z I_z) + \beta_5 \exp(-i2k_z I_z))h^*(k_x, k_y) \\ (\beta_3 \exp(ik_z I_z) + \beta_5 \exp(-i2k_z I_z))h(k_x, k_y) \\ \\ \beta_1 \exp(ik_z I_z) + \beta_2 \exp(i2k_z I_z) \\ (\beta_4 \exp(ik_z I_z) + \beta_5 \exp(-i2k_z I_z))h^*(k_x, k_y) \end{pmatrix}. \tag{11.6}$$

All the significant hopping integrals between two different sublattices on three layers, as clearly indicated in Figure 11.1c, cover $\beta_0 = 3.16$ eV, $\beta_1 = 0.36$ eV, $\beta_2 = -0.02$ eV, $\beta_3 = 0.32$ eV, $\beta_4 = -0.03$ eV, and $\beta_5 = 0.013$ eV [338], in which their definitions are similar to those (Eq. (11.3)) in the AB-stacked graphite.

Using the hexagonal unit cell, the ABC-stacked graphite, as shown in Figure 11.2e and f, has three pairs of valence and conduction bands, mainly owing to the zone-folding effect on the π-electronic structure. For the low-lying electronic structure, the k_z-dependent energy width is less than 10meV (Figure 11.2e), indicating strong competitions among the interlayer atomic interactions in the specific ABC stacking. Band overlap is almost vanishing, so that the 3D density of free electrons and holes is lowest (highest) for the ABC stacking (the AAA stacking). There is one nondegenerate Dirac-cone structure and two degenerate parabolic bands near the K and H points (Figure 11.2f). A similar state degeneracy appears at the middle energies close to the H and L points, in which the energy widths of the interlayer-interaction-induced saddle points are directly reflected in DOS (Figure 11.3c). In fact, only one pair of $2p_z$-dominated valence and conduction bands is revealed under a rhombohedral unit cell [408]. A 3D Dirac-cone structure is composed of tilted anisotropic Dirac cones around the spirally located Dirac points. Moreover, the Dirac points form a nodal spiral in wave vector space due to the accidental degeneracy, which can only be realized in rhombohedral graphite up to now.

The main features of DOS in the ABC-stacked graphite are unique (Figure 11.3c), being thoroughly different from those in the AA- and AB-stacked graphites (Figure 11.3a and b). An almost symmetric V-shape structure (the inset in Figure 11.3a), which purely comes from the 3D Dirac-cone structure, is initiated from the Fermi level [31]. The DOS at E_F is smallest among three kinds of graphites, further illustrating the weakest interlayer atomic interactions after the strong competition. The low-frequency collective excitations are expected to be observable only under sufficiently high temperature. As the state energies gradually increase, a pair of finite-width shoulder structures appear at $0.11 \beta_0$ and $-0.12 \beta_0$. That the anisotropic linear energy dispersions are getting into the parabolic forms is the main reason. The critical points are, respectively, (Dirac points & extreme points) and saddle points at low and middle frequencies. The latter leads to very sharp peaks at $0.959 \beta_0$, $1.010 \beta_0$, and $-0.991 \beta_0$, in which they are similar to the 2D logarithmic-form symmetric structures. The conduction and valence DOSs present slightly splitting peaks and a single peak, respectively. The strong symmetric peaks and the V-shape form, which are, respectively, located at low and middle frequencies, clearly indicate the quasi-2D behaviors [31].

The lower-symmetry rhombohedral graphite has the lowest free electron/hole density among three kinds of well-behaved stacking configurations, so that it is expected to be relatively difficult in creating low-frequency collective excitations. Most of its excitation phenomena are similar to those of the Bernal graphite, as revealed from a detailed comparison between Figures 11.17, 11.18 and 11.11, 11.12. For example, under the parallel transferred momentum, the Landau damping is too prominent to observe 3D optical plasmons. However, even for the perpendicular transferred momentum, the full assistance due to the thermal excitations is necessary to induce the collective charge oscillations along the z-axis (Figure 11.17a–f). At $T = 600$ K, the maximum value of $\text{Im}[\epsilon]/\text{Re}[\epsilon]$ drops from ~ 2000 to ~ 200 as q_z increases from 0.001 to 0.03 Å^{-1} (Figure 11.17a and b). The drastic change in the dielectric function hardly affects the height of plasmon peak (Figure 11.17c), since the latter is determined by the second zero point of $\text{Re}[\epsilon]$ and the value of $\text{Im}[\epsilon]$. The peak positions only weakly depend on the various transferred momenta, directly reflecting the weak energy dispersion along the KH line (Figure 11.2f). Moreover, the bare and screened response functions are very sensitive to the change in temperature. Apparently, the maximum strength of e–h excitations (Figure 11.17d), the frequency corresponding to the second zero point in $\text{Re}[\epsilon]$ (Figure 11.17e), and the height of plasmon peak in the energy loss spectrum (Figure 11.17f) is greatly enhanced by an increase in temperature.

The dispersion relations between the free-carrier-induced plasmon frequency and the transferred momentum/the temperature could provide the first-step information about the experimental examinations. The plasmon frequency lies in the range of $0.02\beta_0 < \omega_p < 0.04\beta_0$ during the variation of momentum or temperature, as clearly shown in Figure 11.18a and b. In general, the q-dependence is weak, directly reflecting a very narrow band width along the KH line (Figure 11.2e and f). The low-frequency optical plasmon modes could survive only at finite temperature even under $q_z \to 0$, being thoroughly different from those in AB-stacked graphite (Figure 11.12a and b). Furthermore, the critical momentum increases with an increase in temperature (Figure 11.18a). The plasmon frequency strongly depends on temperature at very small q's, while the T-dependence becomes negligible for other cases. The temperature-created plasmons in rhombohedral graphite could be verified by

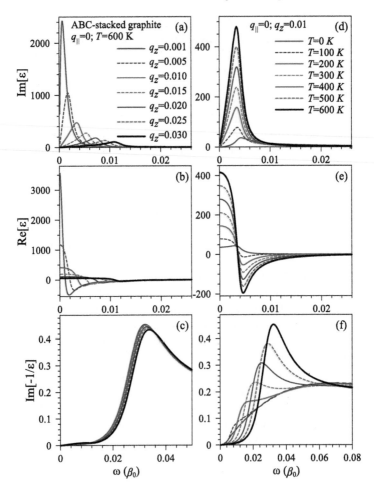

FIGURE 11.17
The imaginary and real parts of the low-frequency dielectric function and the energy loss function for the ABC-stacked graphite, being shown in [(a), (b), (c)]/[(d), (e), (f)], respectively, under the conditions: $T = 600$ K & various q_zs, and $q_z = 0.01$ Å^{-1} and different temperatures.

the high-resolution energy loss spectra [37, 38, 167, 168, 215, 231–233] and the optical measurements, as done for Bernal graphite [231–233, 406].

Concerning the low-frequency single-particle excitations and plasmon modes, it is very interesting to fully explore the stacking-configuration-diversified phenomena. Three kinds of graphites only exhibit optical plasmons, as observed in a 3D electron gas. However, their low-frequency plasmons might sharply contrast to one another in terms of excitation frequencies, main causes, concise relations with the direction of the transferred momentum, and

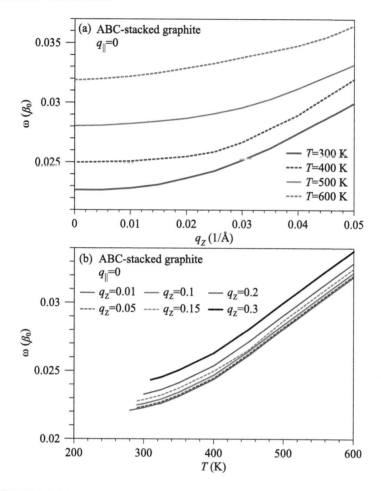

FIGURE 11.18
The dispersion relations of the plasmon frequencies on (a) the transferred momentum and (b) temperature, respectively, under the specific T's and q_z's.

significant temperature dependences. The collective oscillation frequency in the AA-stacked graphite, which mainly arises from the high free e–h density in the Dirac-cone structures, reaches the order of $\omega_p \sim 1$ eV even at zero temperature. This result directly reflects the strongest overlap among three systems. The 3D plasmon modes could survive under parallel and perpendicular transferred momenta, in which the former/latter case experiences serious/light Landau dampings at small momenta; furthermore, the critical momenta are 0.12 and 0.6 Å^{-1}. Apparently, the temperature effects are negligible in simple hexagonal graphite. On the other hand, it is almost impossible to observe the low-frequency optical plasmons in Bernal and rhombohedral graphites, while the transferred momentum is parallel to the graphitic plane. Such phenomenon is attributed to the low free carrier density and the strong e–h dampings. The optical plasmons under the perpendicular momentum are very sensitive to temperature. Specifically, as to the ABC-stacked graphite, the plasmon modes at long wavelength limit are clearly revealed in the energy loss spectra, only when the thermal excitations are required to supply the sufficient free electrons and holes, e.g., the existence of collective excitations at $q \to 0$ only for $T > 100$ K. In general, the higher the temperature (the transferred momentum) is, the larger (the higher) the critical momentum (temperature) is. Up to now, there are no experimental examinations on the screened excitation spectra of AA- and ABC-stacked graphites.

12

1D Electronic Excitations in Metallic and Semiconducting Nanotubes

Carbon nanotubes have attracted a lot of experimental and theoretical researches since their first discovery by Iijima in 1991 [116], mainly owing to the nanoscale radius and hollow cylindrical symmetry. Various experimental methods have successfully synthesized 1D single- and multiwalled carbon nanotubes [116, 117, 409–416] and the 2D/3D array of aligned nanotube systems [417–421]. The achiral and chiral hexagon arrangements relative to the nanotube axis are confirmed by high-resolution scanning tunneling microscopy (STM) measurements [121, 422–425]. Specifically, many 1D energy subbands belong to the parabolic forms, but only few possess linear or weak dispersions. The van Hove singularities, which present the asymmetric square-root forms and the plateau structures, are revealed in the measured density of state (DOS) through scanning tunneling spectroscopy (STS) measurements [424, 425]. The radius and chiral angle, accompanied with the periodical boundary condition, dominate the low-energy physical properties, except for a small carbon nanotube featuring the rather strong sp^3 chemical bondings on a cylindrical surface [118, 119, 426]. In general, (r, θ) and $2p_z$ orbitals are sufficient in exploring the unusual physical phenomena, e.g., three kinds of geometry-dependent energy gaps, and the periodical Aharonov–Bohm effects in the presence of a parallel magnetic field. They are the critical factors in diversifying the fundamental physical properties, such as electronic excitations [19, 39, 48, 173–179, 348, 427–435] and optical absorption spectra [118, 122, 436–447].

Many theoretical predictions exist on the Coulomb excitations of carbon nanotubes [19, 48, 348, 427–434], while most of them require further experimental verifications up to now. The $2p_z$ and $(2s, 2p_x, 2p_y)$ orbitals in a sp^2-bonding carbon nanotube, respectively, induce the π and σ energy bands. The curvature effects, the misorientation of $p\pi$ orbitals and the mixing of $p\pi$ and sp$^2\sigma$ orbitals, have rather strong effects on the low-lying band structures and the low-frequency electronic excitations [118, 426]. According to random-phase approximation (RPA) calculations, each cylindrical nanotube will exhibit momentum- and angular-momentum-dependent (q- and L-dependent) single-particle excitations and plasmons. Only the metallic (armchair) carbon nanotubes are expected at the low-energy plasmon of $L = 0$; that is, such plasmon is absent in narrow- and middle-gap carbon nanotubes

[19, 348, 428–434]. Furthermore, the momentum dependence of 1D plasmon is predicted to be thoroughly different from that of the 2D acoustic mode. The elementary electronic excitations are greatly diversified by the magnetic field (**B**) [448–452] and temperature [435, 453–456], in which the number and mechanism of magnetoplasmons and the range of Landau dampings are sensitive in the magnitude and direction of **B**, T, nanotube geometry, and Zeeman splitting. Moreover, all carbon nanotubes exhibit π plasmons arising from the collective oscillations of whole valence π electrons under the middle frequency of > 5 eV. Such plasmons, which are characteristic of the sp^2-bonding carbon-related systems, are predicted to survive in 2D nanotube arrays [432, 433] and 3D nanotube bundles [430]. However, discrete angular-momentum modes are absent. The 3D transferred momenta ($\mathbf{q} = [\mathbf{q_x}, \mathbf{q_y}, \mathbf{q_z}]$) will play an important role in defining the independent excitation modes.

The high-resolution electron energy loss spectroscopy (EELS) measurements could be utilized to examine the low-frequency, inter-π-band and π plasmons in carbon nanotubes. There is no experimental evidence on the first kind of plasmon modes, being attributed to the difficulties in producing a uniform sample. Pichler et al. have identified several inter-π-band plasmon modes with frequencies of $\sim$1–4 eV [177]. They exhibit very weak momentum dependences; furthermore, their frequencies are close to the inter-π-band excitations associated with the valence and conduction van Hove singularities. These results clearly illustrate that such plasmons only originate from part of valence π electrons, being consistent with the theoretical predictions [48, 428]. A plenty of experimental loss spectra are done for the π plasmons in various nanotube-related systems. For example, Kuzuo et al. show that multiwalled carbon nanotubes, with 21–44 layers, exhibit a pronounced π-plasmon at ω_p $\sim$5.1–5.4 eV or $\sim$6.2–6.4 eV [178]. A $\sim$ 5.8 eV π plasmon is revealed in the finite-size bundle composed of 600 single-walled carbon nanotubes. Friedlein et al. [457] measure the momentum-dependent energy loss spectra of multiwalled carbon nanotubes with $\sim$ 12 layers, single-walled nanotube bundles with $\sim$10–100 layers, and graphite layers. The π-plasmon frequencies linearly rely on the transferred momentum at large q; furthermore, the q-dependence is stronger for the first/third systems, when compared with the second one.

To fully explore the curvature effects on Coulomb excitations, the sp^3 tight-binding model is more effective in understanding the unusual band structures in the presence/absence of a magnetic field [118, 119, 426]. The rich features of electronic properties are expected to clearly present in the chiral/achiral single-walled carbon nanotubes, such as a lot of angular-momentum-defined 1D energy subbands with the significant dependences of the axial wave vector, the simple relation between energy gaps and (r, θ), the periodical Aharonov–Bohm effect due to the coupling of angular momentum and magnetic flux, and the different van Hove singularities in DOS (the square-root asymmetric peaks and plateau structures) arising from the parabolic and linear dispersions. The sensitive dependences of electronic excitations on the geometric structures, momenta & angular momenta, temperature, strength & direction of **B**, and Zeeman splitting are investigated in detail, especially for low-frequency plasmon modes and single-particle excitations. Most importantly, plasmon frequencies are examined for the dispersion relations with the transferred momenta, so that the important roles of 1D bare Coulomb interactions and the low-lying energy dispersions could be understood thoroughly. **B** will create magnetoplasmons in any carbon nanotubes, where the critical mechanisms are proposed to explain how many such modes. The inter-π-band and π plasmons are other focuses of this work. It is worthy of doing a detailed comparison between a single-walled carbon nanotube and a monolayer graphene in three kinds of plasmon modes.

12.1 Rich Electronic Properties in the Absence/Presence of B

Each single-walled carbon nanotube could be regarded as a rolled-up cylindrical tubule of monolayer graphene. Its geometric structure, as clearly shown in Figure 12.1a, is characterized by a specific translation vector of the latter. It is formed by rolling a graphitic sheet from the origin to the vector

$$\mathbf{R_x} = m\mathbf{a_1} + n\mathbf{a_2}. \tag{12.1}$$

The pair of parameters (m, n) will be used to represent a carbon nanotube. The chiral angle of the (m, n) nanotube, $\theta = tan^{-1}[-\sqrt{3}\,n/(2m_n)]$, is the angle between R_x and the original $\mathbf{e'_x}$ (Figure 12.1a), and the radius is $r = [b\sqrt{3(m^2 + mn + n^2)}]/(2\pi)$. It is sufficient to confine the chiral angle to $0 \le \theta \le 30°$, e.g., armchair (m,m) and zigzag (m,0), respectively, corresponding to $\theta = 30°$ and $0°$ (Figure 12.1b and c). Also displayed in Figure 12.1a is the primitive vector $\mathbf{R_y} = p\mathbf{a_1} + q\mathbf{a_2}$ along the tubular axis. The primitive unit cell, being enclosed by $\mathbf{R_x}$ and $\mathbf{R_y}$, covers the carbon-atom number of $N_u = 4\sqrt{(p^2 + pq + q^2)(m^2 + mn + n^2)/3}$. For example, there are $2m$ carbon atoms in the achiral (m,m) and (m,0) nanotubes.

The sp^3 tight-binding model is taken to fully explore the curvatures due to $(2s, 2p_x, 2p_y, 2p_z)$ orbitals on a cylindrical surface. As to monolayer graphene, the zero-field Hamiltonian is described by a 8×8 Hermitian matrix. According to the equivalent A and B sublattices, it could be decomposed into four block matrices:

$$H_{A_i,A_j}(\mathbf{k}) = H_{B_i,B_j}(\mathbf{k}) = E_i \delta_{i,j},$$
$$H_{A_i,B_j}(\mathbf{k}) = \sum_{l=1,2,3} h_{ij}^{(l)} exp[i\mathbf{k} \cdot (r_l - r_A)],$$
$$H_{B_i,A_j}(\mathbf{k}) = \sum_{l'=1,2,3} h_{ij}^{(l')} exp[i\mathbf{k} \cdot (r_{l'} - r_A)]. \tag{12.2}$$

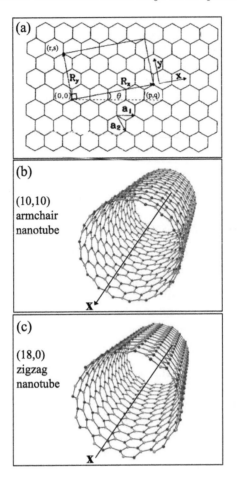

FIGURE 12.1
(a) A rolled-up carbon nanotube from a graphitic sheet, (b) a (10,10) armchair
nanotube and (c) a (18,0) zigzag system.

Each block matrix is a 4×4 matrix, i/j is the four orbitals, and $\mathbf{r_A}/\mathbf{r_B}$
represents the position vector of A/B atom. The cylindrical coordinates,
(r, Ψ, z), are more convenient for calculations. It should be noticed that the
y-direction in monolayer is changed into the z-axis in carbon nanotubes Φ_l's
and z_i's for the three nearest-neighbor A atoms are $\Phi_1 = -b\cos(\pi/6 - \theta)/r$,
$\Phi_2 = b\cos(\pi/6 + \theta)/r$, $\Phi_3 = b\cos(\pi/2 - \theta)/r$, $z_1 = -b\sin(\pi/6 - \theta)$,
$z_2 = -b\sin(\pi/3 - \theta)$ and $z_3 = b\cos(\theta)$.

Similar results are revealed in the B sublattice. The position- and orbital-
dependent hopping integrals are expressed as

$$h_{rr}^{(l)} = V_{pp\pi}\cos\Phi_l + 4(V_{pp\pi} - V_{pp\sigma})\sin^4(\Phi_l/2)r^2/b^2,$$
$$h_{r\Phi}^{(l)} = V_{pp\pi}\sin\Phi_l - 4(V_{pp\pi} - V_{pp\sigma})\sin^3(\Phi_l/2)\cos(\Phi_l/2)r^2/b^2,$$

$$h_{\Phi\Phi}^{(l)} = V_{pp\pi}\cos\Phi_l - (V_{pp\pi} - V_{pp\sigma})\sin^2(\Phi_l)r^2/b^2,$$

$$h_{rz}^{(l)} = -2(V_{pp\pi} - V_{pp\sigma})\sin^2(\Phi_l/2)rz_l/b^2,$$

$$h_{\Phi z}^{(l)} = -(V_{pp\pi} - V_{pp\sigma})\sin(\Phi_l)rz_l/b^2,$$

$$h_{zz}^{(l)} = V_{pp\pi} - (V_{pp\pi} - V_{pp\sigma})z_l^2/b^2,$$

$$h_{sr}^{(l)} = -2V_{sp\sigma}\sin^2(\Phi_l/2)r/b,$$

$$h_{s\Phi}^{(l)} = V_{sp\sigma}\sin(\Phi_l)r/b,$$

$$h_{sz}^{(l)} = V_{sp\sigma}z_l/b,$$

$$h_{ss}^{(l)} = V_{ss\sigma}. \tag{12.3}$$

$h_{z\Phi}^{(l)} = h_{\Phi z}^{(l)}, h_{\Phi r}^{(l)} = -h_{r\Phi}^{(l)}, h_{zr}^{(l)} = -h_{rz}^{(l)}, h_{rs}^{(l)} = h_{sr}^{(l)}, h_{\Phi s}^{(l)} = -h_{s\Phi}^{(l)}; h_{zs}^{(l)} = -h_{sz}^{(l)}$. The subscripts s, r, Φ, and z, respectively, correspond to s, p_π (p_z), p_{σ_1}, and p_{σ_2}. $h_{rr}^{(l)}$ ($h_{rs}^{(l)}$, $h_{r\Phi}^{(l)}$ & $h_{rz}^{(l)}$ is associated with the misorientation of $p\pi$ orbitals (the mixing of $p\pi$ and sp$^2\sigma$). The Slater–Koster hopping parameters and orbital-dependent site energies are as follows: [458] $V_{ss\sigma} = -4.30$ eVAV$_{sp\sigma} = 4.98$ eV, $V_{pp\sigma} = 6.38$ eV, $V_{pp\pi} = -2.66$ eV, $E_s = -7.3$ eV and $E_p = 0$. As a result of the periodical boundary condition, electronic states are defined by the discrete angular momentum J ($k_x r = 1, 2, KN_u/2$) and the longitudinal wave vector k_z ($-\pi \leq k_z(b\sqrt{3(p^2, +pq + q^2)}) \leq \pi$).

Any magnetic field, which have components parallel ($B_\parallel$) and perpendicular to the nanotube axis ($B_\perp$), can create dramatic changes of electron structures and wave functions. When a cylindrical carbon nanotube is threaded by a uniform magnetic flux of $\phi = \pi r^2 B_\parallel$, the angular momentum varies from J into $J + \phi/\phi_0$, leading to the periodical Aharonov–Bohm effect. The angular momentum remains decoupled in the presence of ϕ; that is, J is still a good quantum number. On the other side, the distinct J's can couple one another under $B_\perp$, so that electronic states do not correspond to the well-behaved standing waves. The magnetic field is assumed to deviate from the tubular axis by an angle of α, i.e., $\mathbf{B} = |\mathbf{B}|\cos\alpha\,\hat{z} + |\mathbf{B}|\sin\alpha\,\hat{\Phi} = B_\parallel\hat{z} + B_\perp\hat{\Phi}$. With $B_\perp$, the total carbon atoms in a primitive unit need to be included in the calculations of magnetoelectronic band structures. Each 4×4 block matrix becomes a $2N_u \times 2N_u$ one. The vector potential is chosen as $\mathbf{A} = rB_\perp\sin(x/r)\hat{z}$, where $x = r\Phi$. $\mathbf{A}$ is independent of the z-coordinate; therefore, k_z keeps a good quantum number. However, the obvious dependence on the x-coordinate clearly indicates that the different J's are no longer decoupled. This vector potential creates a Peierls phase of $(2\pi/\phi_0)\int_R^r \mathbf{A}\cdot dl$ in each tight-binding function, leading to a drastic change in the hopping integral. Through detailed derivations, the magnetic Hamiltonian between site A with k_x state and site B with k_x' state is expressed as

$$\langle\Phi_{k_x'}^{B_j}|H|\Phi_{k_x}^{A_i}\rangle = \frac{2h_{ij}}{N_u}\sum_{R^A}\sum_{R^B}e^{-i\Delta k_x x}e^{-i(k_x' + \phi\cos\alpha/\Phi_0 r)\Delta x}$$

$$\times\, e^{ik_z\Delta z}e^{i(e/\hbar)\Delta G}, \tag{12.4}$$

where the phase difference arising from $B_\perp$ is

$$\Delta G = G_{R^A} - G_{R^B} = \frac{\phi \Delta z \sin \alpha}{\pi \Delta x} \left(\cos \frac{x}{r} - \cos \frac{x + \Delta x}{r} \right), \Delta x \neq 0,$$

$$= \frac{\phi \Delta z \sin \alpha}{\pi r} \sin \frac{x}{r}, \Delta x = 0. \qquad (12.5)$$

$\mathbf{R^A} = (x, z)$, $\mathbf{R^B} = (x', z')$, and $\Delta \mathbf{R^A} = \mathbf{R^A} - \mathbf{R^B} = (\Delta x, \Delta z)$.

By diagonalizing the $4N_u \times 4N_u$ magnetic Hamiltonian, we could explore energy dispersion $E^{c,v}(J, k_z, \phi)$ and wave function $\Psi^{c,v}(J, k_z, \phi)$. As to $\alpha \neq 0°$ ($\alpha = 0°$), the magnetic wave function is the linear superposition of the $4N_u$ (8) tight-binding functions, and it consists of different J's (the same J). Although the coupling of angular momenta cannot be ignored, any electronic state is still dominated by a specific J at $\phi < \phi_0/3$. Energy dispersions and wave functions are denoted as a function of J. The magnetoelectronic state energy is the sum of the band energy and spin–B interaction, i.e., $E^{c,v}(J, k_z, \sigma\phi) = E^{c,v}(J, k_z, \phi) + E(\sigma\phi)$. $E(\sigma\phi) = g\sigma\phi/m^*r^2\phi$ The g factor is chosen to be the same with that (~ 2) of Bernal graphite [113], $\sigma = \pm1/2$ is the magnitude of electron spin, and m^* is the bare electronic mass. The Zeeman splitting leads to a rigid separation of spin-up and spin-down states. In general, this interaction could be ignored, except that it needs to be specially emphasized in a certain physical phenomenon.

The 1D carbon nanotubes exhibit rich and unique electronic & magnetoelectronic properties, in which they are further classified into three types according to the concise relation between nanotube geometry and energy gap. Without magnetic field, there are a lot of 1D energy subbands, and their number is very sensitive to the chiral angle, as clearly indicated in Figure 12.2. In general, the achiral nanotubes have less energy subbands in the larger first Brillouin zones, when compared with chiral systems, e.g., (10,10) & (18,0) nanotubes (Figure 12.2a and b). The (m, m) armchair nanotubes present a pair of linearly intersecting valence and conduction subbands of $J = m$, where the Fermi momenta are close to/just located at $k_z = \pm2\pi/3$ in the presence/absence of curvature effects. Most of the higher and deeper energy subbands belong to parabolic dispersions; furthermore, they might be doubly degenerate. Obviously, the DOS is finite at the Fermi level, so that all armchair carbon nanotubes (the first type) are 1D gapless metals. A narrow gap is created in the $2m + n = 3I$ and $m \neq n$ carbon nanotube (I an integer; the second type), e.g., the (18,0) and (17,2) systems (Figure 12.2b and c). The linear energy subbands are slightly changed into parabolic dispersions near $k_z = 0$ or $\pm2\pi/3$. A systematic study has been conducted for the dependence of energy gap on radius under a specific chiral angle [118]. The previous works show that E_g is inversely proportional to r, being verified by the high-resolution STS measurements on a pair of square-root asymmetric peaks nearest to the Fermi level. The nonuniform nearest hopping integrals on a cylindrical surface are the main mechanisms [424, 425]. The low-energy electronic states of narrow-gap systems are almost identical to those sampling from monolayer graphene,

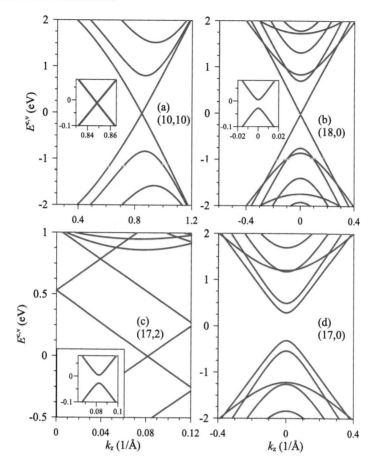

FIGURE 12.2
The low-lying electronic structures for the (a) metallic (10,10), (b) narrow-gap (18,0), (c) narrow-gap (17,2), and (d) middle-gap (17,0) carbon nanotubes.

while they do not correspond to the K/K′ points because of the curvature effects. The other carbon nanotubes of $2m + n \neq 3I$ exhibit the third type of energy gap, being proportional to the inverse of radius, such as $E_g = 0.60$ eV in the (17,0) nanotube (Figure 12.2d). It is impossible to sample Dirac points under the specific boundary condition. As to the tight-binding functions, all the carbon nanotubes behave as perfect standing waves (the details in [350]), as observed in 1D electron gas. They will dominate the dipole matrix elements in the optical absorption spectra (Coulomb matrix elements in bare response functions). As a result, they clearly show the well-known optical selection rule [118].

A uniform magnetic field can diversify the electronic structures of carbon nanotubes. The parallel field of $B_\parallel$ strongly affect the dispersion, gap, and state degeneracy of energy subbands, mainly owing to the dramatic transformation of $J \rightarrow J + \phi/\phi_0$. For example, nondegenerate linear valence and conduction subbands (Figure 12.3a), which are characterized by $J = N_u/4 = 10$, are changed into parabolic dispersions (Figure 12.3b and c). An obvious bandgap is created during the variation of magnetic flux. E_g presents a well-known Aharonov–Bohm oscillation in the absence of Zeeman splitting, i.e., it increases, reaches the maximum at $\phi = \phi_0/2$, declines, and

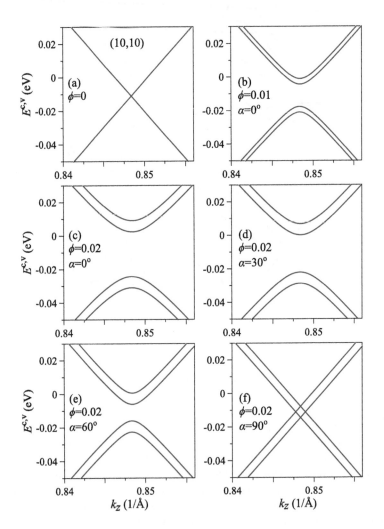

FIGURE 12.3
One or two pairs of valence and conduction bands near to the Fermi level for the $(10, 10)$ nanotube in the presence of Zeeman effect under (a) $\phi = 0$, (b) $\phi = 0.01 \ \phi_0$ & $\alpha = 0°$, (c) $\phi = 0.02 \ \phi_0$ & $\alpha = 0°$, (d) $\phi = 0.02 \ \phi_0$ & $\alpha = 30°$, (e) $\phi = 0.02 \ \phi_0$ & $\alpha = 60°$, (f) $\phi = 0.02 \ \phi_0$ & $\phi = 90°$.

then vanishes at $\phi = \phi_0$ during the variation of magnetic flux [118]. Moreover, the effects of ϕ are distinct on the doubly degenerate energy subbands of J and $N_u/2 - J$. These lead to the destruction of the double degeneracy, as observed in the higher/deeper energy subbands. The magnetic field at $\alpha \neq 0°$ creates the coupling of different J, being strong only at the large ϕ and α. It is relatively weak for type-I $(10, 10)$ nanotube at $\phi = 0.02\ \phi_0$, as clearly shown in Figure 12.3c–f. So that each energy subband is approximately characterized by the decoupled angular momentum. In addition, the magnetic wave functions at $\alpha \neq 0°$ become irregular standing waves. As the direction of magnetic field gradually deviates, this will lead to a decline of energy gap, e.g., $\alpha = 30°$ & $60°$ in Figure 12.3d and e, respectively. The curvatures of energy bands are also reduced with the increase of α. It should be noticed that armchair nanotubes must be metals even in the presence of curvature/Zeeman effects; furthermore, the linear bands are almost identical to those without magnetic flux.

The pairs of valence and conduction bands of the type-II $(18, 0)$ zigzag nanotube, which are nearest to the Fermi level, are also chosen for a model study of the diversified magnetic phenomena. They belong to doubly degenerate energy bands of $J = 2N_u/3 = 12$ and $J = 4N_u = 24$ at zero field, e.g., those at $\phi = 0$ in Figure 12.4a. They have an approximately symmetric energy spectrum about $E_F = 0$. Their energy dispersions are parabolic at small k_zs and linear for others. Energy gap, being inversely proportional to the square of radius, purely come from curvature effects. The double degeneracy vanishes under a parallel magnetic field, e.g., the magnetoband structure at $\phi = 0.01\ \phi_0$ in Figure 12.4b. Energy gap is reduced by increasing the ϕ. The Zeeman effect results in the splitting of spin-up and spin-down states. The lowest/highest conduction/valence band remains at $k_z = 0$ even for $\alpha \neq 0°$. The Zeeman effect results in the splitting of spin-up and spin-down states. Specifically, the spin-up valence band overlaps and the spin-down conduction bands start to touch each other at $\phi = 0.0167\ \phi_0$ and overlap at $\phi = 0.02\ \phi_0$ (Figure 12.4c). The band overlap only comes from Zeeman spitting. The maximum overlap happens at $\phi = 0.0205\ \phi_0$. And then, such overlap vanishes and thus an energy gap is recovered at $\phi = 0.026\ \phi_0$. When the magnetic field deviates from the nanotube axis, the decrease of E_g is more slow and the overlapping of valence and conduction bands occurs at higher critical flux (ϕ_{sm}). For example, semiconductor–metal transitions, corresponding to $\alpha = 30°$, $60°$ & $90°$, are, respectively, revealed under $0.031\phi_0/$ $0.042\phi_0$, $0.031\phi_0/\ 0.042\phi_0$, & $0.087\phi_0$. The narrow- and zero-gap behaviors, which originate from magnetic field and Zeeman effects, are expected to create unusual single- and many-particle magnetoelectronic excitations. In addition, an experimental magnetic field cannot induce such behaviors in type-III carbon nanotubes, except that $|\mathbf{B}|$ is more than 500 T. For example, the $(17, 0)$ nanotube exhibits semiconductor–metal transition at $\phi \sim \phi_0/3$. These systems could be neglected in the discussions of magneto-Coulomb excitations.

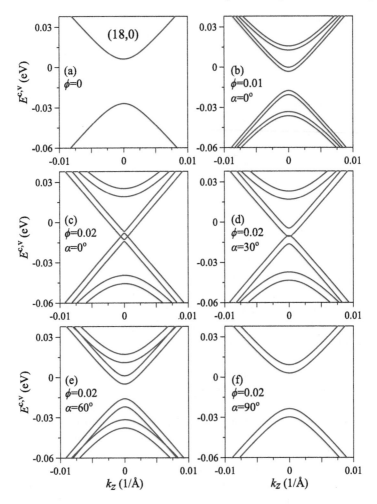

FIGURE 12.4
Similar plots as Figure 12.3, but displayed for the type-II (18,0) carbon nanotube.

12.2   The Low-Frequency Plasmons and Magnetoplasmons

The low-frequency electronic excitations are greatly diversified by three types of carbon nanotubes. The metallic carbon nanotubes exhibit a very strong symmetric peak in the imaginary part of the $L = 0$ dielectric function, as clearly shown in Figure 12.5a by the open circles. This significant structure appears at $\omega_{ex} = E^c(J = 10, k_F + q) - E^v(J = 10, k_F) =$

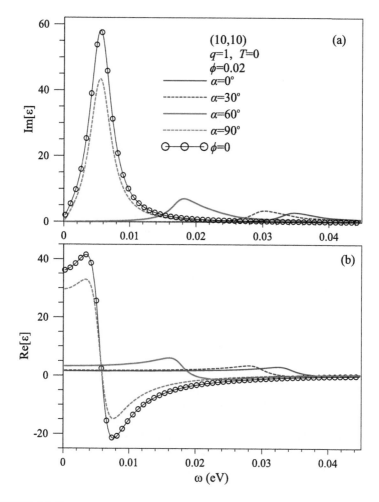

FIGURE 12.5
(a) The imaginary and (b) real parts of the low-frequency $L = 0$ dielectric function for the $(10, 10)$ nanotube are, respectively; displayed in (a) and (b) under $q = 1$, $T = 0$, $\phi = 0.02$ ϕ_0 and distinct α's. Also shown are those in the absence of magnetic flux.

$E^c(J = 10, k_F) - E^v(J = 10, k_F - q)$ (Figure 12.2a), where the Fermi-momentum state is the linearly intersecting point in the absence of doping. Apparently, it is due to the linear valence and conduction bands. Furthermore, the real part of dielectric function, being indicated in Figure 12.5b), illustrates a pair of divergently asymmetric peaks. A vanishing zero point is revealed at ω much higher than ω_{ex}, where $\text{Im}[\epsilon]$ also approaches zero. That is, the low-frequency collective excitations exist at very weak Landau dampings, leading to a strong plasmon peak in the energy loss spectra (Figure 12.7a). The carriers in the linear valence band (Figure 12.2a) can create rather pronounced collective charge oscillations almost without electron–hole (e–h) pairs. The frequency and strength of the interband/first plasmon mode strongly depends on the transferred momentum and angular momentum. With the increase of q, the plasmon frequency increases, but the spectral intensity declines (Figure 12.7a). Moreover, the prominent plasmon peak only survives under the specific $L = 0$ mode, and it is absent for the $L \neq 0$ modes (Figure 12.7b). On the other hand, the narrow-gap carbon nanotubes present the thoroughly different peaks structures in $\text{Im}[\epsilon]$ and $\text{Re}[\epsilon]$, e.g., the dielectric function of the $(18, 0)$ nanotube in Figure 12.6a. The square-root asymmetric peaks only reflect the DOS's in the parabolic dispersions of band-edge states at $k_z = 0$ (Figure 12.4a), i.e., the $k_z = 0$ states of the $J = 12$ & $J = 24$ valence and conduction bands play a critical role in the single-particle and collective excitations. It is very difficult to observe a zero point in $\text{Re}[\epsilon]$ and a very small $\text{Im}[\epsilon]$ simultaneously, so the plasmon peak is only observable in the energy loss spectrum (Figure 12.8a). This plasmon, which is due to the carriers in the $J = 12$ & 24 valence bands, is accompanied by significant interband e–h excitations. In addition, the middle-gap nanotube systems display similar asymmetric peak structures in the dielectric function; furthermore, the strength of plasmon peak is very low in the screened response function (Figure 12.7a) because of serious Landau dampings.

The temperature dependence of Coulomb excitations in pristine carbon nanotubes is worthy of a closer examination. It is very important only for the narrow-gap systems. As for the metallic armchair nanotubes, their single-particle excitations and the low-frequency plasmon of $L = 0$ are almost independent of the various temperatures, as shown in Figures 12.5a and 12.7a for the $(10, 10)$ system. Temperature can create some free conduction electrons, but reduce the same density of valence electrons. Consequently, the extra intraband excitations are generated with the decrease of original interband ones. The dielectric functions and loss functions approximately remain the same, since the bare response functions due to the linear valence and conduction bands are almost identical for these two kinds of Coulomb excitations. That is to say, the intensity, frequency, and critical momentum of the $L = 0$ plasmon in type-I systems are hardly affected by temperature (Figure 12.9a). However, the opposite is true for type-II carbon nanotubes. The T-induced intraband excitations could largely enhance the height of plasmon peak and the critical momentum, as clearly revealed in Figures 12.8a and 12.10a for the $(18, 0)$ nanotube. These results are attributed to the great increment of free conduction electrons and valence holes during the variation of temperature.

Apparently, the Coulomb excitations at a uniform magnetic field display the diverse phenomena. The $L = 0$ single-particle excitations of the type-I carbon nanotubes are strongly modified by **B**, and so does the specific magneto-plasmon mode. The $\phi = 0$ symmetric peak structure in $\text{Im}[\epsilon]$ (the open circles in Figure 12.5a) is getting into the weaker asymmetric peak as α increases at

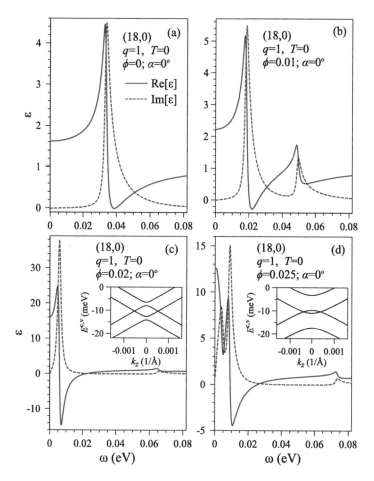

FIGURE 12.6
The real and imaginary parts of magnetodielectric functions in the $(18,0)$
zigzag nanotube at $q = 1$, $L = 0$, $\alpha = 0°$, and various magnetic fluxes:
(a) $\phi = 0$, (b) $\phi = 0.01 \phi_0$, (c) $\phi = 0.02 \phi_0$, & (d) $\phi = 0.025 \phi_0$. The insets in
(c) and (d) also show the low-lying magnetoenergy bands nearest to E_F.

a fixed magnetic flux, such as those of the $(10, 10)$ nanotube at $\phi = 0.02 \phi_0$ for
various field directions in Figure 12.5a. The drastic changes are also revealed
in Re$[\epsilon]$, in which the prominent double-side asymmetric peaks become the
observable single-side structure, as shown in Figure 12.5b. However, a perpendicular magnetic field hardly affects the bare response function (the lightly
dashed curves). i.e., the dielectric functions are identical at $\alpha = 90°$ & $0°$.
These results directly reflect the main features of magnetoband structures
(Figure 12.3) and mainly determine the energy loss spectra.

The $(10, 10)$ armchair nanotube exhibits a ϕ-dependent plasmon peak of
the $L = 0$ mode at $\alpha = 0°$ and $T = 0$, as obviously displayed in Figure 12.7c.

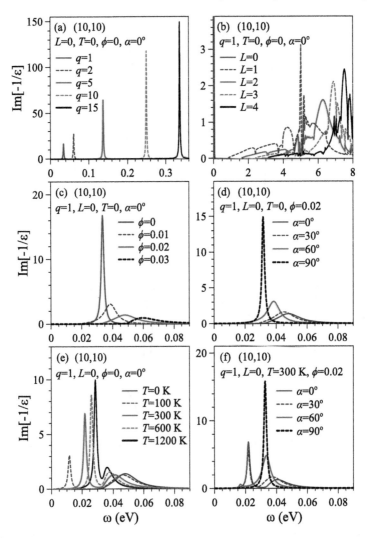

FIGURE 12.7
The energy loss spectra of the $(10, 10)$ carbon nanotube under various conditions: (a) $L = 0$, $T = 0$, $\phi = 0$ and distinct q's, (b) $q = 1$, $T = 0$, $\phi = 0$ and distinct L's, (c) $q = 1$, $L = 0$, $T = 0$, $\alpha = 0°$ and distinct ϕ's, (d) $q = 1$, $L = 0$, $T = 0$, $\phi = 0.02 \, \phi_0$ and distinct α's, (e) $q = 1$, $L = 0$, $\phi = 0.02 \, \phi_0$, $\alpha = 0°$ and distinct T's, and (f) $q = 1$, $L = 0$, $T = 300$ K, $\phi = 0.02 \, \phi_0$ and distinct α's.

The plasmon frequency increases with an increase of ϕ, while the strength of collective excitations shows an opposite behavior. Such results are closely related to the ϕ-created energy gaps and the enhanced Landau dampings (Figure 12.3a and b). A similar phenomena exist with an increase in α (Figure 12.7d). However, a very strong plasmon peak, being identical to that

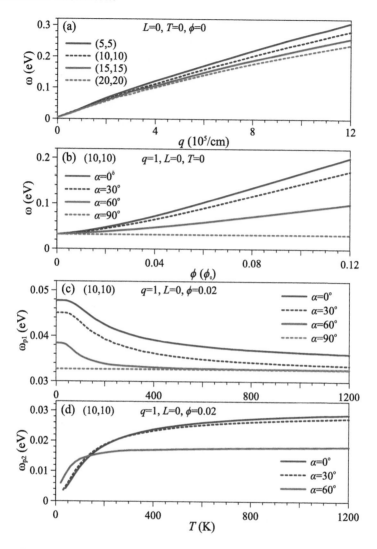

FIGURE 12.8
(a) The plasmon frequencies of the armchair carbon nanotubes under the q-dependence for different systems. The $(10,10)$ nanotube is evaluated at $q = 1$ under various α's: (b) the ϕ-dependent frequency of the first magnetoplasmon at $T = 0$; the T-dependent frequencies of the first and second magnetoplasmons at $\phi = 0.02\ \phi_0$.

at zero flux (the solid curve in Figure 12.7c), appears at $\alpha = 90°$. This further/clearly illustrates that the decrease of interband excitations is fully compensated by the increase of intraband excitations, since the bare response functions are almost the same for them. One plasmon peak is dramatically changed into two plasmon peaks at a sufficiently high temperature, except for $\alpha = 90°$. With the increase of temperature, there are more free electrons/holes in conduction/valence bands. In addition to the $v \to c$ interband excitations, temperature induces the $v \to v$ & $c \to c$ intraband excitations. The latter further creates a lower-frequency plasmon peak. The intensity and frequency of the second plasmon mode are greatly enhanced with an increase in temperature, as obviously indicated at $\alpha = 0°$ and various T's in Figure 12.7e. However, the carriers, which take part in the interband single-particle and collective excitations, are less than those at $T = 0$. This is responsible for the reduced frequency of the first/interband magnetoplasmon mode. The effects of temperature are getting weak when the direction of **B** gradually approaches the radial direction. For example, at $T = 300$ K and $\phi = 0.02 \phi_0$, two plasmon peaks are changed into one plasmon peak during the variation of $\alpha = 0° \to 90°$ (Figure 12.7f). That is, both intraband magnetoplasmon and interband magnetoplasmon are replaced one intraband and interband magnetoplasmon.

The type-I carbon nanotubes exhibit significant dispersion relations between the $L = 0$ plasmon/magnetoplasmon frequency and the critical factors of (r, q, ϕ, T). An obviously monotonous dependence of ω_p on q, as shown in Figure 12.8a, directly reflects the low-lying π-band characteristic (Figure 12.2a), the strong wave-vector relation. This clearly indicates that plasma oscillations of the $L = 0$ mode propagate along the nanotube axis in the wavelength of $2\pi/q$ without the azimuthal variation. The interband plasmon frequency, being independent of temperature, increases quickly at small q's. By the detailed calculations [19, 48, 428, 429], one can show that $\omega \propto q|ln(qr)|^{1/2}$ at small q's, as revealed in 1D electron gas [269]. The analytic q-dependence is not affected by the linear or parabolic energy dispersions, since the low-frequency excitation energy is essentially linear. Such plasmon modes in the small-radius nanotubes could survive at long critical momenta, e.g., q_c more than 0.310^5/cm in the $(10, 10)$ nanotube. With the increase of radius, their frequencies and q_c's decrease simultaneously, being attributed to reduced bare Coulomb interactions. As for the magnetoplasmons in the $(10, 10)$ carbon nanotube, there is only one interband mode under $T = 0$, being clearly illustrated in Figure 12.8b. The interband magnetoplasmon frequency displays a very prominent ϕ dependence except for $\alpha = 90°$, i.e., ω_{p1} stationarily increases with ϕ at $\alpha \neq 90°$. This result is associated with the enhancement of excitation energy/bandgap in the increment of ϕ. But for $\alpha = 90°$ (the light dashed curve), energy dispersions nearest to E_F keep the linear forms with almost same slopes and band overlaps during the variation of ϕ, so that the frequency and intensity of the interband & intraband magnetoplasmons hardly depend on ϕ. The first magnetoplasmon is also strongly affected by the increasing temperature, except for $\alpha = 90°$ (Figure 12.8c). The reduction of ω_{p1} by T directly reflects the decreased magnetointerband Coulomb excitations, since there are less (more) electrons in the valence (conduction) bands. Specifically, at $\alpha = 90°$, the total electronic excitations remain unchanged (Figure 12.7c) even under a very high temperature (e.g., T=1000 K), being attributed to the overlapping linear valence and conduction bands (Figure 12.3f). On the other side, the second magnetoplasmon shows an opposite behavior in the T-dependent frequency (Figure 12.8d). ω_{p2} is enhanced with an increase in temperature. This mode could survive only under a sufficiently high temperature, e.g., $T > 20$ K at $\alpha = 60°$ (the light solid curve). The intraband magnetoplasmon might die out at high temperatures for large deviation angles, e.g., $T > 700$ K for $\alpha = 60°$.

The magnetoelectronic excitations are very different between the type-II and type-I carbon nanotubes. For example, the $(18,0)$ carbon nanotube exhibits a destruction of state degeneracy and an overlap of valence and conduction bands because of the parallel magnetic field (Figure 12.4b and c). These results dominate the single-particle excitation. At $\phi = 0.01 \ \phi_0$ (Figure 12.6b), the splitting of the $J = 12$ and 24 energy band creates two distinct interband excitation channels. Consequently, there are two asymmetric square-root peaks of the right-hand forms in $\text{Im}[\epsilon]$ (two pairs of similar structures in $\text{Re}[\epsilon]$). It is also noticed that the spin-up and spin-down states in a narrow-gap nanotube makes an identical contribution to ϵ, i.e., such states do not alter the number of excitation channels. But for a metallic nanotube, the overlap of spin-up $J = 12$ valence band and spin-down $J = 24$ conduction band (insets in Figure 12.6c and d) induces a new intraband excitation channel. At $\phi = 0.02 \ \phi_0$ and $0.025 \ \phi_0$, $\text{Im}[\epsilon]$ clearly displays an asymmetric prominent peak in the left-hand form at the lowest excitation energy (Figure 12.6c and d). This peak comes from the Fermi-momentum state, but not the $k_z = 0$ band-edge state. Apparently, the mechanisms are different for intraband and interband electronic excitations. It should be noticed that the threshold excitation energies at $\phi = 0.02 \ \phi_0$ are almost identical for these kinds of excitation channels; therefore, a very sharp symmetric peak appears in $\text{Im}[\epsilon]$ (Figure 12.6c). Similar special structures are revealed in α-dependent magnetodielectric functions.

The energy loss spectra of type-II carbon nanotubes, as clearly illustrated in Figure 12.9a–f exhibit one, two, or three magnetoplasmon/plasmon peaks, in which their existence, frequency, and strength are very sensitive to $(\phi,q,\alpha,T,r,\theta)$. The significant interband plasmon peak in the absence of magnetic flux (the heavy solid curve) is purely due to a narrow energy gap in a 1D system (Figure 12.4a). One plasmon peak is changed into two magnetoplasmon peaks as ϕ gradually increases from zero under the specific $\alpha = 0°$. e.g., $\phi = 0.005 \ \phi_0$ and $0.01 \ \phi_0$. The $J = 24$ and 12 energy bands, respectively, create the first interband magnetoplasmon with the higher frequency (ω_{p1}) and the second interband magnetoplasmon with the lower frequency (ω_{p2}). Here, ω_{p1} obviously increases with an increase of ϕ, but the opposite is true for ω_{p2}. This behavior directly reflects the enhanced and reduced interband excitation energies, respectively, arising from the $J = 24$ and $J = 12$ energy bands (Figure 12.4b). With a further increase of magnetic flux, the ϕ-induced intraband excitations in a metallic carbon nanotube alter the characteristics

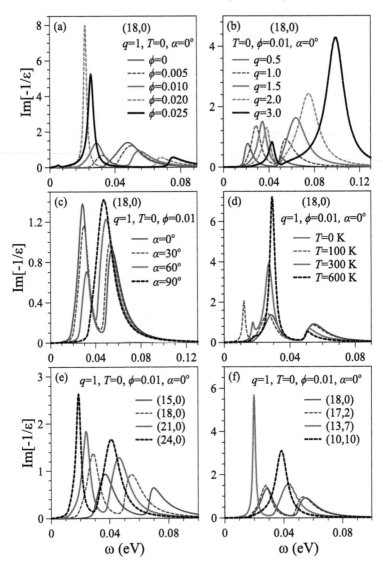

FIGURE 12.9
The energy loss spectra of the $(18,0)$ zigzag carbon nanotube under (a) various ϕ's & $(q = 1, T = 0, \alpha = 0°)$, (b) various q's & $(T = 0, \phi = 0.01, \alpha = 0°)$, (c) various α's & $(q = 1, T = 0, \phi = 0.01)$, and (d) various T's & $(q = 1, \phi = 0.01, \alpha = 0°)$. Also shown are those for (e) different zigzag nanotubes and (f) metallic and narrow-gap nanotubes at $(q = 1, T = 0, \phi = 0.01 \alpha = 0°)$.

of lower-frequency magnetoplasmon peak, such as $\phi = 0.02 \; \phi_0$ and 0.025 ϕ_0. That is to say, this peak should be attributed to the combination of intraband and interband collective excitations, namely, the intraband and interband magnetoplasmons (the second kind of magnetoplasmon). Specifically, at $\phi = 0.025 \; \phi_0$ (the red curve), there exists a weak peak at the lowest frequency. This is named as the intraband magnetoplasmon (the third kind of magnetoplasmon), since the threshold intraband and interband excitations have a wide enough energy spacing (Figure 12.6d). As to the momentum dependence, the loss functions display an unusual behavior, as indicated in Figure 12.9b. In general, the single-particle excitation energy increases with q, and so does the magnetoplasmon frequency. The plasmon peaks become more pronounced as q increases from zero. And then they decline with a further increase of q. But when q is too small or large, the magnetoplasmons are almost replaced by single-particle excitations. The magnetocollective excitations hardly survive below and beyond the critical transferred momenta (Figure 12.10).

The direction of the magnetic field, as clearly displayed in Figure 12.9c, has very strong effects on the number, frequency, and intensity of magneto-plasmon modes. For a narrow-gap $(18, 0)$ nanotube at $\phi = 0.01 \; \phi_0$, the two interband magnetoplasmons are getting into one interband magnetoplasmon during the variation of $\alpha = 0° \rightarrow 90°$. The frequency of the second interband magnetoplasmon becomes high, but its strength gets weak. However, the first interband magnetoplasmon exhibits the opposite behavior. The gradually reduced splitting of $J = 24$ and 12 energy bands is the main mechanism (Figure 12.4c–f). Moreover, the temperature strongly modifies the screened response functions. It can induce extra intraband excitations in a semiconducting nanotube. Electrons/hole will occupy the conduction/valence bands at $T \neq 0$, $\phi = 0.01 \; \phi_0$, and $\alpha = 0°$ (Figure 12.4b). These two types of free carriers generate an intraband magnetoplasmon with the lowest frequency (Figure 12.9d). They even affect the first and second interband magnetoplasmons by means of reducing the excitation strength. The temperature needs to high enough in the creation of a significant intraband magnetoplasmon peak. On the other hand, the intraband magnetoplasmon might be mixed up with the second magnetoplasmon at very high temperatures ($T \geq 600$ K), leading to a prominent peak of the latter.

The nanotube geometries, radii and chiral angles, principally determine the low-energy magnetoelectronic properties and thus the magnetoplasmons. Figure 12.9e presents the radius-dependent energy loss spectra in the narrow-gap zigzag carbon nanotubes at $\phi = 0.01 \; \phi_0$ and $\alpha = 0°$. Apparently, there are two interband magnetoplasmon modes, in which their frequencies and intensities, respectively, decrease and increase with the increase of zigzag nanotube radius. Both energy gap and band curvature decrease with an increase in radius, and so do the single-particle excitation energies and the magnetoplasmon frequencies. Since such magnetocollective excitations are damped by the interband e–h pair excitations, their strong competitions dominate the

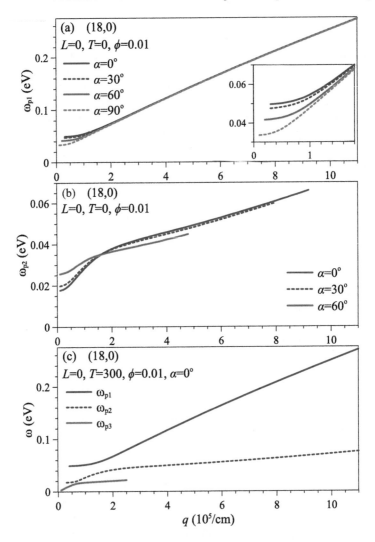

FIGURE 12.10
The momentum-dependent magnetoplasmon frequencies of the $(18,0)$ zigzag nanotube calculated at $\phi = 0.01\ \phi_0$, (a) $T = 0$ and various α's for the first mode, (b) $T = 0$ and various α's for the second mode, and (c) $T = 300$ K and $\alpha = 0°$ for three distinct modes. The inset in (a) shows the details of ω_{p1} at small q's.

strengths of magnetoplasmon peaks. The reduced bare Coulomb interactions have a stronger effect on the Landau damping, leading to enhanced magnetoplasmon peaks. As a result, it is relatively easy to observe the low-frequency magnetoplasmons in large zigzag carbon nanotubes. As to the chiral-angle-dependent energy loss functions, all narrow-gap nonarmchair carbon nanotubes exhibit two interband magnetoplasmon modes, as clearly shown in Figure 12.9f. However, a (m, m) armchair carbon nanotube display one interband magnetoplasmon mode. Owing to the singlet valence and conduction bands of $J = m$, there are no simple relations between the chiral angles and the magnetoelectronic properties (energy gaps and band curvatures). So the magnetoplasmon frequencies and intensities strongly depend on the detailed magnetoband structures. The main features of magnetoplasmon modes do not vary with the chiral angles monotonously.

The dispersion relations of magnetoplasmon frequency with momentum provide unusual characteristics for magneto-Coulomb excitations. For the narrow-gap $(18, 0)$ zigzag nanotube under $\phi = 0.01 \ \phi_0$ and $T = 0$, Figure 12.10a and b, respectively, show the significantly q-dependent frequencies of the first and the second magnetoplasmons, respectively. They approach finite values at small q's for any field directions; hence, the interband magnetoplasmons belong to optical modes. Also noticed that such magnetoplasmons hardly survive at $q \to 0$. The important differences between these two magnetoplasmons lie in the fact that the first mode exhibits a stronger q-dependence, a larger critical momentum, and a simple relation of ω_{p1} with q, and the second one is absent at $\alpha = 90°$. Their frequencies are somewhat affected by temperature, as indicated in Figure 12.10c under $T = 300$ K. Specially, the T-induced intraband magnetoplasmon presents a vanishing frequency at $q \to 0$, so it is an acoustic mode. At room temperature, this mode only exists below the critical momentum of $q \leq 2.4$. It is hybridized or replaced by the second magnetoplasmon at larger momenta, in which the latter could display a very prominent plasmon peak.

12.3   Doping Effects, Inter-π-Band Plasmons, and π Plasmons

Under the electron/hole doping, all carbon nanotubes can dramatically change the single-particle excitations and exhibit low-frequency plasmon modes. Apparently, the new/distinct e–h excitations are created/strongly modified, and so do the collective excitations. The more complicated (momentum, frequency)-phase diagrams exist during the variation of the Fermi level, such as the relations of the $L = 0$ and $L \neq 0$ plasmon modes with E_F, the q-dependent frequencies, and the critical momenta. This is attributed to the enhanced band-structure asymmetry. Experimental measurements on the electrical resistivities [459] and Raman Spectra [460] clearly show that electrons are transferred from intercalants to carbon atoms or vice versa. Charge transfer also occurs, while carbon nanotube lies on some substrates (e.g., gold [121, 461]). It is worthy of a detailed examination on the energy loss spectra of doped carbon nanotubes.

The type-I carbon nanotubes, as indicated in Figure 12.11a and b for the $10, 10$ armchair nanotube, clearly show strong doping effects under any Fermi level. At $E_F = 0$ (the solid red curve), the symmetric peak and a pair of asymmetric peaks, which are, respectively, revealed in $\text{Im}[\epsilon]$ and $\text{Re}[\epsilon]$, are purely associated with the Fermi-momentum states of interband excitations. However, with the gradual increase of E_F, such excitations decline quickly and even disappear at a very small $E_F = 3V_{pp\pi}bq/2$, in which the E_F-created intraband excitations increase rapidly and fully replace them at $\omega \leq 0.2$ eV

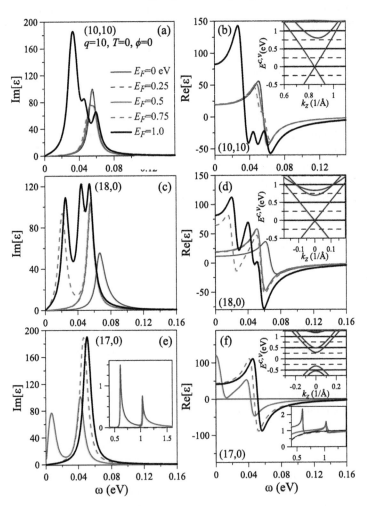

FIGURE 12.11
The low-frequency dielectric functions of three types of carbon nanotubes: $\text{Im}[\epsilon]$ of the (a) type-I $(10, 10)$, type-II $(18, 0)$, and type-III $(17, 0)$ nanotubes; similarly for $\text{Re}[\epsilon]$ in (b), (d) and (f). The insets in (b), (d) & (f) show the E_F-dependent band structures.

(not shown). For $E_F \leq 0.25$ eV (the dashed blue curves), the special structures in Im[ϵ] and Re[ϵ] almost keep the same within the range of $\omega \leq 0.2$ eV, since the great reduction of the former is compensated by the doping-induced large enhancement of the latter. The full replacement is attributed to the linear low-energy dispersions (inset in Figure 12.11b). It should be noticed that the interband excitations could survive at $\omega \geq 2E_F = 3V_{pp\pi}bq/2$. But their strengths are very weak. On the other hand, the dielectric function presents strong modifications for the sufficiently high E_F, e.g., the splitting of special structures at $E_F = 0.5$ eV (the green curve) and 0.75 eV (the dashed curve curve) arises from the different slopes in the left- and right-hand linear dispersions. Specifically, at a high Fermi level, e.g., $E_F = 1.0$ eV (the solid black curve), the originally occupied conduction band of $J = 10$ exhibits two obvious splitting structures. Furthermore, the partial occupation of the higher-energy conduction bands of $J = 9$ & $J = 11$ creates a rather prominent structure in Im[ϵ] and Re[ϵ]. The E_F-enriched bare response functions are directly reflected in the energy loss spectra, leading to a strong low-frequency plasmon, e.g., the E_F-dependent plasmon peaks of the $(10, 10)$ nanotube. The plasmon frequency is somewhat affected by the Fermi level, but without a simple relation between ω_p and E_F. The intraband plasmon of $E_F = 0$ is changed into the intraband plasmon at $E_F \neq 0$. Only at $E_F = 1$ eV, the frequency and intensity of the low-frequency plasmon mode are greatly enhanced by the more free conduction electrons. Such carriers can create a low-frequency interband plasmon of the $L = 1$ mode, with $\omega_p > 1.3$ eV, as obviously observed in the inset of Figure 12.12a by the dashed brown curve. The $L \geq 2$ modes do not display strong plasmon peaks at low frequency.

Apparently, the type-II carbon nanotubes present the doping-diversified Coulomb excitation phenomena, being directly illustrated in Figure 12.11c and d. The $(18, 0)$ nanotube, with a narrow gap of $E_g \sim 10$ meV, has the initial interband excitations from the $k_z = 0$ band-edge valence state (discussed earlier), so that they induce a special structure in Im[ϵ] and Re[ϵ] at $\omega \sim E_g + 3V_{pp\pi}bq/2$ (66 meV; the almost linear dispersions). The similar special structures are revealed before the occupation of the higher-energy conduction bands, e.g., those at $E_F = 0.25$ eV (the dashed blue curve) and 0.5 eV (the solid green curve). However, the intraband excitations are getting into the intraband ones, and the latter appears at $\omega \sim 3V_{pp\pi}bq/2$ (56 meV) under the stronger response. The number of intraband excitation channels increases with a further increase of the Fermi level. For example, the $J = 11$ and 25 conduction bands, with the smaller curvature, are simultaneously occupied under $E_F = 0.75$ eV (the inset in Figure 12.11b), and their intraband contributions generate new special structures at the lower frequency (~ 21 meV in the dashed red curve). Concerning $E_F = 1.0$ eV, there are more occupied conduction bands, and so do the intraband excitation channels and special structures in ϵ. The initial intraband excitation energies, which, respectively, originate from the Fermi-momentum states of the $(12, 24)$, $(11, 25)$ and $(10, 26)$ conduction bands, correspond to $\sim$55, 45, and 25 meV. Although the single-particle

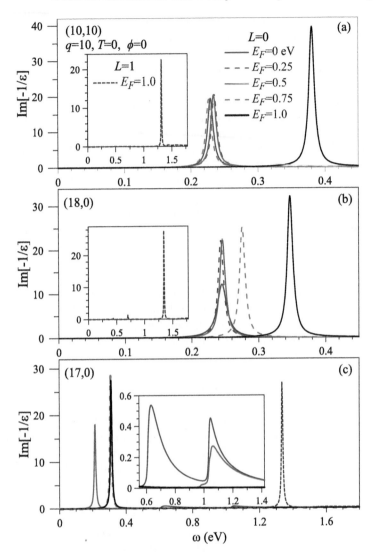

FIGURE 12.12
The energy loss spectra under various Fermi levels for the (a) type-I $(10, 10)$, type-II $(18, 0)$, and type-III $(17, 0)$ carbon nanotubes.

response functions directly reflect the E_F-dependent characteristics, there is only one prominent low-frequency plasmon peak in the energy loss spectra, as shown in Figure 12.12b). This mode should represent the $L = 0$ collective excitations of all free conductions in distinct energy bands.

To dramatically change the excitation properties, the Fermi levels of type-III carbon nanotubes need to be higher than half of the energy gap, as clearly shown in Figures 12.11e,f and 12.12c. For example, the threshold Fermi level is $E_F \sim 0.303$ eV for the $(17,0)$ moderate-gap nanotube (the inset in Figure 12.11f). In general, the bare response functions, without the intraband excitations, do not exhibit low-frequency excitations and thus the special initial structures, e.g., those at $E_F = 0$ and 0.25 eV. As to $E_F = 0.5$ eV, it crosses two doubly degenerate conduction bands, covering the lowest energy bands of $J = 11$ & 22 and the next higher ones of $J = 12$ & 22. Their free conduction electrons induce two roughly symmetric peaks in $\text{Im}[\epsilon]$/two pairs of asymmetric peaks in $\text{Re}[\epsilon]$ (Figure 12.11e and f). At $\omega = 8$ and 45 meV (the green solid curve), such distinct structures are merged together with a further increase of Fermi level, e.g., those generated at $E_F = 0.75$ eV and 1.0 eV (the dashed red curve and the solid black curve). The main reason is that energy dispersions are most identical for the larger-k_z states. On the other hand, the interband excitations of the $L = 0$ mode only survives at a specific frequency range of $\omega \geq 2E_F$. Within 0.4 eV $\leq \omega \leq 1.6$ eV (the insets in Figure 12.11e and f), the square-root asymmetric peaks are observable at $E_F = 0$, 0.25 and 0.50 eV, in which they disappear at higher E_F's. Specially, the $L = 1$ interband plasmons due to the free carriers hardly depend on the type of carbon nanotubes with almost the same radii, e.g., the roughly identical frequencies and intensities of the $(10,10)$, $(18,0)$, and $(17,0)$ nanotubes (the inset in Figure 12.12a, the inset in Figure 12.12b and c).

Obviously, the doped carbon nanotubes exhibit geometry-diversified plasmon modes, as obviously revealed in Figure 12.13a–c. The low-frequency plasmons of the $L = 0$ mode, which present the $q|ln(qr)|^{1/2}$-dependent frequencies at long wavelengths, belong to acoustic modes. They can survive in type-I and type-II nanotubes (Figure 12.13a and b) with any Fermi levels, but for type-III systems only under $E_F \geq E_g/2$ (Figure 12.13c). In general, such plasmons arise from the intraband excitations of free conduction electrons. However, at $E_F = 0$, they are related to the interband excitations of valence electrons (the red curves in Figure 12.14a and b). Their frequencies hardly depend on low doping, while the opposite is true for the higher Fermi level with the extra occupation of conduction bands, e.g., the strong Fermi-level dependence in the $(10,10)/(18,0)$ nanotubes at $E_F > 0.75$ eV, and the $(17,0)$ system at $E_F > 0.5$ eV. On the other hand, the $L = 0$ interband plasmons, being revealed in the $(17,0)$ nanotube, appear at a high enough frequency (> 0.5 eV) and sufficiently large q's and, e.g., two distinct plasmon modes at $E_F = 0$ (the red curves in Figure 12.14c) and a single optical mode at $E_F = 0.5$ eV (the solid green curve). Specifically, the $L = 1$ interband plasmons due to free carriers exist only at high Fermi levels, e.g., those under $E_F = 1.0$ eV (the dashed brown curves). The critical momentum is largest in the $(17,0)$ nanotube. The aforementioned results directly reflect the main features of energy bands.

The strong competitions between low-frequency plasmons and e–h excitations are clearly illustrated by the (q,ω)-phase diagram of the type-I $(10,10)$ carbon nanotube. A pristine system at zero temperature, as observed in Figure 12.14a, exhibits the $L = 0$ interband transitions. The acoustic plasmon

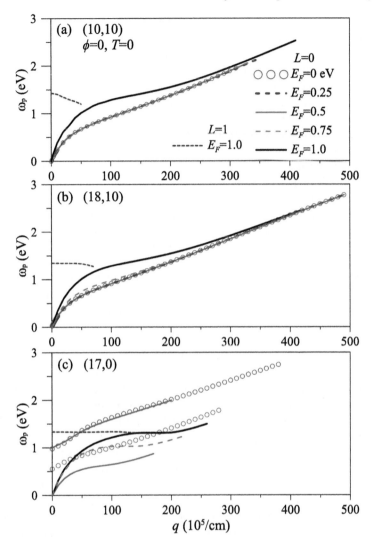

FIGURE 12.13
The momentum-dependent frequencies of the $L = 0$ and $L = 1$ modes under various Fermi levels for the (a) type-I (10,10), (b) type-II (18,0), and (c) type-III (17,0) carbon nanotubes.

experiences e–h dampings at any transferred momenta. This mode has a strong intensity at small q's, while it will disappear under $q > 0.4$ Å^{-1}. During the increase of doping level (Figure 12.14b–e), dramatic changes are revealed in the interband and intraband single-particle excitation regions (the dashed and solid white curves), such as the variation of boundary and more excitation

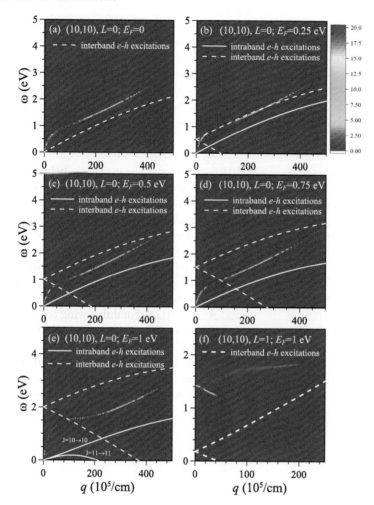

FIGURE 12.14

The (momentum, frequency)-phase diagrams of the $L = 0$ mode for $(10, 10)$ nanotube under (a) $E_F = 0$, (b) $E_F = 0.25$ eV, (c) $E_F = 0.5$ eV (d) $E_F = 0.75$ eV, and (e) $E_F = 1.0$ eV. (f) The $L = 1$ mode is also shown at $E_F = 1.0$ eV.

channels. At long wavelength limit, the collective excitations appear in the absence of Landau dampings (the left-lower corners by black circles), being different from the undoped case. However, the plasmon frequency and critical momentum are hardly affected by Fermi energies. Moreover, the $L = 1$ optical plasmon is seriously damped by the interband single-particle excitations (Figure 12.14f), leading to a small critical momentum. The unusual doping effects are attributed to the special Coulomb matrix elements and energy dispersions of the linearly intersecting band structure.

All carbon nanotubes exhibit higher-frequency inter-π-band plasmons and π plasmons, as clearly revealed in energy loss spectra of $L = 0$ for three types of systems (Figure 12.15a–c). The frequencies of the former and latter lie, respectively, in the range of 1.5 eV$\leq \omega_p \leq$ 5 eV and $\omega_p >$ 5.3 eV. There are several inter-π-band modes, being identified from the significant plasmon peaks. They arise from the specific valence$\rightarrow$conduction band transitions under the same J. Apparently, the number of plasmon modes is determined by the geometric

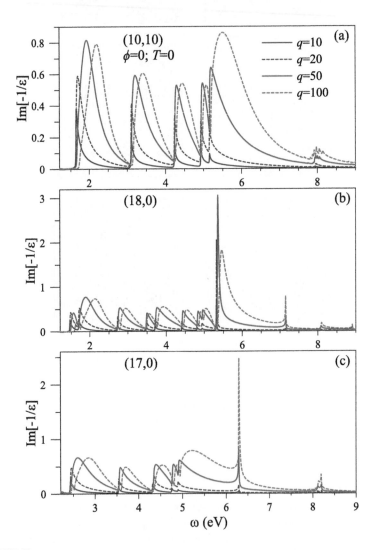

FIGURE 12.15
The energy loss spectra, with a broadening factor of 0.02 eV, under the various transferred momenta for the (a) (10,10), (b) (18,0), and (17,0) carbon nanotubes.

structures/the J-dependent band structure. For example, the $(10, 10)$, $(18, 0)$, and $(17, 0)$ carbon nanotubes, respectively, present 4, 8, and 5 inter-π-band modes. In general, the intensity and frequency of inter-π-band peak increase with an increase of transferred momentum [428]. Moreover, the π plasmons, which are due to all valence π electrons, exhibit prominent peaks at the highest frequency. They are relatively easy to be observed at large transferred momenta. It should be noticed that few weak peaks at frequency higher than 7 eV originate from the σ bands. In addition to the q-dependence, both inter-π-band and π plasmons strongly depend on the discrete L modes, and they are sensitive to the nanotube radius only for $r < 40$ Å. The larger the L is, the higher (stronger) the frequency (intensity) of the π plasmon is.

Figure 12.16a–c clearly shows the dispersion relations of the $L = 0$ inter-π-band and π plasmon frequencies with the transferred momenta, being useful in understanding the collective carrier oscillations. Among three types of carbon nanotubes, the armchair and narrow-gap zigzag systems (Figure 12.16a and b), respectively, exhibit the smallest and largest mode numbers, reflecting the lowest and highest available channels under various inter-π-band transitions. The q-dependence might be strong or weak, regardless of the plasmon frequencies. Furthermore, the critical momenta for some plasmon modes behave so. These two results indicate that it will be very difficult to identify which plasmon modes have unusual behaviors. However, most of the plasmon modes could survive at $q > 0.5$ Å^{-1}. Specifically, for the $(17, 0)$ nanotube (Figure 12.16c), the highest-frequency inter-π-band plasmon is larger than the π plasmon in excitation frequencies, since their loss peaks, respectively, correspond to the (thin & high) and (broad & middle) ones. The aforementioned plasmon modes are verified from the composite bundle of single-walled carbon nanotubes using accurate EELS measurements [177].

The angular-momentum-decoupled π plasmons, which only appear in condensed-matter systems with the cylindrical symmetries, are worthy of a closer examination. All carbon nanotubes exhibit L-dependent collective excitations due to the whole valence π electrons, as clearly shown in Figure 12.17a–c. Such π plasmons, belonging to optical modes, could survive at very large transferred momenta. The discrete modes are absent in other sp^2- and sp^3-bonding carbon-related systems [177]. The π-plasmon frequencies increase with an increase in q or L, and so do the single-particle excitation energies. Among the discrete π plasmons, the $L = 0$ modes display the strongest momentum dispersion relations, owing to the almost symmetric valence and conduction energy bands with the optimal q- and k_z-dependences. Most important, the $L = 0$ and $L \neq 0$ π-plasmon modes, respectively, correspond the collective π-electron oscillations along the nanotube axis and the longitudinal & transverse directions (the axial & azimuthal ones). For example, the $L = 1$ mode behaves as a propagating wave along the cylindrical surface, as well as the dipole-like standing wave along the azimuthal direction. Up to now, there are no experimental measurements in distinguishing the various π plasmons, except that the measured optical spectra could identify the $L = 0$

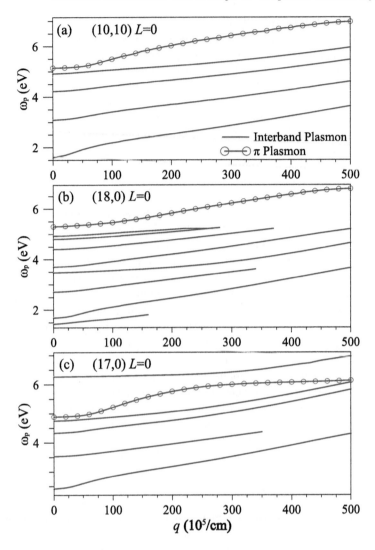

FIGURE 12.16
The dispersion relations of the inter-π-band plasmons and the π plasmon (red circles) for the (a) (10,10), (b) (18,0), and (17,0) carbon nanotubes.

and $L = 1$ single-particle excitation channels [177]. The high-frequency π plasmons might be suitable in testing the L decoupled modes through the high-resolution energy loss spectra [39,173–179]. This will open a new experimental research category.

In addition to a single-walled carbon nanotube, the π-electronic excitations have been thoroughly investigated for a multiwall carbon nanotube, single-walled carbon nanotube bundle with a finite number, and the graphitic

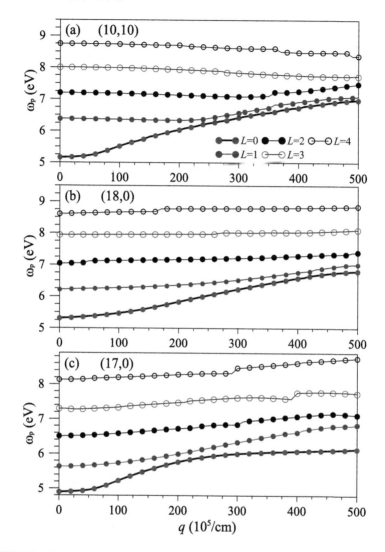

FIGURE 12.17
The dependences of the L-decoupled π plasmon modes for the (a) (10,10), (b) (18,0), and (17,0) carbon nanotubes.

systems with any layers [462]. The superlattice model is utilized to explore the higher-frequency Coulomb excitations, in which the low-energy band-structure effects due to the intertube/interlayer atomic interactions are negligible. The energy loss spectra of nanotube systems exhibit several plasmon peaks. The most prominent one corresponds to the π plasmon with $\omega_p > 2V_{pp\pi}$, and the others are the inter-π-band plasmons (> 1 eV). Apparently, the distinct inter-π-band plasmons disappear in the layered graphitic systems, owing to the absence of J-characterized energy subbands, The π plasmons are very sensitive to the changes in the number (N) of carbon nanotubes or graphitic layers, the transferred momentum, and the transferred angular momentum. The intertube/interlayer Coulomb interactions greatly enhance the frequency and oscillator strength of π plasmons during the increase of N. A multiwall carbon nanotube can present L-decoupled π plasmons, as observed in a single-walled system. For the $L = 0$ π plasmon, the multiwall carbon nanotube behaves as a layered graphite, but not like a single-walled nanotube bundle. The radius dependence could be ignored for a multiwall carbon nanotube; however, it is very strong in a single-walled nanotube bundle. Among three systems, the first one shows stronger collective excitations, higher oscillation frequency, and a relatively rapid increase of ω_p with q. The calculated results are responsible for the up-to-date experimental measurements, such as two kinds of plasmon modes in a multiwall carbon nanotube [462], the 5.8 eV-π plasmon in a large single-walled nanotube bundle [462], and the similarities and differences between a single-walled carbon nanotube, and a multiwall carbon nanotube. The EELS measurements are required to examine the predicted results: the q- and N-dependent π plasmons in the graphite-related systems and the L-decoupled modes in a multiwall carbon nanotube.

12.4   Significant Differences between 1D Carbon Nanotubes and 2D Planar Graphenes

In addition to the similar Coulomb excitation phenomena, there are certain important differences between 1D cylindrical carbon nanotubes and 2D layered graphenes. These two systems exhibit low frequency, π, and $\pi + \sigma$ plasmon modes, in which they, respectively, appear in the loss energy spectra at < 1, > 5, and > 20 eV. Such collective excitations mainly arise from the free carriers, the whole valence π electrons, and all the $(2s, 2p_x, 2p_y, 2p_z)$ orbitals, directly reflecting the characteristics of sp^2-bonding carbon-related systems. However, the cylindrical and planar symmetries further diversify the Coulomb excitations, since the distinct periodical boundary conditions will determine the main characteristics of electronic states and Coulomb interactions.

A carbon nanotube possesses the translation symmetry along the tubular and azimuthal rotation symmetry, while a planar graphene only has a 2D translation symmetry. Apparently, the electronic states of the former and the latter are, respectively, characterized by (k_z, J) and 2D $\mathbf{k}$, and so do the conservation of the transferred (momentum, angular momentum) and momentum $[(q, L)]$ and momentum $[\mathbf{q}]$. A cylindrical system can create a lot of discrete excitation modes coming from the distinct L's, being never observed in any other condensed-matter systems. However, the experimental verifications on the L-decoupled energy loss spectra are difficult to achieve up to now [177]. This is the main reason why there exist many inter-π-band plasmons of $\sim$1–4 eV's, being absent in the layer graphene systems [198,201,202]. On the other hand, the 2D transferred momentum is associated with the significant direction-dependent energy loss spectra; that is, the planar graphene systems might exhibit anisotropic screened response functions.

The bare Coulomb interactions, being determined by the geometric symmetries, are very different for carbon nanotubes and layered graphenes. As a result, the transferred momentum dependences are greatly diversified. Only the metallic armchair nanotubes present the acoustic plasmon of $L = 0$, while this collective mode cannot survive in type-II and type-III ones. It purely originates from the 1D linearly intersecting valence and conduction bands (the interband excitations). The strength of the interband $L = 0$ plasmon is reduced at a finite temperature, and an extra intraband $L = 0$ plasmon is generated under a sufficiently high temperature. At a long wavelength limit, the plasmon frequency is roughly proportional to $q|ln(qr)|^{1/2}$, as revealed in 1D electron gas. [269] Of course, the low-frequency plasmon is absent in a pristine monolayer graphene. After the e–h dopings, all carbon-related systems will exhibit free carrier-induced collective excitations. The low-frequency plasmons in any system belong to the acoustic mode. However, the momentum dependence is approximately described by $\sqrt{q}$. This significant difference is attributed to the distinct form in the 1D and 2D bare Coulomb interactions, respectively, corresponding to $I_0(qr)K_0(qr)$ and $v_q = 2\pi e^2/q$. It should be noticed that low doping hardly affects the low-frequency plasmons in type-I carbon nanotubes, and the high doping will create $L = 1$ plasmon in these systems, but not type-I and type-III ones.

Any external magnetic field cannot create highly degenerate Landau levels (LLs) in a cylindrical carbon nanotube, except under a giant strength. The main mechanism is that the net magnetic flux is vanishing at $\alpha \neq 0°$, and the parallel field only leads to decoupled angular momentum of $J + \phi/\phi_0$. The 1D energy dispersions and energy gaps are strongly modified by **B**. The van Hove singularities in DOSs consist of square-root asymmetric peaks, accompanied with one plateau structure across the Fermi level in a metallic nanotube. They exhibit the well-known Aharonov–Bohm effect in the absence of Zeeman splitting. The magnetic wave functions, which are due to the parallel and perpendicular fields, respectively, present the well-behaved and abnormal standing waves with the specific zero point and irregular oscillation. On the other side, a uniform perpendicular magnetic field in layered graphenes can induce dispersionless LLs. Their spatial distributions present symmetric/antisymmetric behaviors about the -dependent localization centers. In general, the LLs clearly show unusual dependences on quantum number and field strength, being sensitive to the stacking configuration and number of layers. There are only delta-function-like symmetric peaks in DOSs. The diverse magnetoelectronic properties in these two systems are directly reflected in other essential physical properties, e.g., the distinct magneto-optical selection rules [118] and electrical magnetoconductances [463,464].

The magneto-Coulomb excitations are rich and unique in 1D nanotube and 2D graphene systems. Each armchair carbon nanotube exhibits one interband magnetoplasmon at low temperature, when the magnetic field is not perpendicular to the tubular axis. The plasmon frequency decreases/increases with the increase of α/ϕ, while the opposite is true for its strength. Specifically, one intraband and interband magnetoplasmon appears at $\alpha = 90^{c}irc$. Furthermore, it is almost the same with that in the absence of ϕ, regardless of the temperature. The nonarmchair narrow-gap carbon nanotubes, which are present in a **B**-field, clearly reveal two interband magnetoplasmons. However, the nonarmchair metallic systems exhibit one interband magnetoplasmon and one intraband & interband magnetoplasmon The important differences among such plasmon modes become relatively obvious if the magnetic field is more close to the nanotube axis. The transferred momenta will determine the frequency and existence of magnetoplasmons, in which the dispersion relation is a monotonous dependence. Moreover, the temperature can create an intraband magnetoplasmon or change an interband magnetoplasmon into an intraband & interband magnetoplasmon. As to magnetoplasmons in monolayer graphene (discussed earlier in Chapter 10), they originate from the interband inter-LL transitions at $T = 0$. Such plasmon modes are purely induced by a perpendicular field even without any free carriers. In general, the magnetoplasmons cannot survive at large q's, and their frequencies present a nonmonotonous q-dependence. The main reason is the strong competition between the longitudinal Coulomb interactions and the transverse cyclotron forces. At finite temperatures, there are more magnetoplasmons, in which the extra modes correspond to the intraband inter-LL ones.

13

Electronic Excitations in Monolayer Silicene and Germanene

The layered silicene and germanene have stirred a lot of theoretical and experimental studies, mainly owing to the buckled structure, important spin–orbital interactions, the hexagonal symmetry, and the nanoscaled thickness, especially for the first two critical factors. These 2D systems are composed of dominant sp^2 bondings and the nonnegligible sp^3 ones, in which their strong competitions determine the optimal bucklings with the different heights of A and B sublattices. Recently, the buckled and hexagonal geometries, with the distinct stacking configurations, are verified by the high-resolution scanning tunneling microscopy (STM) and transmission electron microscopy (TEM) measurements [145, 465, 466], such as the recent experimental identifications on the AB- and AA-stacked bilayer silicene [467]. It should be noticed that both bilayer/few-layer silicene (germanene) and graphene quite differ from each other in essential properties because of very strong interlayer hopping integrals (even larger than the intralayer nearest-neighbor one) [387]. The significant buckling structures and spin–orbital couplings clearly indicate that the fundamental properties of monolayer silicene and germanene are easily modulated by external electric and magnetic fields, since the equivalence of A and B sublattices and the spin degree of freedom are greatly modified by them. As to the theoretical predictions, the tight-binding model [34, 151, 152], the effective-mass approximation [137, 138, 142, 142], and the first-principles calculations [134, 468] are used to study electronic properties in the absence/presence of electric field, and they are approximately consistent with one another in the low-lying energy bands. Furthermore, the former two methods are available in understanding the magnetoelectronic properties [34, 138, 142] and thus magnetoabsorption optical spectra [137, 142] of the highly degenerate Landau levels (LLs). On the experimental side, for monolayer silicene, the angle resolved photoemission spectroscopy (ARPES) measurements show the separated Dirac-cone structure [469], and the STS examinations display the magnetic quantization phenomena [127, 470]. Similar experimental measurements are done for monolayer germanene, identifying the parabolic conduction band [471] and the V-shape-like density of state (DOS) near the Fermi level [472]. The former measurements indicate the non-Dirac-cone band structure, since there exist strong hybridizations between germanene and substrate. In addition, no evidences of long-range

Friedel oscillations are revealed in the STM images [472], mainly owing to the semiconducting behavior of monolayer germanene.

The significant differences among monolayer silicene, germanene, and graphene in Coulomb excitations are worthy of a systematic investigation. The generalized tight-binding model, which covers the intrinsic geometric structures, orbital hybridizations, spin–orbital interactions, and external electric and magnetic fields, is combined with the 2D random-phase approximation (RPA) to fully explore the many-particle properties induced by the electron–electron Coulomb interactions. The $3p_z/4p_z$-orbital chemical bondings of Si/Ge atoms are sufficient for studying the low-energy essential properties. Temperature, electric field, magnetic field, & doping, and even their composite effects are included in the detailed calculations. The strong competitions of cooperation between the intrinsic and extrinsic factors are expected to greatly diversify the electronic excitation phenomena. It is well known that the energy gap of monolayer silicene (germanene), being associated with the slightly separated Dirac-cone structure, is mainly determined by the important spin–orbital coupling. It will modify the boundary of the interband electron–hole (e–h) excitations, and its cooperation with temperature further leads to the separation of interband and intraband excitation boundaries. The T-induced acoustic plasmon modes are predicted to exhibit diverse behaviors. Apparently, Coulomb excitations are largely enriched by the electric and magnetic fields, since they, respectively, create the semiconductor–semimetal transition & the splitting of energy bands (the destruction of the z-direction mirror symmetry) and the spin-dependent magnetic quantization. The complex effects due to the combination of temperature, electric field, magnetic field, and doping are studied thoroughly. Moreover, a detailed comparison is also made for monolayer graphene, silicene, and germanene.

13.1   Temperature-Induced Electronic Excitations in Narrow-Gap Systems

Both monolayer silicene and germanene have buckled structures, as clearly shown in Figure 13.1a–c. The A and B sublattices are situated at distinct heights with a difference of $2l$, such as $l = 0.23$ and 0.33 Å for Si and Ge, respectively (the top view in Figure 13.1b). Similar to graphene, their honeycomb lattices consist of two atoms in a primitive unit cell. The tight-binding mode, which is built from the $3p_z/4p_z$ orbitals, is suitable for describing the low-lying energy bands, when compared with the first-principles calculations [34, 151, 152]. With the inclusion of significant spin–orbital couplings, the nearest-neighbor Hamiltonian in the presence of π bonding is expressed as

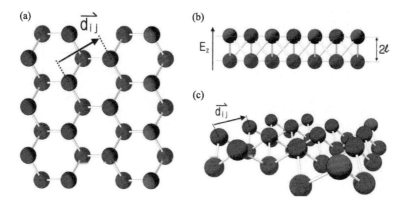

FIGURE 13.1
The buckled structure of silicene and germanene: (a) the top view, (b) the side view under a uniform perpendicular electric field, and (c) the 3D geometry.

$$H = -t \sum_{\langle ij \rangle \alpha} c_{i\alpha}^\dagger c_{j\alpha} + i\frac{\lambda_{SO}}{3\sqrt{3}} \sum_{\langle\langle ij \rangle\rangle \alpha\beta} v_{ij} c_{i\alpha}^\dagger \sigma_{\alpha\beta}^z c_{j\beta} - i\frac{2}{3}\lambda_{R2}$$
$$\sum_{\langle\langle ij \rangle\rangle \alpha\beta} u_{ij} c_{i\alpha}^\dagger (\vec{\sigma} \times \vec{d}_{ij}^{\,0})_{\alpha\beta}^z c_{j\beta} + \ell \sum_{i\alpha} \mu_i E_z c_{i\alpha}^\dagger c_{i\alpha}, \qquad (13.1)$$

where $c_{i\alpha}^\dagger$ ($c_{j\beta}$) creates (destroys) an electronic state with spin polarization of α (β) at the i-th (j-th) site. The summation is taken for all pairs of nearest neighbors $\langle ij \rangle$ or those of the next-nearest neighbors $\langle\langle ij \rangle\rangle$. The first term represents the nearest-neighbor hopping integrals, and t is, respectively, 1.1 and 0.86 eV for monolayer silicene and germanene. The second term, the effective spin-orbital (SO) coupling, has the strength of $\lambda_{SO} = 3.9$ meV for silicene (46.3 meV for germanene). $\sigma = [\sigma_x \, \sigma_y \, \sigma_z]$ is the Pauli matrix vector. v_{ij} is mainly determined by the orientation of the two nearest bonds connecting the next-nearest neighbors, i.e., $v_{ij} = +1(-1)$ if the next-nearest-neighbor hopping is anticlockwise (clockwise) with respect to the positive z-axis. The third term is the so-called intrinsic Rashba SO coupling with $\lambda_R = 0.7$ and 10.7 meV for silicene and germanene, respectively, where $\vec{d}_{ij} = \vec{d}_{ij}/|\vec{d}_{ij}|$ with $\vec{d}_{ij}$ connecting the i-th and j-th sites in the same sublattice, and $u_{ij} = +1(-1)$ denotes the A (B) sublattice. Finally, a uniform perpendicular electric field creates a strong effect on the sublattice-dependent Coulomb potential energies, where $\mu_i = 1$ and -1, respectively, correspond to the A and B sites.

Both monolayer silicene and germanene, as clearly shown in Figure 13.2a and b, exhibit unusual low-lying band structures, strongly depending on the spin–orbit coupling and the wave vector. The unoccupied conduction states are symmetric to the occupied valence one about the Fermi level, as observed

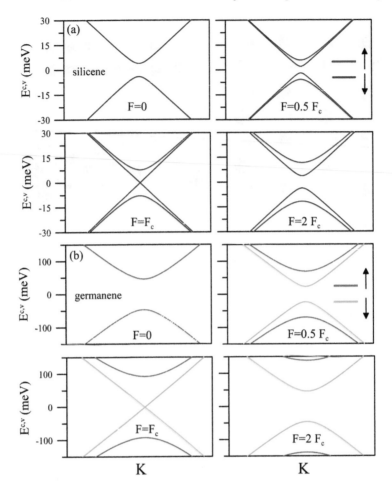

FIGURE 13.2
The low-lying band structures for monolayer (a) silicene and (b) germanene in the presence of various electric fields.

in monolayer graphene (Figure 3.1 in Chapter 3). There also exist equivalent valleys, the K and K′ ones, mainly the unchanged hexagonal symmetry. Electronic states in the presence of spin–orbital couplings are doubly degenerate for the spin freedom (Figure 13.2a), although they have the up- and down-dominated spin configurations simultaneously. These two distinct spin configurations can make the same contribution to the Coulomb excitations under zero fields. Moreover, the spin–orbital interactions lead to the separation of Dirac points and thus induce an energy spacing of $E_D = 2\lambda_{SO}$ (just E_g in the pristine case) near the K point (the blue curves). For example, silicene and germanene, respectively, show $E_g = 7.8$ ad 93 meV. The distorted energy bands possess parabolic dispersions near the band-edge states and then gradually are getting into linear ones in the increment of wave vector. However, at higher energy ($|E^{c,v}| > 1$) eV, they recover into parabolic ones and have a saddle point at the M point with very high DOSs. It is noticed that only the low-lying states, $|E^{c,v}| \leq 0.3$ eV for silicene (0.2 for germanene) display an isotropic energy spectra.

The single-particle and collective excitations of monolayer silicene are chosen for a model study. Its dielectric function, being in the absence of external electric and magnetic fields, is similar to that of monolayer graphene, since electronic states are double degenerate for the spin degree of freedom (Figure 13.2a). At zero temperature, Im[ϵ] and Re[ϵ] exhibit different bare responses under the vanishing and finite spin–orbital couplings, as clearly shown in Figure 13.3a and b. For the former case, the linearly intersecting Dirac cone, as revealed in monolayer graphene, only induce interband excitations. The threshold transition channels are due to the Dirac point (the Fermi momentum or the band-edge state); therefore, the excitation frequency is expressed as $\omega_{th} = v_f q$ ($v_F \sim 5.52 \times 10^5$ m/s). These make Im[ϵ] and Re[ϵ], respectively, divergent in the forms of $\sqrt{\omega - \omega_{th}}$ and $\sqrt{\omega_{th} - \omega}$ [132], e.g., those shown by the black curves under $q = 1$ (in unit of 10^5/cm). On the other hand, the slightly separated Dirac cone, which is generated by the spin–orbital interactions (Figure 13.2a), leads to the shoulder and logarithmically divergent peak in the imaginary and real parts, respectively (the red curves). Such structures are located at the threshold frequency $\omega_{th} = \sqrt{E_g^2 + v_F^2 q^2}$. Apparently, the zero point of Re[ϵ] at a small Im[ϵ] is absent under both cases. This means that no prominent plasmon mode could survive at zero temperature.

Temperature can induce available excitation channels and thus dramatically change the number and form of van Hove singularities in bare polarization functions. When temperature gradually starts from zero, the intraband transitions are created and quickly enhanced. For example, Im[ϵ] (Re[ϵ]) with the spin–orbital coupling, as shown for $q = 1$ and $T = 50$ K in Figure 13.3c and d exhibits a strong square-root peak and a weak shoulder structure (a prominent asymmetric peak and an observable logarithmic peak) at $\omega = v_F q$ and $\sqrt{E_g^2 + v_F^2 q^2}$, respectively. The second Re[ϵ] zero point is located in the Im[ϵ] gap, where an undamped plasmon mode exists. In addition, only the former pronounced structure could survive in the absence of spin–orbital couplings because of the same e–h excitation boundary. Apparently, the intraband transition channels strongly suppress the interband ones with the increase of T. As a result, the latter is almost vanishing at $T = 100$ K, being clearly illustrated in Figure 13.3e and f. Furthermore, a zero point in Re[ϵ] appears at small Im[ϵ] if temperature is sufficiently high. This suggests the existence of T-induced collective excitations. As to monolayer germanene, it needs to have a higher temperature to initiate single-particle excitations ($T = 300$ K in Figure 13.3g and h), since this system has the largest energy gap [154].

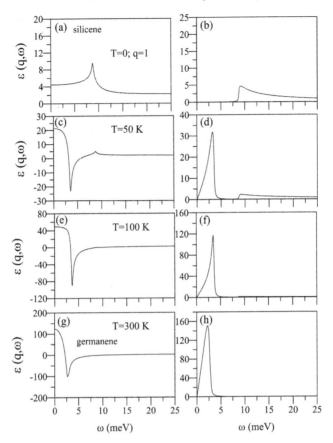

FIGURE 13.3
The real and imaginary parts of the dielectric function for monolayer silicene at $q = 1$ (10^5/cm) under [(a), (b)] $T = 0$ K, [(c), (d)] $T = 50$ K, [(e), (f)] at $T = 100$ K, and the similar results for monolayer germanene under [(g), (h)] $T = 100$ K. Also displayed in (a–f) for comparison are those without spin–orbital couplings.

The momentum- and temperature-dependent single-particle excitations, which are clearly characterized by the imaginary part of dielectric functions in Figure 13.4a–c, are very useful in understanding the available transition channels and Landau dampings. At $T = 0$, the interband e–h excitations only survive above the threshold curve defined by $\sqrt{E_g^2 + v_F^2 q^2}$. (the dashed curve in Figure 13.4a). There exists the strongest interband response only at larger momenta near the boundary. The T-induced intraband transitions create a lower boundary corresponding to $v_F q$ [the maximum intraband excitation frequency due to the conduction or valence Dirac point], e.g., Im[$\epsilon(q, \omega)$] at $T = 50$ K in Figure 13.4b. Apparently, an excitation gap is generated

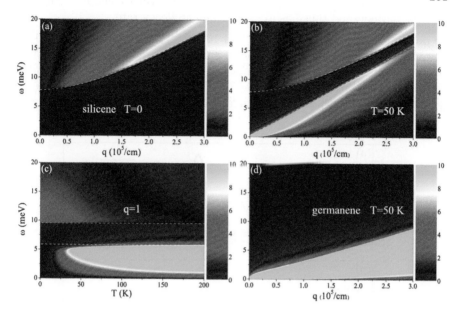

FIGURE 13.4

The dependences of the e–h excitation strength $(\mathrm{Im}[\epsilon])$ on the frequency & momentum for monolayer silicene under (a) $T = 0$ K and (b) $T = 50$ K, and (c) another variations on the temperature & frequency under the specific $q = 1$. (d) The similar relations with (ω, q) are shown for monolayer germanene under $T = 50$ K. The white dashed curves indicate the boundaries of single-particle excitations.

between the maximum intraband excitation frequencies and the threshold interband ones, strongest in which it is larger at small transferred momenta with a prominent intraband response. The coexistence of two excitation categories is sensitive to temperature. For example, at $q = 1$ (Figure 13.4c), such behavior is observable in the range of 25 K$\leq T \leq 100$ K. Under lower and higher temperatures, the e–h excitations are fully dominated by the interband and intraband transition channels. Also, the coexistent region depends on the transferred momentum.

The energy loss spectra of monolayer silicene and germanene are mainly determined by the intrinsic interactions, temperatures, transferred momenta, as clearly illustrated in Figure 13.5a–d. With or without spin–orbital couplings, only a weak peak appears at zero temperature (the black curves in Figure 13.5a and b). When temperature is high enough, an observable and broad peak, being regarded as the collective excitation mode, appears in the absence of intrinsic interactions, e.g., the plasmon peaks at $T = 50$, 100, & 200 K in Figure 13.5a. They are damped by interband single-particle excitations (Figure 13.3a–f). However, the spin–orbital couplings make such peaks become very strong and sharp for a certain range of temperature, such as the prominent peaks at $T = 50$, 75, and 100 K in Figure 13.5b. In addition, they induce another weak and higher-frequency plasmon peak, corresponding to the interband excitation channels (Figures 13.3d and 13.4c). Under the higher-temperature case, only the high and wide plasmon peaks appear, e.g., those at $T = 200$ K. The bandgap is much smaller than the thermal

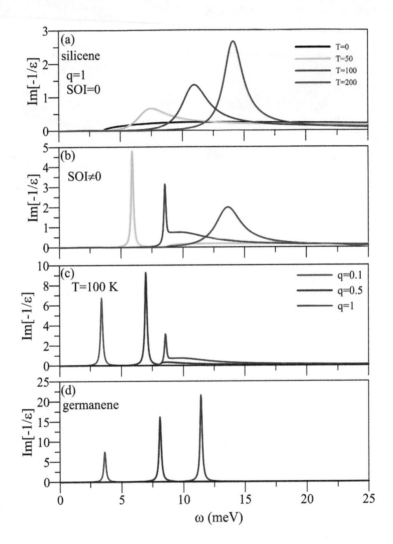

FIGURE 13.5
The energy loss spectra of monolayer silicene (a)/(b) at various T's under a specific $q = 1$ in the absence/presence of spin–orbital interactions, and (c) at distinct q's for a fixed $T = 100$ K with intrinsic interactions. Similar momentum dependences are also shown for monolayer germanene in (e)/(f).

energy, so that the former does not play a critical role in Coulomb excitations, i.e., the E_g-dependent interband transitions are almost fully suppressed by the intraband ones. The frequency of plasmon peak increases monotonously with an increase in temperature, while its intensity displays a nonmonotonous behavior because of the complex T- and q-induced e–h dampings (Figure 13.4a–c). Similar dependences are revealed in the q-dependent energy loss spectra (Figure 13.5c). However, the second plasmon is absent at very small q's, e.g., the screened response function at $q = 0.1$ and $T = 100$ K [the black curve]. That the second mode is observable depends on T and q. As to monolayer germanene, only the sufficiently high temperature can create a prominent plasmon peak, e.g., the loss spectra at $T = 300$ K (Figure 13.5d). Obviously, it is relatively easy to verify the T-induced intraband single-particle excitations and plasmons in monolayer silicene.

The (q, ω)- and (T, ω)-phase diagrams of monolayer silicene and germanene exhibit a rich and unique Coulomb excitation phenomena. Figure 13.6a–c, with the momentum dispersion relations under the specific temperature, clearly illustrates three kinds of excitation phases in terms of diverse plasmon behaviors. Apparently, the energy gap results in the full separation of intraband and interband e–h boundaries [the white dashed curves]. When the thermal energy is comparable and less than the bandgap

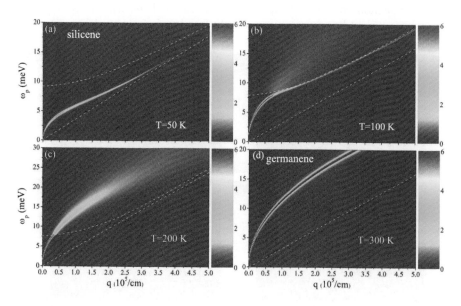

FIGURE 13.6

The momentum-dependent plasmon spectra of monolayer silicene under (a) $T = 50$ K, (b) 100 K & (c) 300 K, and the similar dependence for monolayer germanene under (d) $T = 200$ K. The upper and lower dashed curves, respectively, correspond to the interband and intraband excitation boundaries.

($T = 50$ K in Figure 13.6a), the plasmon mode remains undamped before the critical momentum ($q_c \sim 2.5$), in which it experiences very strong intraband e–h dampings near q_c. However, if $k_B T$ is higher and close to E_q ($T = 100$ K in Figure 13.6b), the strength of plasmon is greatly enhanced. Its existence range can reach $q_c \sim 5$, in which the Landau dampings arise from interband e–h excitations. Specifically, another higher-frequency plasmon, being seriously suppressed by the similar dampings, exists in the range of $0.5 \leq q \leq 1.5$. On the other hand, the drastic change in the Fermi–Dirac distribution near the Fermi level appears under $k_B T >> E_g$ ($T = 200$ K in Figure 13.6c), so that the interband excitations dominate the phase diagram. This undamped plasmon only survives within the narrow range of $0 \leq q \leq 0.25$, and then the pronounced interband e–h dampings make a sharp and strong peak transform into a wide and significant structure with the critical momentum of $q_c \sim 4$. In addition, the second plasmon mode is absent at very high temperatures, corresponding to the singly dominating interband transition channels.

The (T, ω) phase diagram under a specific q, as clearly illustrated in Figure 13.7a–d, could provide the evolution of screened excitation spectra. When temperature starts to rise from zero, it must be high enough in creating the dominating intraband transition channels and thus the observable plasmon peak, such as $T = 15$, 25, and 35 K for $q = 0.2$, 1 and 2, respectively (the spectral curves in Figure 13.7a–c). And then, the intraband-induced plasmon, which is located between two e–h excitation boundaries (two white dashed lines), experiences slight Landau dampings within a certain

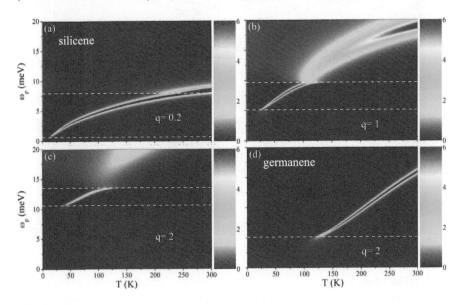

FIGURE 13.7
The frequency-temperature phase diagrams for monolayer silicene at (a) $q = 0.2$, (b) 1 & (c) 2, and (d) for germanene at $q = 2$.

temperature range, e.g., 15 K$\leq T \leq$ 250 K, 25 K$\leq T \leq$ 110 K, and 35 K$\leq T \leq$ 115 K, respectively, corresponding to $q = 0.2$, 1, and 2. This unusual T-range becomes narrow as q increases. With the further increase of temperature, a wide and significant plasmon peak, being closely related to the interband e–h Landau dampings, appears under any transferred momenta. Obviously, the critical temperature in determining the intraband plasmon is higher for monolayer germanene with a larger bandgap (Figure 13.7d). Moreover, it is quite different among silicene, germanene, and graphene (details in Sections 13.1 and 4.1). For example, at $q = 1$, its value is, respectively, 24, 125, and 40 K (Figure 13.7b).

13.2   Electric-Field-Enriched Coulomb Excitations

The main features of low-energy electronic properties in buckled systems are easily modulated by a gate voltage. Apparently, the Coulomb potential energy difference, which corresponds to different-height A and B sublattices (Figure 13.8b), induces the destruction of mirror symmetry about the $z = 0$ plane. The spin-dependent electronic states are split into spin-up- and spin-down-dominated ones; that is, one pair of valence and conduction bands changes into two pairs of spin-dependent ones (Figure 13.2). These valence and conduction bands are, respectively, denoted by $1^{(c,v)}$ and $2^{(c,v)}$, such as $E^{c,v}$ at $V_z = 2eFI$ in unit of λ_{SO} (the red curves), where the effective spin–orbital coupling is λ_{SO} ($\sim E_D$). Concerning electronic states near the K point, the first pair of energy bands, with the spin-down-dominated configuration, is relatively close to the Fermi level. However, the opposite is true for those near the K' point (not shown). When the gate voltage starts to increase from zero, the first pair gradually approaches to E_F and E_D is getting smaller. The normal Dirac-cone structure exists under $V_z = \lambda_{SO}$ (the green curves), where the electronic spectrum has a pair of linearly intersecting valence and conduction bands. With the further increase of V_z (the dashed orange curves), energy spacing (just E_g) of parabolic valence and conduction bands is opened and enlarged. On the other hand, the second pair of energy bands for the spin-up-dominated configuration is away from the Fermi energy. Most important, two splitting spin-dependent configurations are expected to greatly diversify single-particle excitations and plasmon modes.

The bare response function exhibits the dramatic transformation of the joint van Hove singularities during the variation of temperature and electric field, especially for the latter. The special structures in Im[ϵ], as clearly shown by the red curves in Figure 13.8a–e, have a diverse excitation phenomena. At zero T and F, it only has interband transition channels in a narrow-gap system, leading to a specific shoulder structure (Figure 13.8a). With a suitable temperature (e.g., $T = 50$ K in Figure 13.8b), the dominating intraband

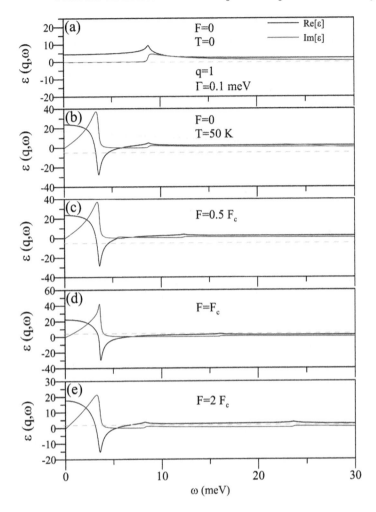

FIGURE 13.8
The dielectric functions of monolayer silicene at various temperatures and
electric fields: (a) $T = 0$ & $F = 0$, (b) $T = 50$ K & $F = 0$, (c) $T = 50$ K &
$F = 0.5F_c$, (d) $T = 50$ K & $F = F_c$, and (e) $T = 50$ K & $F = 2F_c$.

transitions and the significant interband ones could survive together, in which
they, respectively, induce the left-hand asymmetric peak and the shoulder
structure at the lower and upper boundary frequencies. When an electric field
begins to rise from zero under a fixed temperature, an extra shoulder exists,
e.g., Im[ε] for $F = 0.5F_c$ in Figure 13.8c. That is, there are three coexistent special structures, the asymmetric peak, shoulder, and shoulder (three arrows), in
which the first and second ones mainly come from the spin-down-dominated
valence and conduction bands (Figure 13.2b), and the third one is determined

by the spin-up-dominated energy bands. Apparently, the energy gap, energy spacing, and thermal energy are responsible for such structures. However, only two special structures (Figure 13.8d) appear in a zero-gap system just under $F = F_c$ (Figure 13.2c). The interband transitions are almost thoroughly suppressed by the intraband ones for the linearly intersecting energy bands, leading to the absence of the second structure. With the further increase of field strength, three special structures are recovered at higher frequencies, e.g., $\text{Im}[\epsilon]$ at $F = 2F_c$. That energy gap is getting into a larger value is the main reason. In addition, the corresponding structures in the imaginary and real parts of dielectric functions satisfy the Kramers–Kronig relations (the red and black curves).

The energy loss spectra strongly depend on the composite effects due to the temperature and electric field, as clearly illustrated by Figure 13.9a–d. The dependence on F is prominent, being revealed by the number, intensity, and frequency of plasmon peaks. When the temperature is fixed at $T = 50$. The intraband excitations dominate the bare and screened response functions and thus account for a significant plasmon peak. Its strength is strongest at $F = 0$ (Figure 13.9a), reduces in the range of $F \leq F_c$ (Figure 13.9b and c), and then becomes more pronounced at $F \geq 2F_c$ (Figure 13.9d). It is very difficult to observe another higher-frequency plasmon peak (Figure 13.9a and c), respectively, owing to the larger bandgap and energy spacing between the spin-down and spin-up split energy bands (Figure 13.2). This mode is observable at $F = 0.5F_c$ (Figure 13.9b) for a smaller-gap silicene. As to the frequency of the most prominent plasmon, a simple relation with electric field is absent.

The (q, ω)-, (T, ω)-, and (F, ω)-phase diagrams, as clearly indicated in Figures 13.10–13.12, is useful in understanding the complex excitation relations due to the bandgap, energy spacing, energy dispersions (or the intrinsic atomic and spin interactions), temperature, and electric field; that is, these critical factors are responsible for the diverse single-particle and collective excitations. Under the different electric fields, the e–h excitation boundaries and plasmon mode/modes in (q, ω) space have four types of strong relations with the transferred momentum, namely, types (I)–(III) in Figures 13.9a and 13.10c at $T = 50$ K and type (IV) in Figure 13.10d at $T = 200$ K. At zero field ((I) in Figure 13.10a), there exists an excitation gap between the lower and upper boundary curves (the region within the dashed white curves), in which they, respectively, correspond to the highest intraband transitions and the initial interband ones. The T-induced intraband plasmon, which belongs to an acoustic mode, survives within this gap. Its strength gradually declines and then disappears at larger q's near the intraband boundaries. But for $F = 0.5F_c$ ((II) in Figure 13.10b), three distinct e–h boundaries exist; furthermore, the lower, middle, and upper ones, respectively, originate from the intraband excitations of the spin-down-dominated valence and conduction bands, the interband excitations from this pair of energy bands, and the interband excitations due to the spin-up-dominated pair of energy bands. By the approximate expansion near the K point, the middle and upper boundaries are described by

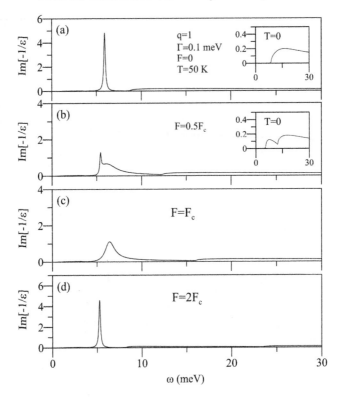

FIGURE 13.9
The energy loss spectra for monolayer silicene at $q = 1$, $T = 50$ K and distinct electric fields: $F = 0$, (b) $F = 0.5F_c$, (c) $F = F_c$ and (d) $F = 2F_c$.

$$\omega = 2\sqrt{\left(\ell E_z - s\sqrt{\lambda_{SO}^2 + a^2\lambda_R^2 q^2/4}\right)^2 + \hbar^2 v_F^2 q^2/4}, \qquad (13.2)$$

where $s = +1$ and -1, respectively, correspond to the former and the latter. The intraband plasmon exists in the gap between the lower and middle boundaries, and it is damped by the interband e–h excitations and then the intraband excitations with the increase of q. Furthermore, another weak plasmon mode arising from the interband excitations of the spin-down-dominated pair is seriously suppressed by similar transitions. This plasmon is absent at long wavelength limit. Specifically, at the critical electric field ((III) at $F = F_c$ in Figure 13.10c), the linearly intersecting energy bands (Figure 13.2c) only create the same boundary curve for the intraband and interband excitations, being characterized by $v_F q$. The upper e–h boundary is determined by the spin-up-dominated pair of energy bands. The excitation gap cannot survive for the zero bandgap system, so that the T-created intraband plasmon is frequently accompanied with the significant interband e–h dampings. When the

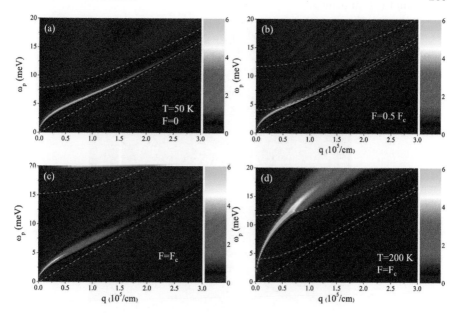

FIGURE 13.10

The (q, ω) phase diagram in monolayer silicene at various temperatures and electric fields: (a) $T = 50$ K & $F = 0$, (b) $T = 50$ K & $F = 0.5F_c$, (c) $T = 50$ K & $F = F_c$, and (d) $T = 200$ K & $F = 0.5F_c$.

electric field is fixed under $F = F_c$ and the temperature is enhanced to 200 K, an extra phase diagram (IV) appears in Figure 13.10d. A prominent T-induced plasmon is undamped very close to $q \to 0$ and quickly enters into the interband e–h dampings. Such mode is observable at larger momenta (the critical value ~ 6. The aforementioned q-dependent phase diagrams are never revealed in graphene systems (Figure 3.4).

The temperature-dependent phase diagrams could provide the full information about the coexistent intraband and interband plasmon modes due to the first pair of energy bands, as clearly shown in Figure 13.11a–c. Under a specific $q = 1$, the critical temperatures in determining the dramatic transformation in the intensity of energy loss spectrum strongly depend on the electric field strengths; that is, they are greatly diversified by the various F's. The initial temperature for the creation of intraband plasmon is, respectively, 30, 25, and 22 K under $F = 0$, $F = 0.5F_c$, and $F = F_c$. This result mainly arises from the strong competition among the bandgap, threshold excitation frequency, and thermal energy. However, for such fields, the second critical temperatures are in sharp contrast with one another, in which they are 100, 50, and 250 K. The discontinuous transformation of the screened response function means the emergence of higher-frequency interband transitions. These e–h excitations are, respectively, the interband excitations due to the nondegenerate first pair

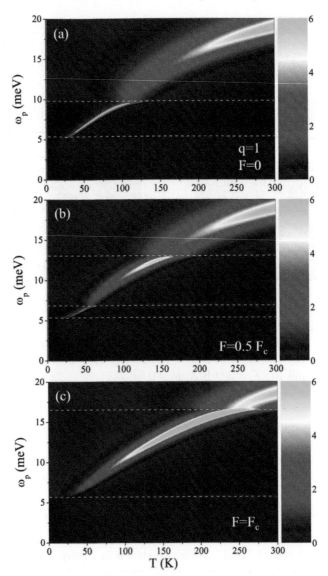

FIGURE 13.11
The (T, ω) phase diagram of monolayer silicene at $q = 1$ and distinct electric fields: (a) $F = 0$, (b) $F = 0.5F_c$, and (c) $F = F_c$.

of valence and conduction bands, the spin-down-dominated pair, and the spin-up-dominated pair (Figure 13.2). Specifically, concerning $F = 0.5F_c$, the third critical temperature occurs at 175 K, since the interband excitations associated with the spin-up-dominated energy bands become important there. It should be noticed that two plasmon peaks (strong and weak modes) might

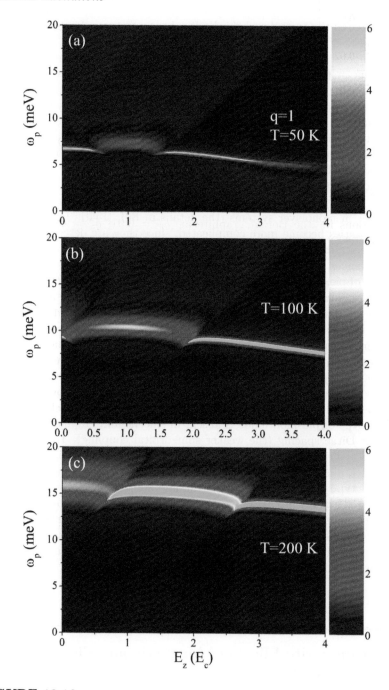

FIGURE 13.12
The (F, ω) phase diagram for monolayer silicene at $q = 1$ and different temperatures: (a) $T = 50$ K, (b) $T = 100$ K, and (c) $T = 200$ K.

coexist close to the second and third critical temperatures. In short, the composite effects of the nearest-neighbor hopping integral, the spin–orbital coupling, temperature, and electric field can create diverse Coulomb excitation phenomena in (q, ω)-, (T, ω)-, and (F, ω)-phase diagrams. They cover four types of single-particle and collective excitations, being characterized by the different e–h boundaries/the available transition channels, one or two plasmon modes, and the existence or absence of Landau dampings. Moreover, the different critical momenta, temperatures, and field strengths are revealed under various cases. Such critical factors represent the emergence or disappearance of plasmon mode and significant e–h dampings.

In addition, the composite effects, which principally originate from the electric field and electron doping, have been thoroughly explored for the Coulomb excitations in monolayer systems (the details in Ref. [154]). The e–h excitation boundaries become more complicated, since they are determined by the band-edge states and the doping Fermi momenta. There exist certain significant differences among germanene, silicene, and graphene in terms of diverse behaviors of plasmon modes, mainly owing to the strength of spin–orbit interactions and buckled structures. For the extrinsic monolayer systems, all of them clearly display an acoustic plasmon mode at long wavelength limit, as observed in 2D electron gas. That is, the effects due to the spin–orbital-induced energy spacing are fully suppressed. However, for larger momenta, they might change into another kind of undamped plasmons (the first type), experience the seriously suppressed mode during heavy intraband eVh excitations (the second type), keep the same undamped plasmons (the third type) or decline, and then vanish within the strong interband e–h excitations (the fourth type). The first type of plasmon mode is absent in silicene and graphene as a result of smaller Dirac spacing (the weaker spin–orbital coupling). To observe the azimuthal-angle-dependent plasmon modes, the required Fermi energies are, respectively, about 1.0, 0.4, and 0.2 eV for graphene, silicene, and germanene. The third type is revealed in germanene and silicene, with the requirement of extremely low Fermi energy (0.01 eV) for the latter. The fourth type for germanene and silicene is purely generated by the electric field, while it frequently appears in few-layer extrinsic graphenes without external fields [23]. In short, germanene possesses four kinds of plasmon modes only under the small variation in Fermi energy.

13.3   Composite Effects of Magnetic and Electric Fields and Doping

For monolayer silicene, the magnetoelectronic Coulomb excitations, which are in the presence of an applied electric field and a tunable Fermi level, are thoroughly investigated using suitable and delicate calculations. The generalized tight-binding model is directly combined with the 2D RPA; furthermore, they simultaneously incorporate all meaningful interactions, including the nearest-neighbor hopping integral, the important spin–orbit coupling, the significant interactions between charge carriers and external fields, and the strong electron–electron interactions. Since all low-energy states are included in numerical evaluations, the predicted results are reliable over a wide range of excitation frequencies, magnetic-/electric-field strength, and Fermi energies. The dispersion relation with the transferred momentum for the magnetoplasmons could be characterized by two categories. One is a propagating mode, while another belongs to a localized mode. The former is mainly driven by the longitudinal Coulomb interactions, while the latter is principally governed by magnetic forces. The electric field in the buckled structure creates localized plasmon modes, mainly owing to the lifting of spin and valley degeneracy. This section is focused on the B_z-dependent plasmon spectrum and the dramatic changes in the main features of plasmon modes as a result of occupation transformation. The observable modulation of plasmon excitations by the electric and magnetic fields provides a possible way to design an active plasmon device in low-buckled materials.

The magnetic Hamiltonian for monolayer silicene/germanene becomes more complicated, when compared with that in graphene. It needs to cover the spin–orbital coupling and the different on-site energies for A and B sublattices arising from uniform perpendicular electric field. This method has been discussed thoroughly in Chapter 10, while the spin degree of freedom is included in the calculations. The enlarged rectangular unit cell (Figure 13.13c) [387], which corresponds to a period of the Peierls phase, includes $4R_B$ Si atoms. The Hamiltonian matrix is a $8R_B \times 8R_B$ Hermitian matrix, as illustrated in Figure 13.13a. With $B_z = 4$ T, where $R_B = 8,000$, the Hamiltonian possesses a dimension of $64,000 \times 64,000$. Only the nearest- and next-nearest-neighbor hopping integrals contribute to the nonvanishing matrix elements. Furthermore, the well-behaved Landau wave functions, which are almost identical to those of a harmonic oscillator (the product of the Hermite polynomials and the exponential decay), exhibit localized distributions near the 1/6, 2/6, 4/6, and 5/6 positions of the enlarged unit cell. The localization centers are determined by the effective momentum of the magnetic field and the k_y component of two valleys [387]. Consequently, the numerical calculation time associated with the giant Hamiltonian matrix could be largely reduced. The independent Hamiltonian matrix elements could be found in the Appendix of Ref. [34]. The spin–orbital-induced hopping phases are illustrated in Figure 13.13b.

The important differences in the LL energy spectra and magnetic wave functions between silicene and graphene (details in Chapter 10) are that the former has the $n^c = 0$ and $n^v = 0$ LLs split by the significant spin–orbital coupling; furthermore, the low-lying LL degeneracy and energy are sensitive to a perpendicular electric field, mainly owing to an obvious on-site potential difference between the A and B sublattices. These are clearly revealed in the

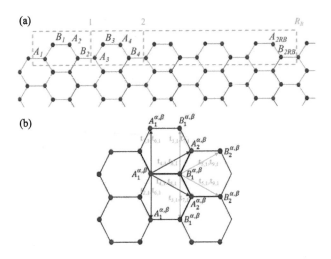

FIGURE 13.13
(a) The enlarged rectangular unit cell of monolayer silicene in the presence
of a uniform perpendicular magnetic field and (b) the various hopping phases
related to the spin-orbital couplings and the external fields (details in the
appendix of Ref. [34]). α and β indicate spin-up and spin-down configurations.

F-dependent LL spectrum shown in Figure 13.14a. From the nodal structure
of the wave functions (Chapter 10), the effective quantum number n^c/n^v for
each conduction/valence LL is mainly determined from the zero-point number
in the dominating sublattice (one of four sublattices: A and B sublattices)
with spin-up and spin-down configurations. The highest occupied LL, which
depends on the doping level, is denoted by n_F , e.g., $n_F = 1$ for two occupied
LLs of $n^c = 0$ and 1 under low electron doping. Except for the $n^{c,v} = 0$ ones,
each LL is eightfold degenerate for each (k_x, k_y) state, being attributed to the
K and K' valleys, the mirror symmetry of $z = 0$ plane, and the spin degree of
freedom. The 2D carrier density for each $n^{c,v} \geq 1$ LL is $8/3\sqrt{3}b^2 R_B$, where
the inverse of denominator is the area of the reduced first Brillouin zone.
The degeneracy of the $n^{c,v} = 0$ LLs is only half that of the others, and the
intrinsic bandgap determined by the former is about 7.9 meV. The splitting of
the spin-up- and spin-down-dominated $n^{c,v} = 0$ LLs exist for any electric-field
strengths, in which the former dominates the magnetoenergy gap between
the conduction and valence LLs (the green arrow). However, the splittings of
$n^{c,v} \geq 1$ LLs only occur at high electric fields and their energies are almost
independent of the electric-field strength, so that such behaviors could be
neglected in the magnetoelectronic excitations.

    The composite effects, which arise from the spin–orbital coupling, elec-
tric field and magnetic field, deserve a closer examination. In general, the
highly degenerate quantized LLs can exhibit rich and unique energy spectra

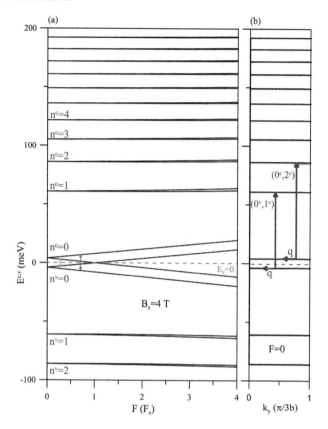

FIGURE 13.14

(a) The electric-field-dependent LL energy spectrum at $B_z = 4$ T for the well-behaved wave functions in monolayer silicene, and (b) the inter-LL transitions with the transferred momentum of q under the zero electric field.

and magnetic wave functions under sufficiently strong spin–orbital interactions, as predicted in monolayer germanene and tinene [133]. For monolayer silicene, with a weak spin–orbital coupling, the prominent magnetic quantization almost dominates the main features of LLs; that is, the effect due to the magnetic field strongly suppresses that associated with the first critical factor. For example, even at low magnetic field ($B_z = 4$ T in Figure 13.14a), the energy spacing between $n^{c,v} = 0$ and 1 LL is much higher than the order of spin–orbital interactions. Here, only the $n^{c,v} = 0$ LLs are very sensitive to three critical mechanisms. Such LLs clearly show the spin- and valley-split behavior. The spin-up-dominated LLs of the 1/6 (4/6) localization center are identical to the spin-down-dominated ones of the 2/6 (5/6) localization center in terms of state energy and wave function. This is responsible for fourfold degeneracy. The crossing behavior only reveals in the $F-$ and B_z-dependent

$n^{c,v} = 0$ energy spectra. On the other hand, for monolayer germanene and tinene, the crossing and anticrossing phenomena frequently appear in the field-related energy spectra, owing to the stronger spin–orbital coupling and the weaker hopping integrals (the smaller Fermi velocities). As a result, it is relatively easy to observe the well-behaved and perturbed LLs, the unusual Coulomb excitations, and the extra magneto-optical selection rules in these 2D systems.

For intrinsic silicene, with $n_F = n^v = 0$, electrons could be excited from the occupied valence LLs to the unoccupied conduction ones. A single-particle mode has an excitation frequency $\omega_{th} = E^c(n^c, \mathbf{k} + \mathbf{q})$-$E^v(n^v, \mathbf{k})$ through the energy and momentum conservation. A pair of numbers (n^c, n^v) is employed to label the specific inter-LL transition channel, being clearly illustrated by the blue solid curve in Figure 13.14b. It belongs to the interband inter-LL transitions. When the monolayer system is under low electron doping, only the $n^c = 0$ LLs at $B_z = 4$ T are fully occupied states, corresponding to $E_F = 8$ meV. The $(0^c, 1^c)$ inter-LL transitions should be classified into intraband ones (the blue dashed curve). Apparently, there exist interband and intraband inter-LL transitions in extrinsic systems. The aforementioned number pair is also utilized to represent a plasmon branch, indicating that the collective excitations are closely related to the single-particle channel. The evidence is identified in the plasmon frequency that approaches the e–h excitation energy in the large or small limit of q (discussed later).

The single-particle excitation spectrum, being characterized by the imaginary part of dielectric function, is very sensitive to the changes in Fermi energies and electric fields, as clearly indicated in Figure 13.15b, d, f and h. Each delta-function-like peak in $\text{Im}[\epsilon]$ represents a major inter-LL transition channel. Its intensity strongly depends on the Coulomb-matrix elements $|\langle n', \mathbf{k} + \mathbf{q} | e^{i\mathbf{q} \cdot \mathbf{r}} | n, \mathbf{k} \rangle|^2$, being closely related to the wave-function overlap between the initial and final LL states. Based on the spatial characteristics of the Hermite polynomials (the well-behaved magnetic wave functions), the inter-LL transitions, with a lower transition order of $\Delta n = |n - n'|$, present the larger Coulomb-matrix elements under small q's, especially for $\Delta n = 1$ & 2. Furthermore, $\Delta n = 0$ is almost forbidden because of the vanishing Coulomb matrix elements. The opposite is true for the larger-Δn inter-LL transitions. For example, at $q = 1$, the intrinsic silicene exhibits three initial excitations peaks $(0^v, 1^c)$ & $(1^v, 0^c)$, $(1^v, 2^c)$, and $(2^v, 3^c)$ from the low- to high-frequency channels, as illustrated by the black curve in Figure 13.15b. All of them arise from the interband transition channels of $\Delta n = 1$. Moreover, the corresponding real part in $\text{Re}[\epsilon]$ shows the neighboring asymmetric peaks (Figure 13.15a), accompanied with the zero points (indicated by the purple arrow) or the smaller values at higher frequencies.

The channels of single-particle excitations are drastically altered under extrinsic condition, as clearly illustrated in Figure 13.15d, f and h for various Fermi energies. If the $n^c = 0$ LLs are fully occupied with $n_F = n^c = 0$ at $E_F = 8$ meV and $B_z = 4$ T (Figure 13.15d), the intensity of the $(0^v, 1^c)$ &

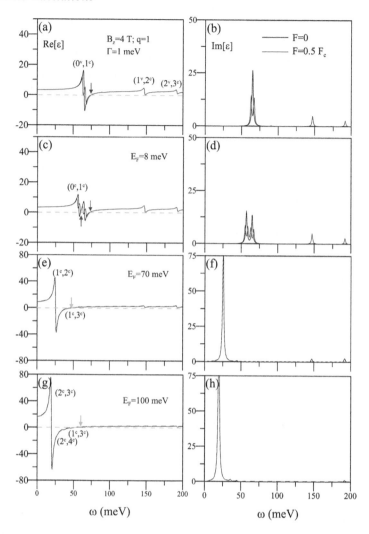

FIGURE 13.15
The real/imaginary parts of the magneto-dielectric functions at $B = 4$, $q = 1$, distinct electric fields and various Fermi energies (a)/(b) $E_F = 0$, (c)/(d) $E_F = 8$ meV, (e)/(f) $E_F = 70$ meV and (g)/(h) $E_F = 100$ meV. $F = 0$ and $F = 0.5F_c$, respectively, correspond to the black and red curves.

$(1^v, 0^c)$ interband transitions is reduced by a factor of 2 as half the spectral weight is shifted to a lower-frequency intraband peak for the $(0^c, 1^c)$ mode. However, the aforementioned two channels $(0^c, 1^c)$ and $(0^v, 1^c)$ are Pauli blocked when the Fermi level crosses the $n^c = 1$ LLs for $n_F = 1$ and $E_F = 70$ meV (Figure 13.15f). The replacement is a more prominent lower-frequency peak due to the intraband channel of $(1^c, 2^c)$. In addition, there exists a very weak channel of $(1^c,3^c)$. The significant changes in peak intensity and frequency mainly originate from the in-phase inter-LL transition and the reduced LL spacing at higher energy. This is even more evident for a larger n_F, e.g., the strongest $(2^c,3^c)$ channel under $n_F = 2$ and $E_F = 100$ meV (Figure 13.15h), accompanied with the emergence of the weak $(2^c,4^c)$ and $(1^c,3^c)$ channels and the absence of the $(1^v,2^c)$ one.

A finite electric field creates the spin- and valley-polarized LLs of $n^{c,v} = 0$ (Figure 13.14a) and thus enriches the single-particle excitation spectra. For an intrinsic silicene, the strong peak of $(0^v,1^c)$ is split into two according to different valley transitions for a spin-related configuration (the red curve in Figure 13.15b under $F = 0.5F_c$); that is, the neighboring half-intensity peaks are, respectively, due to the spin-down- and spin-up-dominated LLs (Figure 13.14a). The two interband peaks, respectively, exhibit the red and blueshifts when F is increased from zero to F_c. Their frequency difference reaches a maximum as the bandgap completely vanishes at $F = F_c$. But for $F > F_c$, both split peaks move to higher frequencies due to the reopening of bandgap (not shown). If both $n^c = 0$ LLs are fully occupied (Figure 13.15d), there are four robust spin- and valley-polarized peaks. They result from two-spin and two-valley transitions. This phenomenon is absent under $n_F = 1$ and $n_F = 2$ (Figure 13.15f and h), since the lowest-frequency prominent peak is independent of $n^{c,v} = 0$ LLs. Moreover, all e–h excitation spectra remain identical.

The energy-loss function is useful for understanding the E_F- and F-diversified magnetoplasmon modes in monolayer silicene, as clearly illustrated in Figure 13.16a–d. Each spectral peak in $\text{Im}[-1/\epsilon(q,\omega)]$ could be interpreted as the screened response of a plasmon mode, with different degrees of Landau damping. Under the intrinsic case, the most intense interband plasmon mode is located between the single-particle excitations of $(0^v,1^c)/(1^v,0^c)$ and $(1^v,2^c)$ (the green vertical dashed lines in Figure 13.16a). The most prominent peak corresponds to the zero point of $\text{Re}[\epsilon]$ at a very small $\text{Im}[\epsilon]$ (the purple arrow in Figure 13.15a). However, the second and third peaks, which are close to the single-particle frequencies of $(1^v,2^c)$ and $(2^v,3^c)$, respectively, are relatively weak as a result of considerable Landau dampings from the higher-frequency e–h excitations. The main features of magnetoplasmon modes, their intensities and frequencies, change noticeably as the Fermi level crosses a specific LL. The spectral weight of the lowest-frequency interband plasmon of $(0^v,1^c)$ is partially transferred to a lower-frequency intraband mode $(0^c,1^c)$ when the $n^c = 0$ LL is completely occupied, which is shown as a black curve in Figure 13.16b. The former experiences serious Landau dampings and is much weaker than the latter. More free conduction electrons produce an enhanced intraband magnetoplasmon, e.g., the most strong plasmon peak in Figure 13.16c for $n_F = 1$. The intraband peak increases in intensity obviously with the increase in N_F, such as the $n_F = 2$ case in Figure 13.16d. This is purely due to the increased (decreased) number of intraband (interband) channels plotted in the blue (green) vertical dashed lines. Changing the broadening parameter does not

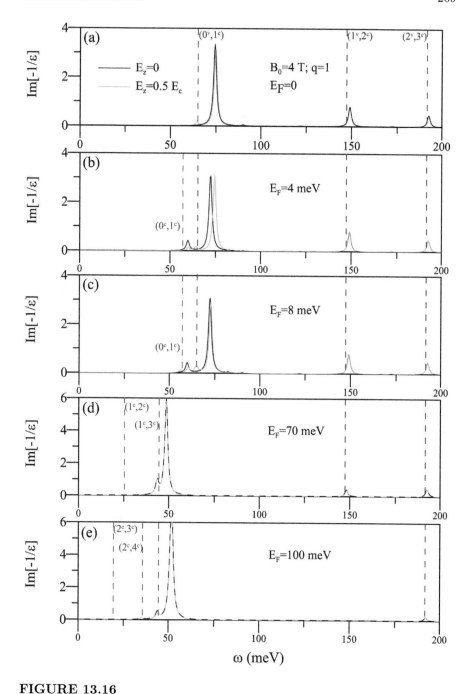

FIGURE 13.16

The magneto-energy loss spectra of monolayer silicene evaluated from Figure 13.15.

change the magnetoplasmon frequencies, but it does affect the spectral widths and intensities. The smaller the parameter value for Γ, the narrower spectral width and the stronger intensities are obtained (the inset of Figure 13.16c). A quite strong peak in energy loss function clearly indicates that this magnetoplasmon mode is due to the collective excitations of all free carriers in the conduction LLs. As to F-enriched loss spectra, an electric field provides a tool for tuning the dispersion relations of collective excitations. It can create new peaks and reduce the threshold frequency, mainly owing due to the separated spin and valley polarizations of the $n^{c,v} = 0$ LLs (the red curves in Figure 13.16a and b) under $E_F = 0$ and $E_F = 8$ meV, respectively. The extra plasmon modes experience rather strong Landau damping, since the splitting energies between any two LLs are small compared with other LL spacings. However, the splitting energies may be increased by making F larger, which results in higher weights of the newly created peaks (not shown here). In addition, with $n_F = 1$ and $n_F = 2$ (Figure 13.16c and d), the F-field effects will be evident only for larger q's, as demonstrated later.

The (q, ω)- and (B_z, ω)-phase diagrams (Figures 13.17–13.20), which are under various Fermi energies and electric fields, could provide the full information on the diverse magnetoelectronic excitation phenomena. Intrinsic silicene (Figure 13.17a) only possesses interband plasmon modes, being strongly confined to lie between two neighboring single-particle inter-LL channels in

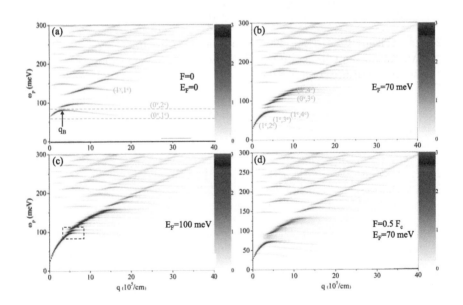

FIGURE 13.17
The intensities of energy loss spectra in the (q, ω) phase diagrams under the various cases: (a) $(F = 0, E_F = 0)$, (b) $(F = 0, E_F = 70$ meV$)$, (c) $(F = 0, E_F = 100$ meV$)$, and (d) $(F = 0.5F_c, E_F = 70$ meV$)$.

a limited q range. For example, the lowest-frequency plasmon branch of $(0^v,1^c)/(1^v,0^c)$ lies between the single-particle energies of $(0^v,1^c)$ and $(0^v,2^c)$, as illustrated by the two horizontal blue-dashed lines in Figure 13.17a. This plasmon is seriously damped in both short and long wavelength limits, with its frequency close to the e–h excitation frequency of $(0^v,1^c)$. All the q-dependent magnetoplasmon frequencies possess humplike features (exhibit the nonmonotonous behaviors), indicating that charge oscillations behave quite differently below and above a critical momentum (q_B), indicated by a purple arrow. q_B is defined by the transferred momentum corresponding to zero group velocity, and it is mainly determined by the comparable characteristic lengths in longitudinal charge oscillations and transverse cyclotron motion. That is, q_B indicates a strong competition between the longitudinal Coulomb interactions and transverse magnetic forces. If $q < q_B$, the group velocity is positive and the plasmon intensity increases with an increase in momentum. However, the opposite is true for $q > q_B$. The value of q_B is enhanced by the stronger magnetic field. For a plasmon-excitation channel, the larger the transition order of Δn, a higher rate of increase in q_B as a function of B_z is achieved [32]. The peculiar dependence of q_B on B_z might lead to a rich B_z-dependent magnetoplasmon spectrum, as demonstrated in Figures 13.18–13.20.

The main features of intraband magnetoplasmons sharply contrast with those arising from interband transitions in (q,ω) space. For $n_F = 1$ and $E_F = 70$ meV, Figure 13.17b clearly shows that the lowest-frequency interband magnetoplasmon branch of $(0^v,1^c)$ no longer exists and is replaced by a combination of three intraband modes, namely, $(1^c,2^c)$, $(1^c,3^c)$, and $(1^c,4^c)$. The three intraband modes consist of a continuous branch that exhibits a longer range of positive group velocity and higher intensity. In addition, the disappearance of the interband magnetoplasmon $(1^v,1^c)$, marked in Figure 13.17a, supports the other interband modes, such as $(0^v,3^c)$ and $(0^v,4^c)$. Although such interband magnetoplasmons are close to each other, they disperse independently under a strong transverse restoring force coming from the magnetic field. With larger $n_F = 2$ when $E_F = 100$ meV in Figure 13.17c, the intraband magnetoplasmons display a more obvious continuous branch, mainly owing to the enhanced number of intraband channels. The gap between the intraband magnetoplasmon and the lowest-frequency interband mode is quickly diminishing (the purple rectangle). Moreover, an electric field creates additional discrete subbranches as a result of $n^{c,v} = 0$ LL splitting, as shown in Figure 13.17d for $E_F = 70$ meV and $F = 0.5F_c$. The newly created modes in the range of 75 meV$\leq \omega \leq$ 100 meV are weakly dispersive due to the combined effects of buckled structure and magnetic field. Their momentum range, frequency, and number could be easily controlled by varying E_F and F. These subbranches, arising from the spin- and valley-polarized LLs, might play critical roles in fine-tuning details of the magnetoplasmon spectra.

The intensity and frequency of magnetoplasmon modes exhibit an unusual B_z-dependence under various transferred momenta and Fermi energies. The interband mode frequencies for intrinsic silicene, as shown in Figure 13.18a–c

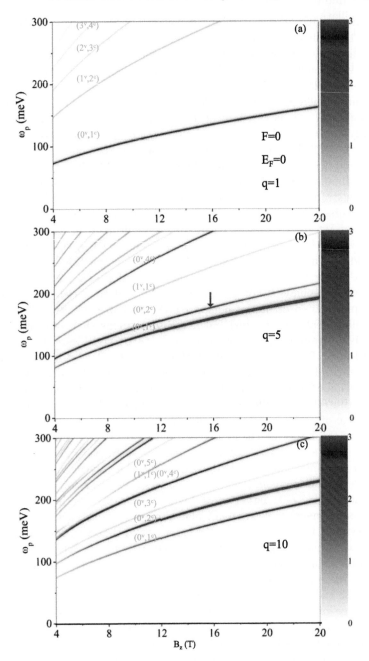

FIGURE 13.18
The energy-loss spectral intensities for the (B_z, ω) phase diagram at $(F = 0, E_F = 0)$ under the various transferred momenta: (a) $q = 1$, (b) $q = 5$, and (c) $q = 10$.

for the distinct q's, increases monotonically with an increase in B_z, mainly owing to the enlarged LL spacings and the absence of changes in occupation number. Specially, the magnetoplasmon intensities are drastically changed during the variation of B_z, depending on the LL state degeneracy, momentum transfer, and q_B. At long wavelength (e.g., $q = 1$ in Figure 13.18a), the interband magnetoplasmons are dominated by the single-particle excitation channels with $\Delta n = 1$. This simple rule is mainly determined by the spatial characteristics of Hermite polynomial functions. Their intensities are purely enhanced by B_z, since there are more states in each LL. When q increases and becomes comparable with the q_B for the $(0^v, 2^c)$ mode (e.g., $q = 5$ in Figure 13.18b), $\Delta n = 1$ and two magnetoplasmon branches, corresponding to the initial two modes, can exist simultaneously, and their intensities are strengthened and weakened by the increase of B_z, respectively. The different B_z dependence mainly comes from the competition between q and q_B. The latter is augmented on account of the enlarged ratio of carrier oscillation wavelength to magnetic length. As to $\Delta n = 1$, magnetoplasmon modes with $q = 5 > q_B$, an increasing B_z causes q_B to move closer to q, enhancing the collective oscillations, e.g., the $(0^v, 1^c)$ mode. However, the opposite holds true for the $(0^v, 2^c)$ mode with $q = 5 < q_B$. This means that an increase in B_z will cause the specific q to deviate from q_B and thus weaken the magnetoplasmon intensity. The aforementioned magnetoplasmons could be enhanced simultaneously when q is larger than q_B of all modes, such as, $q = 10$ in Figure 13.18c.

The magnetoplasmon spectrum, which corresponds to $n_F = 1$ under a weak magnetic field, experiences an abrupt change in intensity, frequency, and bandwidth at a critical magnetic field, e.g., $B_c \sim 5.5$ T for $E_F = 70$ meV shown in Figure 13.19a–c. Above B_c, all free conduction electrons are accumulated in the $n^c = 0$ LL, so that the threshold magnetoplasmon mode is dramatically changed from the intraband $(1^c, 2^c)$ to the stronger interband $(0^v, 1^c)$ and the weaker intraband $(0^c, 1^c)$ at $q = 1$ in Figure 13.19a. In contrast, the other interband magnetoplasmon modes pass the critical field continuously. This unusual behavior for the intraband and interband magnetoplasmons is a key feature that distinguishes them. With an increase in momentum (e.g., $q = 5$ in Figure 13.19b), the obvious changes in the magnetoplasmon spectrum appear near B_c, in which they cover the discontinuous transformation from the lowest-frequency intraband mode $(1^c, 4^c)$ to the interband one $(0^v, 1^c)$ and the newly created interband branch of $(1^v, 1^c)$. For $q = 10$ in Figure 13.19c, there exist more magnetoplasmon modes involving the $n^c = 1$ LL, like $(1^c, 5^c)$ and $(1^c, 8^c)$ under a low B_z. These intraband magnetoplasmons modes disappear when the magnetic field is beyond the critical one, leading to more discontinuities in the magnetoplasmon spectrum. Apparently, the transferred momentum is a critical factor in tuning the B_z-dependent magnetoplasmons.

Starting with $n_F = 2$, there are two critical magnetic fields for which n_F is reduced to 1 and 0, as obviously indicated in Figure 13.20a–c. Concerning $E_F = 100$ meV, the first critical magnetic field appears at $B_{c1} \sim 5.5$ T (identical to that in Figure 13.19), and the second one is at $B_{c2} \sim 10.8$ T.

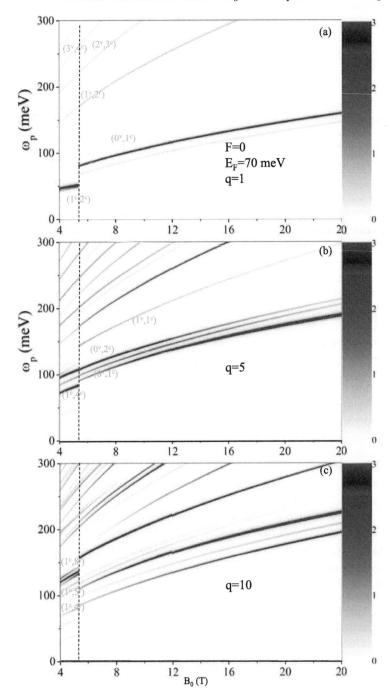

FIGURE 13.19
The similar plot as Figure 13.18, but shown at $E_F = 70$ meV.

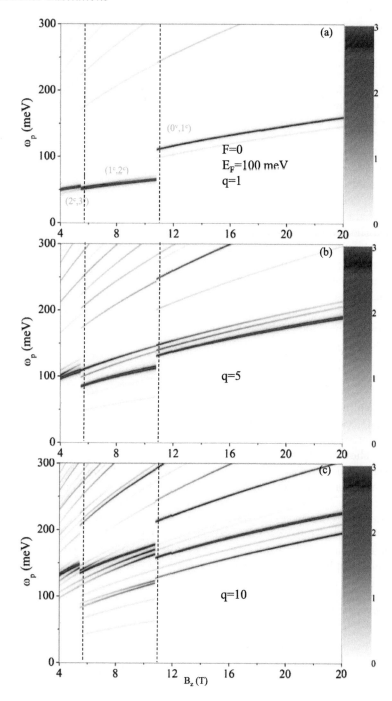

FIGURE 13.20
Same plot as Figure 3.17, except at $E_F = 100$ meV.

For small momentum in Figure 13.20a, the lowest-frequency magnetoplasmon belongs to the intraband of $(2^c, 3^c)$ under $B_z < B_{c1}$ is achieved, while between B_{c1} and B_{c2} corresponds to the $(1^c, 2^c)$ channel. Beyond B_{c2}, the threshold mode is replaced by the interband $(0^v, 1^c)$ and the intraband $(0^c, 1^c)$. The redistributions of strong magnetoplasmon modes are achieved for various transferred momenta, referring to Figure 13.20b and c. In general, the magnetoplasmon spectra are quite different in the three B_z-field ranges characterized by $0 < B_z < B_{c1}$, $B_{c1} < B_z < B_{c2}$, and $B_{c2} < B_z$. These results directly reflect the B_z-dependent occupation number and degeneracy of LLs under $n_F = 2$, $n_F = 1$, and $n_F = 0$. The dramatic transformations of the magnetoplasmon spectra near the critical magnetic fields could be verified by EELS [35, 127, 376, 378, 379] and infrared spectroscopy [375, 376] as well as inelastic light scatterings [473, 474]. The magnetically tunable plasmon spectra, with the strong dependences on the momentum transfer as well as intraband and interband modes, might be useful in the design of magnetoplasmonic components for various applications.

The significant differences between the intraband and interband magnetoplasmons are summarized later. The inter-LL magnetoelectronic excitations are available if electrons are excited from the occupied LLs to the unoccupied ones. They could be further categorized into intraband $(c \rightarrow c)$ and the interband $(v \rightarrow c)$ inter-LL transitions. Their important differences cover the strength, frequency, and number of single-particle excitations and magnetoplasmon modes. The former possesses the stronger e–h excitation spectra (Figure 13.15a–h), since Coulomb matrix elements are associated with the linearly symmetric superposition of the tight-binding functions at two different sublattices. However, the weaker spectra in the latter originate from the linearly antisymmetric superposition. The first few intraband single-particle channels show the closer frequencies, so that they could induce the collective excitations together. This intraband magnetoplasmon, which is due to all free carriers in conduction LLs, displays a continuous dispersion relation with the momentum transfer (Figure 13.17b and c). This clearly indicates that the intraband collective excitations behave as a propagating wave with a continuous wavelength of $2\pi/q$. Compared with the transverse magnetic force, the longitudinal Coulomb interactions dominate the main features of intraband magnetoplasmons. Its strong intensity is effectively suppressed by the interband e–h excitations at large q's. On the other hand, there are several interband magnetoplasmons corresponding to the interband inter-LL transitions. These modes possess weaker intensities because of the serious Landau dampings of the higher-energy inter-LL transitions. They possess very weak dispersion relations, being strongly confined between two single-particle excitations frequencies (Figure 13.17a). That is, the strongly competitive relations of the transverse magnetic quantization and the longitudinal Coulomb interactions exist in the interband magnetoplasmons. Their group velocities are rather small; therefore, the interband magnetoplasmons could be categorized as the localized modes. From the classical point of view, the localized modes show that the charge density oscillations experience a strong restoring force arising from the magnetic field.

13.4 Differences among Graphene, Silicene, and Germanene

Monolayer silicene, germanene, and graphene quite differ from one another in certain geometric structures, electronic properties, and Coulomb excitations. The buckled structures, which mainly originate from the optimal competition of sp^2 and sp^3 chemical bondings, only appear in the former two systems. They become more obvious with an increase in atomic number, and so does the strength of spin–orbital coupling. Such intrinsic interactions can create an important energy gap in the hexagonal condensed-matter systems. As a result, their energy gaps are, respectively, $\sim$8, 90, and 0 meV. The Fermi velocity, being proportional to the nearest-neighbor hopping integral, is smallest/largest for germanene/graphene. Its magnitude directly reflects the uniform bond length. Apparently, the bandgap and Fermi velocity are responsible for the threshold excitation frequency and the critical temperature in determining the available transition channels. The initial excitation energies of the intraband and interband transition channels are characterized by $v_F q$ and $\sqrt{E_g^2 + v_F^2 q^2}$, respectively; that is, there exist obvious excitation gaps in silicene and germanene. To clearly reveal the dominating intraband e–h excitations and the observable plasmon mode, a high enough mode is required for all three systems. For example, under $q = 1$, the critical temperature is, respectively, $T = 40$, 200, and 200 K for these three systems. It is relatively easy to observe the T-induced plasmon in silicene. Silicene and germanene could exhibit three types of collective excitations: (I) the undamped plasmon mode due to the intraband transition in between two gapped boundaries, (II) the similar one and another weaker plasmon mode, and (III) the undamped and damped plasmons, in which (I) & (III) [(II)] are affected by the interband (intraband) Landau dampings. Such behaviors are determined by the range of temperature: (I) $T <$75 K, (II) 75 K$\leq T \leq$ 120 K, and (III) $T >$ 120 K.

The electric field can destroy the mirror symmetry about $z = 0$ plane only for buckled silicene and germanene, but not planar graphene. The close cooperation with the spin–orbital coupling further leads to the split spin-down- and spin-up-dominated energy bands and even the semimetal–semiconductor transition during the variation of field strength. As a result, the Coulomb excitation is greatly diversified by the composite effects due to the bandgap, energy spacing, Fermi velocity, spin–orbital interaction, temperature, and electric field. There exist four types of single-particle and collective excitations, being characterized by the available e–h transition channels and the main features of plasmon modes (the number and Landau dampings). They cover (I) the excitation gap between the intraband and interband transitions of the spin-degenerate valence and conduction bands (without F), and an intraband plasmon undamped at small q's and absent at large q's because of the strong dampings from the intraband boundary; (II) three e–h boundaries, the lower, middle, and upper ones, respectively, corresponding to the intraband transitions of the spin-down-dominated bands, the interband transitions of the same pair, and the interband excitations of the spin-up-dominated pair, where the T-induced intraband plasmon exists between the former two boundaries and experiences an interband e–h damping near the middle boundary, accompanies with another weaker plasmon, and then this mode is damped out at large q's by the intraband e–h excitations (close to the lower boundary); (III) the merged intraband and interband boundaries coming from the linear Dirac cone and the higher-frequency interband boundary related to the spin-up-dominated pair, in which the intraband plasmon shows a strong e–h dampings even at $q \to 0$; (IV) the same e–h boundaries with those in (II), and the plasmon mode undamped only at long wavelength limit and then the distinct interband Landau dampings associated with the first and second pairs of spin-split energy bands. It should be noticed that three e–h boundaries are described by $v_F q$ and threshold excitation frequencies in Eq. (13.2). Moreover, the (q, ω)-, (T, ω)-, and (F, ω)-based phase diagrams clearly illustrate the diverse critical momenta, temperatures, and field strengths under the various cases. Such significant factors could provide the full information about the existence/absence and damping/nondamping behaviors of the prominent T-induced intraband plasmon.

For germanene and silicene, the composite effects due to the electric field and electron doping can enrich the transition channels and diversify the collective excitations with various single-particle Landau dampings. The rich e–h boundaries mainly come from the band-edge states and the Fermi momenta after doping. According to the unique frequency- and momentum-dependent phase diagrams, the collective excitations are classified into four types: a 2D acoustic plasmon mode at the long wavelength limit, but for the larger momenta, changing into another kind of undamped plasmon (the first type), becoming the seriously suppressed mode in the significant intraband e–h transitions (the second type), remaining identical undamped plasmons (the third type), or declining and then vanishing within the prominent interband e–h transitions (the fourth type). Germanene, silicene, and graphene are quite different from one another in the main features of the diverse plasmon modes, e.g., the absence of the first (third) one in silicene and graphene (graphene). Apparently, the diversified Coulomb excitation phenomena are closely related to the strength of spin–orbital coupling and the significant buckled structures.

Apparently, the complex effects, which arise from the spin–orbital interactions, magnetic field, electric field, and electron doping, can greatly diversify the magnetoelectronic properties and Coulomb excitation phenomena [34, 138, 142, 153, 475]. The magnetic quantization, combined with the on-site

Coulomb potential energies in the buckled structures, creates the rich and unique LLs for monolayer silicene and germanene. The splitting of spin configurations and (K,K') valleys is clearly revealed in the LL energy spectra and wave functions under the electric and magnetic fields. For silicene, only the $n^{c,v} = 0$ LLs exhibit this behavior because of the weak spin–orbital coupling [34,138,142]. However, germanene shows the splitting behavior for more LLs and even presents frequent LL anticrossings during the variation of electric/magnetic field, mainly owing to the largest spin–orbital coupling among (C, Si, Ge) systems. As for the magnetoelectronic excitations, there exist two categories: intraband and interband inter LL transitions. The intensities of the delta-function-like peaks in $\text{Im}[\epsilon]$ are reduced to half, and the neighboring double peak exists when the electric-field-induced LL splitting appears. Similar results are directly reflected in magnetoplasmon modes. The intraband and interband magnetoplasmons, with the long and short dispersion relations with the transferred momentum, could be regarded as continuous and localized modes, respectively. In addition, a dramatic transformation of single-particle and collective excitations occurs near the critical magnetic fields, being similar for all layered systems.

14

Coulomb Decay Rates in Graphene

There are some theoretical [239, 240, 258, 259, 476 483] and experimental [41–43, 246–257, 484, 485] studies on the decay rates of quasiparticle states in layered graphenes, Bernal graphite, and carbon nanotubes. The electron–electron (e–e) Coulomb scatterings [237, 239, 240, 476, 478], as well as electron–phonon scatterings [41–43, 237, 239, 240, 246, 484, 485], play critical roles in the decay rates (the mean free paths) of excited electron/hole states, especially for the former. By using the self-energy method (Section 2.6.1) and the Fermi golden rule, the theoretical predictions of Coulomb decay rates in monolayer graphene clearly show that they purely arise from T-created intraband electron–hole (e–h) excitations, and the dependence on temperature and wave vector is very strong [239, 240]. The Coulomb decay rate is much faster than electron–phonon scattering rate. A 2D monolayer graphene sharply contrasts with a 2D electron gas [237] or a 1D carbon nanotube [239] in electronic excitations and deexcitations. The 2D superlattice model (discussed earlier in Chapter 11), corresponding to the effective-mass model of monolayer graphene, is utilized to evaluate the Coulomb decay rates and quasiparticle energies in 2D doped graphene and 3D graphite intercalation compounds [234, 261]. It should be noticed that the state energies of quasiparticles can only be evaluated by self-energies [486]. In addition to interband excitations, doping-induced intraband excitations and acoustic/optical plasmons are effective deexcitation channels [239, 240]. For donor-type systems with $E_F \sim 1$ eV, the excited valence bands present an oscillatory energy dependence, in which the largest energy widths achieve the order of 0.1 eV because of strong plasmon decay channels. The highly anisotropic behavior is revealed using the calculations of the tight-binding model, such as the direction-dependent Coulomb decay rates in doped graphene [132, 238, 240], silicene, and germanene at large wave vectors. There exist certain important differences among these three systems, being attributed to very different hopping integrals.

There are four kinds of experimental methods in measuring the quasiparticle decay rates/lifetimes of carbon-related sp^2-bonding systems. The femtosecond pump–probe spectroscopies are powerful tools in exploring the ultrafast relation of photoexcited electrons. They cover the time-resolved photoemission spectroscopy [41–43, 246–249], absorption/transmission/reflectivity [250–257], and fluorescence spectroscopies [258, 259], in which their measurements are useful in understanding the quasiparticle behavior of excited conduction electrons. Moreover, the energy distributions of angle resolved photoemission spectroscopy (ARPES) measurements could provide inelastic scattering rates of the excited valence holes. From the experimental measurements on graphite by the first equipment [248], the quasiparticles above the Fermi level are identified to exhibit a 3D metallic behavior, being consistent with the layer electron gas theory [487]. The generation, relaxation, and recombination of nonequilibrium electronic carriers are clearly observed by the second tool [257]. Furthermore, the fourth method is used to confirm the pronounced temperature-dependent decay rate for excited electrons near E_F. As to carbon nanotubes, the measurements of the first method show the carrier relaxation of excited electrons in metallic single-walled and multiwalled systems. The measured decay rate is ~ 1 meV for the low-energy excited states [246,247]. The second [250,252] and third [258,259] methods are made on bundled and isolated single-walled carbon nanotubes. The decay rate of the first (second) conduction band is ~ 1–5 meV (~ 0.5–3 meV, which depends on the nanotube [478]. The first geometry, or energy gap, is deduced from the intraband carrier deexcitations. The first conduction band also shows a smaller decay rate about several percentages of the large one [478]. This is attributed to the interband recombination or defect trapping. The temperature-dependent photoluminescence spectra are measured for a very small $(6,4)$ nanotube between 48 and 182 K. The nonexponential behavior might be associated with the nonradiative decay of excitations. The high-resolution ARPES spectra of doped graphene systems (e.g., potassium-adsorbed graphenes) have confirmed the strong and non-monotonous dependence on wave vector/energy, mainly owing to the isotropic Dirac-cone band structure and the prominent Coulomb interactions of free conduction electrons [41–43].

The Coulomb inelastic scattering rates of pristine graphene are derived by the Fermi golden rule in addition to the self-energy method (Section 2.6.1). Their consistence could promote full understanding of the complicated dynamic charge screenings and deexcitation processes. For the low-lying excited conduction electrons and valence holes near the Fermi level, the dependence on wave vector and temperature is investigated completely. The decay rates of a gapless carbon nanotube are also evaluated to thoroughly comprehend the dimensionality effect. The e–e interactions are compared with the electron–phonon ones in the carrier decay rates. It is worthy of detailed comparisons with those of 3D Bernal graphite and 2D electron gas. After the electron/hole doping in monolayer graphene, the deeper valence states and the higher conduction ones are expected to exhibit an unusual deexcitation phenomena. Now, there are three kinds of excited states under a donor-type Fermi level, namely the valence holes, conduction holes, and conduction electrons, according to the increment of state energy. The distinct deexcitation processes and the dependence of decay rate on the direction of wave vector, state energy, and Fermi energy need to be explored in detail, such as the critical roles of the intraband & interband single-particle excitations and plasmon modes, the strong isotropic/anisotropic behavior at lower/higher (deeper) state energies, the monotonous or oscillatory energy dependence, the similarity with 2D electron gas for low-energy conduction electrons and holes, and the equivalence/nonequivalence of valence and conduction Dirac points. The doped graphene, silicene, and germanene might display significant differences in Coulomb decay rates because of the distinct nearest hopping integrals and spin–orbital interactions.

14.1 Temperature-Induced Inelastic Scatterings in Monolayer Graphene

The temperature- and doping-dependent Coulomb decay rates of monolayer graphene/single-walled carbon nanotube could also be calculated from the Fermi golden rule [240]. The bare and screened Coulomb potentials are critical mechanisms in understanding the electronic excitations and deexcitations. One first considers the Coulomb interaction between the $|\mathbf{k}, h\rangle$ and $|\mathbf{p}, h''\rangle$ states. The square of the bare Coulomb interaction, which includes the band-structure effect, is expressed as

$$|V(\mathbf{k}, \mathbf{q}, \mathbf{p}, h, h', h'', h''')|^2 = v_q^2 |\langle \mathbf{k} + \mathbf{q}, h'|e^{i\mathbf{q}\cdot\mathbf{r}}|\mathbf{k}, h\rangle\langle\mathbf{p} - \mathbf{q}, h'''|e^{-i\mathbf{q}\cdot\mathbf{r}}|\mathbf{p}, h''\rangle|^2,$$
$$(14.1)$$

where $|\mathbf{k}, h'\rangle$ and $|\mathbf{p}, h'''\rangle$ are two final states. The Coulomb scattering matrix elements in Eq. (2.4) is evaluated in an analytic form [486]:

$$|\langle \mathbf{k} + \mathbf{q}, h'|e^{i\mathbf{q}\cdot\mathbf{r}}|\mathbf{k}, h\rangle|^2 = \frac{1}{2}I^2(q)\left[1 \pm \frac{k + q\cos\phi}{\sqrt{k^2 + q^2 + 2kq\cos\phi}}\right], \quad (14.2)$$

where $+$ and $-$, respectively, correspond to the intraband and interband scatterings. ϕ is the angle between $\mathbf{k}$ and $\mathbf{q}$. $I(\mathbf{q}) = [1 + (q/6)^2]^{-3}$ is very close to one at small transferred momenta. All the π valence electrons would dynamically screen the bare Coulomb potential by means of various e–e interaction channels. The effective Coulomb potential, being characterized by the dielectric function, is given by

$$|V^{eff}(\mathbf{k}, \mathbf{q}, \mathbf{p}, h, h', h'', h'''; T)|^2 = \left|\frac{V(\mathbf{k}, \mathbf{q}, \mathbf{p}, h, h', h'', h''')}{\epsilon(\mathbf{q}, \omega_{de}(\mathbf{q}); T)}\right|. \quad (14.3)$$

$\omega_{de} = E_{\mathbf{k}}^h - E_{\mathbf{k}+\mathbf{q}}^{h'}$ is the deexcitation energy from the initial $|\mathbf{k}, h\rangle$ state to the final $|\mathbf{k} + \mathbf{q}, h'\rangle$ state through the intraband or interband process. Such potential is sensitive to the direction and magnitude of transferred momentum.

The effective e–e interactions could be utilized to explore the decay rate due to inelastic Coulomb scatterings. From the Fermi golden rule, the quasiparticle decay rates of the excited $|\mathbf{k}, h\rangle\}$ state at any temperature is ($\hbar = 1$)

$$\frac{1}{\tau(\mathbf{k}, h; T)} = 2\pi \sum_{\mathbf{p},\mathbf{q},\sigma} \sum_{h',h'',h'''} f(E_{\mathbf{p}}^{h''})(1 - f(E_{\mathbf{k+q}}^{h'}))(1 - f(E_{\mathbf{p-q}}^{h'''}))$$

$$\times |V^{eff}(\mathbf{k}, \mathbf{p}, \mathbf{q}; T)|^2 \delta(E_{\mathbf{k+q}}^{h'} + E_{\mathbf{p-q}}^{h'''} - E_{\mathbf{q}}^{h''} - E_{\mathbf{k}}^{h}). \qquad (14.4)$$

Equation (14.2) has covered all the intraband and interband Coulomb excitations. By using the two relations

$$\delta(E_{\mathbf{k+q}}^{h'} + E_{\mathbf{p-q}}^{h'''} - E_{\mathbf{q}}^{h''} - E_{\mathbf{k}}^{h})$$

$$= \int_{\infty}^{-\infty} \delta(E_{\mathbf{k+q}}^{h'} - E_{\mathbf{k}}^{h} + \omega)\delta(E_{\mathbf{p-q}}^{h'''} - E_{\mathbf{p}}^{h''} - \omega)d\omega \qquad (14.5)$$

and

$$\frac{Im[\chi(\mathbf{q}, \omega; T)]}{\pi[exp(-\omega/k_B T) - 1]} = \sum_{\mathbf{p},\sigma,h'',h'''} f(E_{\mathbf{p}}^{h''})f(1 - E_{\mathbf{p-q}}^{h'''})$$

$$\times |\langle \mathbf{p} - \mathbf{q}, h'''|e^{-i\mathbf{q}\cdot\mathbf{r}}|\mathbf{p}, h''\rangle|^2 \delta(E_{\mathbf{p-q}}^{h'''} - E_{\mathbf{p}}^{h''} - \omega). \qquad (14.6)$$

Equation (4.6) is further reduced

$$\frac{1}{\tau(\mathbf{k}, h; T)} = 2\sum_{h'} \int \int \frac{d\phi dq}{2\pi^2}$$

$$\times \frac{coth[\omega_{de}(\mathbf{q})/2k_B T] - tanh[\omega_{de}(\mathbf{q}) - E_{\mathbf{k}}^{h}]/2k_B T]}{exp(-E_{\mathbf{k}}^{h}) + 1} \qquad (14.7)$$

$$\times v_q |\langle \mathbf{k} + \mathbf{q}, h'|e^{i\mathbf{q}\cdot\mathbf{r}}|\mathbf{k}, h\rangle|^2 Im\left[\frac{-1}{\epsilon(\mathbf{q}, \omega_{de}(\mathbf{q}); T)}\right].$$

The Coulomb decay rate, as clearly indicated in Eq. (14.5), is determined by the Fermi–Dirac distribution of the final electronic state (the first term) and the energy loss function (the second term). Equation (14.5) is identical to Eq. (2.19) (discussed in Section 2.6.1).

As to the specific Fermi-momentum state (the Dirac point; $k_F = 0$), its deexcitation energy is isotropic as a result of $\omega_{de}(|\mathbf{q}|)_{k_F=0} = \pm v_F q. + (-)$ comes from the final valence (conduction) band sate. Specially, the negative energy transfer is allowed at finite temperatures. The azimuthal-angle-part integration in Eq. (14.5) could be performed thoroughly:

$$\frac{1}{\tau(k_F; T)} = \frac{1}{2\pi} \sum_{h'} \int dq \left[coth\left(\frac{\pm v_f q}{2k_B T}\right) - tanh\left(\frac{\pm v_f q}{2k_B T}\right)\right]$$

$$\times v_q Im\left[\frac{-1}{\epsilon(q, \pm v_f q; T)}\right]. \qquad (14.8)$$

Temperature significantly affects the free carrier density, as clearly discussed in Section 3.1 and thus the low-frequency Coulomb excitations (the intraband, interband, and collective excitations) and decay rates. For monolayer graphene, the Fermi–Dirac distribution presents obvious changes near the Fermi level only in the energy range of room temperature. Apparently, the linear and isotropic Dirac-cone band structure is suitable for understanding temperature-induced excitations and deexcitations; that is, all the T-dependent dielectric function, energy loss function, and Coulomb scattering rate hardly depend on the direction of transferred momenta/wave vectors (Figures 14.1–14.3). With an increase in temperature (Figure 14.1a), the interband single-particle excitations at $T = 0$ are quickly transformed into intraband ones, since the right-hand square-root peak in the imaginary-part dielectric function dramatically becomes the left-hand form (red curve). The sufficiently high temperature (the enough small momentum), there exists a prominent acoustic plasmon peak in the screened response function (Figure 14.1b). The T-induced plasmon mode is easily damped by the interband e–h excitations, so that it cannot survive at long transferred momenta.

In general, the low-energy quasiparticle states, the excited conduction electronic and valence holes, can decay by utilizing the intraband excitations, the T-dependent acoustic plasmon, and the interband excitations. Concerning the specific Fermi-momentum state, its Coulomb inelastic scatterings come from the upper boundary of the first channel (Figure 14.2a and b) because of comparable deexcitation energies. Furthermore, the screened response function, which is due to the available intraband excitations, exhibits a low-intensity peak, as clearly indicated in Figure 14.1c. At zero temperature, the $k_F = 0$ state does not have effective final states as a result of phase-space restrictions, so that its decay rate is vanishing (Figure 14.2c). With an increase in temperature, the Coulomb decay rate, $1/\tau(k_F)$, increases quickly and monotonously. The magnitude at room temperature is $1/\tau(k_F, 300 \text{ K}) \sim 4.2$ meV. The Fermi-momentum state could decay into unoccupied valence-band (conduction-band) states under a positive (negative) deexcitation energy of $\omega_{de} = v_F q > 0$ $(-v_F q < 0)$. The response function of the screened intraband excitations, which characterizes the effective Coulomb interaction between the Fermi-momentum state and other unoccupied states, makes important contributions to the decay rate at smaller and larger transferred momenta (Figure 14.1c). The larger-(q, ω_{de}) electronic excitations are nonnegligible during the inelastic Coulomb scatterings. Consequently, it is impossible to evaluate the temperature dependence for the final-state distribution function (the first term of the integral function in Eq. (14.6)) and the energy loss function (the second term). The analytic formula for the temperature-dependent Coulomb decay rates is absent; however, $1/\tau(k_F, T)$ is approximately fitted by $\sqrt{T}$. On the other hand, 2D electron gas [269] and graphite intercalation compounds [234] are verified to exhibit a simple relation of $T^2 ln(T)$, being attributed to the smaller-q dominating intraband deexcitation channels.

The temperature-dependent Coulomb scattering rates are very sensitive to wave vector or state energy, as clearly revealed in Figure 14.3b. There are

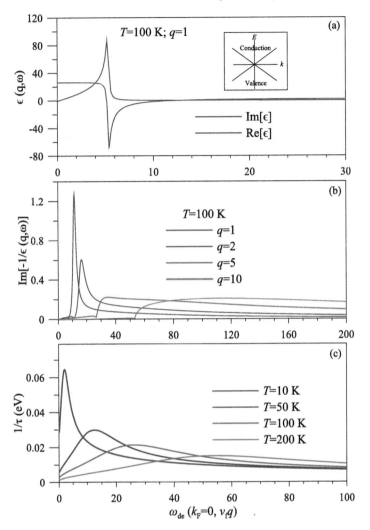

FIGURE 14.1
For monolayer graphene, (a) the real and imaginary parts of the dielectric function at $T = 100$ K and $q = 1$, (b) the energy loss spectra under different q's and $T = 100$ K, and (c) the excitation spectra of the Fermi-momentum state for distinct Ts. Also shown in the inset of (a) is the isotropic Dirac-cone structure.

excited electrons and holes in conduction band, in which these two kinds of carriers are, respectively, similar to holes and electrons in valence band. That is to say, the deexcitation phenomena due to the excited electrons in conduction and valence bands can provide the full information. Such excited carriers have almost vanishing scattering rates at zero temperature (not shown). That

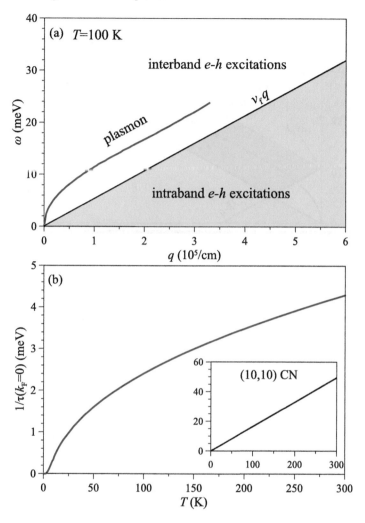

FIGURE 14.2
As to the specific Fermi-momentum state, the available intraband excitation channels, with the various intensities, at (a) $T = 100$ K and (b) $T = 300$ K, and (c) the temperature-dependent Coulomb decay rates. Compared with that of the $(10, 10)$ metallic carbon nanotube. The upper boundary of intraband excitations corresponds to the only deexcitation channels.

the available interband e–h excitations, being consistent with the conservation of momentum and energy, hardly survive is responsible for this unique property. Under a finite temperature, the total Coulomb decay rate present an oscillatory wave-vector dependence centered at the Dirac point, e.g., $1/\tau(k)$ in Figure 14.3b at $T = 100$ K by the heavy solid curve. As for the conduction-band (valence-band) states, $1/\tau$ increases rapidly as k gradually increases

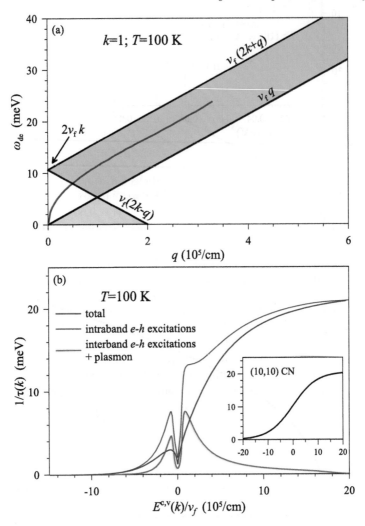

FIGURE 14.3
(a) For the excited conduction $k = 1$ state at $T = 100$ K, its available deexcitation channels cover the intraband e–h excitations (the circled region), the plasmon mode and the interband e–h excitations (the cross-hatched region). (b) The wave-vector-dependent Coulomb decay rates are contributed by the three kinds of electronic excitations.

from the Dirac point. It obviously displays a shoulder structure (reaches a maximum) and then continues to increase (declines and quickly vanishes) with a further increment of k.

It is worthy of a detailed examination on the effective scattering mechanisms of the unusual k-dependence in $1/\tau(k, T)$. For small k's, the conduction-

and valence-band excited states can decay using interband e–h excitations, the acoustic plasmon mode, and intraband e–h excitations, e.g., the Coulomb excitations, respectively, indicated by the green region, the red curve, and the blue region in Figure 14.3a for the $k = 1$ state under $T = 100$ K. The former two kinds of excitations make important contributions to the decay rate (green curve in Fig, 14-3(b)), which, thus, results in the special shoulder or peak structure. However, they are not involved in the inelastic Coulomb scatterings of larger-k states. The main reason is that the deexcitation energy is smaller than the threshold frequency of interband e–h excitations and plasmon modes. On the other side, the intraband e–h excitations are getting important for the conduction-band states (the right-hand blue curve). Furthermore, their contributions to the valence states are seriously suppressed by Fermi–Dirac distributions, i.e., a strong competition between two critical factors (the Fermion distribution and the energy loss spectrum) induces a peak structure in the valence-band decay rate (the left-hand blue curve). Up to now, for the wave-vector-dependent decay rate, only the Dirac-cone band structure presents a double-peak form centered at Fermi momentum, being absent in other energy bands.

It is interesting to understand the dimensionality-diversified deexcitation phenomena, such as a detailed comparison between a single-walled carbon nanotube and monolayer graphene. The dimensionalities dominate the strength of bare Coulomb interaction and density of states. Equation (4.9)/(4.10) could be further generalized to Coulomb decay rates of type-I, type-II, and type-III carbon nanotubes [239, 477, 478]. Their decay behaviors quite differ from one another, revealing the rich and unique features. Only the linear energy bands in a metallic (type-I) armchair carbon nanotube are similar to the Dirac-cone structure. Such energy dispersions are expected to exhibit intraband e–h excitations, the acoustic plasmon, and interband excitations, in which only the $L = 0$ mode is the effective deexcitation channel because of the too high deexcitation energies for the other $L \neq 0$ modes. It should be noticed that the frequency and intensity of the $L = 0$ plasmon hardly depend on temperature (discussed in Section 12.2); furthermore, ω_p is much higher than the intraband e–h excitation energy. Concerning the Fermi-momentum state of armchair carbon nanotube, its decay rate presents a linear temperature dependence (the inset in Figure 14.2c for the (10,10) nanotube), as seen in a 1D electron gas [237]. Those states exhibit a nonlinear T-dependance (not shown). Moreover, the wave-vector-dependence is monotonous (the inset in Figure 14.3b). The Coulomb decay rates of the conduction-band states increase rapidly as the wave vector increases from $K_F = 0$, while the opposite is true for those of the valence-band states as a result of the Pauli principle. The intraband e–h excitations are the critical channels of inelastic Coulomb scatterings, being responsible for the monotonous relation between $1/\tau$ and k. Specifically, the bare Coulomb interaction in a 1D carbon nanotube is much stronger than that in a 2D monolayer graphene, and so is the Coulomb decay rate (~ 10 times).

The significant differences between monolayer graphene and 2D electron gas in Coulomb excitations and scattering rates clearly illustrate the band-structure/geometric effects. The latter exhibits an acoustic plasmon at any temperatures, and there are no interband e–h excitations in the absence of valence band. This collective mode is damped by the intraband e–h excitations at very large transferred momenta [240]. The intraband e–h excitations are the only effective channels, because the deexcitation energies of the excited states are smaller than the plasmon frequencies. As for the Fermi-momentum state, its deexcitation channels mainly arise from the small-(q, ω_{de}) intraband e–h excitations, but not the whole ones. Consequently, the final-state distribution and the energy loss spectrum are proportional to temperature; furthermore, $1/\tau(k_F, T)$ is derived to be in the analytic form of $T^2 ln(T)$ [240]. Such excitations are also responsible for the decay rates of low-lying excited states; therefore, the Coulomb decay rates show a specific energy dependence: $(E - E_F)^2 ln|E - E_F|$. Obviously, different temperature- and wave vector-/energy-dependences lie in the existence/absence of the acoustic plasmon and the interband e–h excitations as efficient deexcitation channels.

The Coulomb decay rates in layered graphite are worthy of further theoretical and experimental studies. From the electron energy loss spectroscopy (EELS) and optical measurement on the AB-stacked (Bernal) graphite [37,38,167,168,215,231–233], this semimetallic system could exhibit an optical plasmon mode due to the band-overlap-induced free electrons and holes, with frequency ~ 50 meV. The intraband e–h excitations, the optical plasmon, and the interband e–h excitations survive under any temperature simultaneously. Up to now, the theoretical works on the essential properties of low-lying quasiparticle states are absent, in which they are required to examine the many-particle effects in detail. The anisotropic energy bands will induce a multidimensional integration, and this creates very high numerical barriers in fully exploring the temperature- and wave-vector-dependent inelastic Coulomb scatterings. On the experimental side, the femtosecond photoelectron spectra display [41–43, 246, 247] that the decay rate (the lifetime) of the Fermi-momentum state is higher than 8 meV ($\tau < 0.5$ ps) at room temperature, being close to that in monolayer graphene ($1/\tau(k_F) \sim 4.3$ meV or $\tau(k_F) \sim 0.9$ ps at $T = 300$ K in Figure 14.2c).

In addition to the e–e Coulomb interactions, the inelastic electron–phonon in inelastic scatterings is another deexcitation in monolayer graphene. From the previous theoretical calculations [484], the electron–phonon scattering rate is predicted to present the linear energy dependence, and it almost vanishes at room temperature. The Coulomb decay rate is much faster than the electron–phonon scattering rate. That is to say, the e–e Coulomb interactions are more efficient in carrier deexcitations. However, these two kinds of typical inelastic scatterings might play critical roles on other physical properties, e.g., transport properties and electronic excitations (the coupling of e–e and electron–phonon interactions).

14.2 Doping-Enriched Coulomb Decay Rates and Differences among Graphene, Silicene, and Germanene

Monolayer graphene exhibits a feature-rich band structure, mainly owing to the hexagonal symmetry and nanoscaled thickness (details in Chapter 3). The linear Dirac-cone bands are isotropic for low doping ($E_F \leq 1$ eV), while they gradually become parabolic at higher/deeper state energies (Figure 14.4a). The Fermi energy/free carrier density determines the main features of electronic excitations and thus dominate the Coulomb decay channels. When the Fermi level is situated at gapless Dirac points, the excited electrons/holes at zero temperature could decay into conduction/valence band states by utilizing interband single-particle excitations. The increment of E_F creates intraband e–h excitations & plasmon modes and induces drastic changes in interband single-particle excitations. Such E_F-induced Coulomb excitations greatly diversify the decay channels. As for the excited conduction electrons above the Fermi level, the final states during Coulomb deexcitations only lie between the initial states and Fermi momentum (a red arrow in Figure 14.4b), according to the Pauli exclusion principle and the conservation of energy and momentum. The available deexcitation channels, the intraband e–h excitations, make the most important contribution to the Coulomb decay rates for the low-lying conduction electrons, corresponding to the orange part in Figure 14.5a. But when the initial state energy is high, the interband single-particle excitations might become effective deexcitation mechanisms (discussed later in Figure 14.5b). Concerning the excited holes in the conduction band, they could be deexcited to the conduction states (a blue arrow in Figure 14.4c) through the intraband e–h excitations because of the low deexcitation energies and transferred momenta. On the other hand, the valence holes present two kinds of decay processes: $v \rightarrow v$ and $v \rightarrow c$ in Figure 14.4d and e, respectively. Their available decay channels, respectively, cover (intraband & interband e–h excitations) and (interband single-particle excitations & acoustic plasmon modes), corresponding to the blue arrow in Figure 14.4d and red arrow in Figure 14.4e. Specifically, the latter has large deexcitation energies at small momenta and is thus expected to exhibit efficient and unusual Coulomb decay rates.

The Coulomb decay rates strongly depend on the quasiparticle states of $(\mathbf{k}, h)$. Concerning the excited conduction electrons near the Fermi momenta, the $c \rightarrow c$ intraband decay processes are the only deexcitation channels; that is, the intraband single-particle excitations make the main contributions to such processes (the orange part in Figure 14.5); therefore, the Coulomb decay rates monotonously increase with $E^c - E_F$, as indicated in Figure 14.6a and b by the orange curves. Apparently, the zero decay rates (the infinite lifetimes) appear at the Fermi-momentum states because of the steplike Fermion distribution function at zero temperature. Furthermore,

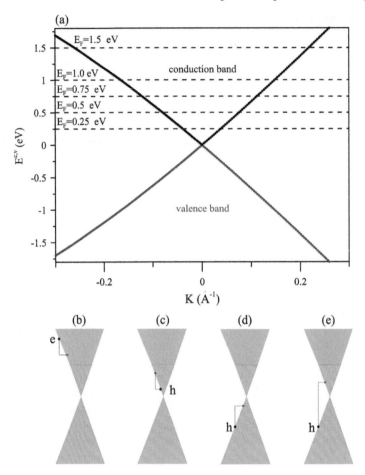

FIGURE 14.4
(a) Band structure of monolayer graphene with the various Fermi energies
($E_F = 0.25$, 0.5, 0.75, 1.0, and 1.5 eV by the dashed curves), and the available deexcitation channels of the specific excited states: (b) conduction electrons, (c) conduction holes, and valence holes with the (d) intraband and (e) interband decays.

those of the neighboring excited states $[|E^{c,v} - E_F| < 0.5E_F]$ are roughly proportional to $(E^{c,v} - E_F)^2 ln|E^{c,v} - E_F|$, purely according to the numerical fitting. Also, analytic derivations have been done for graphite intercalation compounds [18]. Such an energy dependence is characteristic of a 2D electron gas [237, 488]. This is not surprising, since, when $E^{c,v} \to E_F$, the deexcitation energy is essentially linear in q whether the energy band has a linear quadratic energy dispersion. Furthermore, the low momentum-frequency intraband single-particles are the only deexcitation channels. It is for such reasons that the widths of the doped graphene and electron gas near the

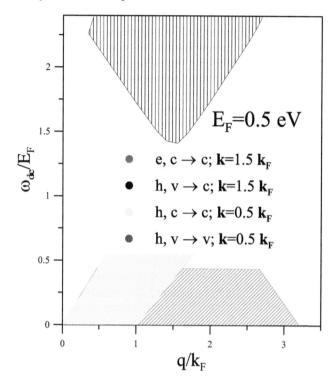

FIGURE 14.5
The relations between the deexcitation energies and transferred momenta for the specific excited states; conduction electrons (Figure 14.4b), conduction holes (Figure 14.4c)], and valence holes (Figure 14.4d and e).

Fermi level share a common character. The similar results are revealed in doped silicene and germanene [238]. In addition, for the Fermi-momentum states, the temperature-dependent Coulomb decay rates display the $T^2 ln[T]$ behavior, as observed in a 2D electron gas [489].

The Coulomb decay rates of the higher-energy conduction electrons are sensitive to anisotropic energy bands (Figure 14.4a). Along both KM and KΓ directions (Figure 14.6a and b), $[1/\tau]_{e,c \to c}$ increases with an increase of $|E^c - E^F|$. The energy dependence on the two high-symmetry directions lies in whether the interband single-particle excitations become effective deexcitation channels. The higher-energy electronic states have stronger energy dispersions along KΓ (Figure 14.4a); therefore, their deexcitation energies at large transferred momenta are consistent with those of the interband e–h excitations. For example, the conduction state of $E^c = 3E_F$ along KΓ presents a plenty of deexcitation channels, as indicated by the orange curves in Figure 14.7a at $\theta_q = 0°$. Similar results are revealed in different momentum directions, e.g.,

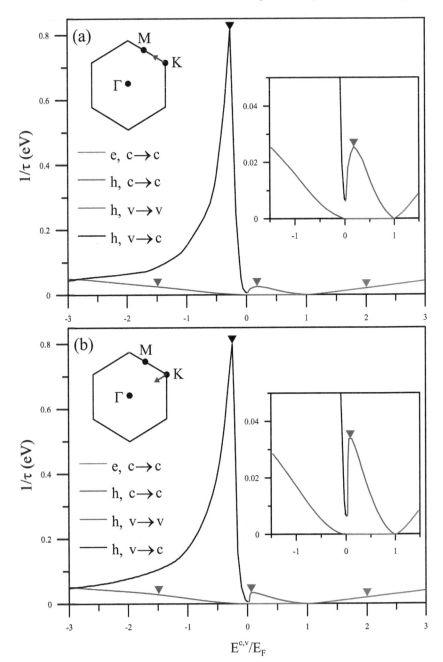

FIGURE 14.6
The energy-dependent Coulomb decay rates of the quasiparticle states along
the high-symmetry directions: (a) KM and (b) KΓ under $E_F = 0.5$ eV. The
insets show the first Brillouin zone.

$\theta_q = 30°$ in Figure 14.7c. The effective deexcitation channels cover the intra-band and interband e–h excitations. The latter is responsible for enhanced Coulomb decay rates in the high-energy conduction states.

The deexcitation behaviors of excited holes strongly depend on whether they belong to the conduction or valence states. Concerning the conduction holes, the Coulomb decay rates are almost isotropic, as indicated by the approximately identical $[1/\tau]_{h,c}$ along KM and KΓ (green curves in Figure 14.6a and b). Furthermore, the energy dependence is similar to that of the low-lying conduction electrons (2D electron gas). Such results directly reflect the fact that the intraband single-particle excitations are the only available deexcitation channels, e.g., the pink curves related to the states very close to the conduction Dirac point (Figure 14.7a–d). As a result of distorted linear dispersion, the finite decay rate at the Dirac point is different from the zero decay rate derived from the effective-mass model [18]. Specifically, the Coulomb decay rate is local minimum at the K point, indicating that the Dirac-point state is the most stable among all excited conduction holes (green curves in Figure 14.6a and b).

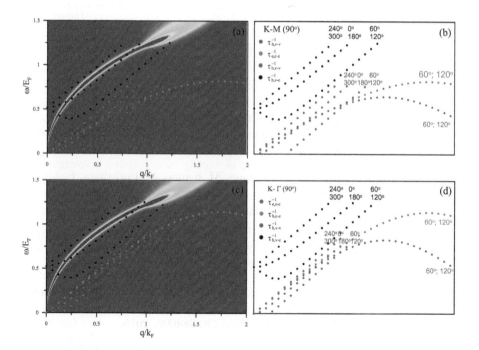

FIGURE 14.7
The available deexcitation spectra due to the specific states indicated by the arrows in Figs. 14.6a and b are displayed for (a) $\theta_q = 0°$ and (c) $\theta_q = 30°$. (b) and (d) The details of the θ_q-dependent deexcitation energies. The curves are defined by the conservation of energy and momentum.

On the other hand, the decay rates of valence holes display unique **k** dependences. The valence states slightly below the Dirac point have significant decay rates (purple arrow in Figure 14.6b), approaching to those from above the conduction ones. They present only the $v \to c$ decay process, in which the deexcitation channels mainly arise from interband single-particle excitations and undamped acoustic plasmon modes, as indicated by the black and brown symbols in Figure 14.7a and c. They create important difference above and below the Dirac point. With the increasing valence-state energy, two decay channels, $v \to c$ and $v \to v$, contribute to Coulomb decay rates simultaneously. Concerning the former, the available range of strong acoustic plasmon increases and then decreases quickly for the low-lying valence holes, leading to an unusual peak structure in $[1/\tau]_{h, v \to c}$ at small E^v's (the red curve in Figure 14.6a and b). For example, the $E^v = 0.9 E_F$ valence state along KM possesses the widest plasmon-decay range, associated with the black and brown curves in Figure 14.7a and c, so it could show a fast Coulomb decay (blue arrow in Figure 14.6a). The plasmon-induced deexcitations almost disappear for the deeper valence states, e.g., their absence under $E^v < -1.5 E_F$ along the KM direction The interband e–h excitations also make part of contributions to $[1/\tau]_{h, v \to c}$, and they dominate the Coulomb decay rates of the deeper-energy valence holes, e.g., the red curves along KM and KΓ at Ev < -1.5 E_F . Specifically, for the $v \to v$ process, the excited valence holes (the blue curves in Figure 14.6a and b) behave as excited conduction electrons (the orange curves) in terms of k dependence and deexcitation channels. The intraband single-particle excitations are the dominating mechanisms in determining $[1/\tau]_{h, v \to v}$ of the low-lying valence states. They are substituted by the intraband and interband e–h excitations for the deeper valence holes along KΓ. This is responsible for the anisotropic Coulomb decay rates under the specific KΓ and KM directions.

The effective deexcitation channels is worthy of a detailed investigation. Each excited state experiences inelastic e–e Coulomb scatterings along any direction, as clearly indicated by the summation of **q** in Eq. (14.2), where the transferred momentum is a function of q (magnitude) and θ_q (azimuthal angle in the range of 2π). Through the specific excitation spectra, it might exhibit several dispersion relations (less than six) in the q-dependent deexcitation energies for a fixed θ_q . The main reason is that both Coulomb excitations and energy bands possess a hexagonal symmetry; that is, the excitation spectra are identical for θ_q, $\theta_q + \pi/3$, $\theta_q + 2\pi/3$, $\theta_q + \pi$, $\theta_q + 4\pi/3$, and $\theta_q + 5\pi/3$. For example, the excited valence hole state, with the highest Coulomb decay rate along KM (KΓ), displays three (four) independent dispersive functions (blue curves) for $\theta_q = 0°$ ($\theta_q = 30°$). The other excited states in Figure 14.7a and c exhibit similar behaviors. The total deexcitation regions cover the θ_q-dependent dispersion relations; that is, they strongly depend on the direction and magnitude of **q**, as expected from the basic scattering pictures.

Figures 14.8–14.11 clearly illustrate the wave-vector- and Fermi-level-dependent Coulomb scattering rates. The decay rates of the valence holes

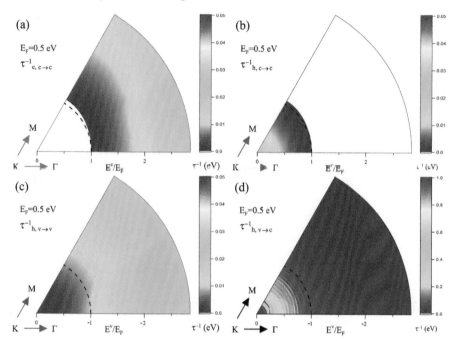

FIGURE 14.8
The wave-vector-dependent Coulomb scattering rates of electrons and holes via available deexcitation channels at $E_F = 0.5$ eV.

exhibit a nonmonotonous energy dependence along any wave-vector direction, since composite deexcitation channels cover intraband & interband e–h excitations, and the damped & undamped acoustic plasmons. The strongest Coulomb scatterings, which are dominated by undamped/damped collective excitations, exist for the valence states below the Dirac point (Figures 14.8d, 14.9d, and 14.10d). The valence-state decay rate strongly depends on the direction of $\mathbf{k}$; the anisotropic decay is more obvious for the Fermi states. This is closely related to the strong anisotropy of the deeper valence band (Figure 14.4c). As for conduction holes and electrons, the Coulomb scattering rates, as measured from that of the Fermi-momentum states, present different behaviors. The former possesses nearly isotropic Coulomb decay rate in the phase diagram due to the low-energy isotropic Dirac cones. However, the latter exhibits monotonous energy dependences, and the anisotropic deexcitations appear only for the high doping cases.

It would be relatively easy to observe the oscillatory energy dependence and the anisotropic behavior under higher Fermi energies. Electronic excitations and Coulomb decay rates are very sensitive to the changes in free carrier densities with the variation of E_F. The (q, ω)-phase diagram of $1/\tau$ is drastically altered, as clearly shown in Figures 14.8–14.11. For example, the less-damped acoustic plasmon and the almost isotropic excitations

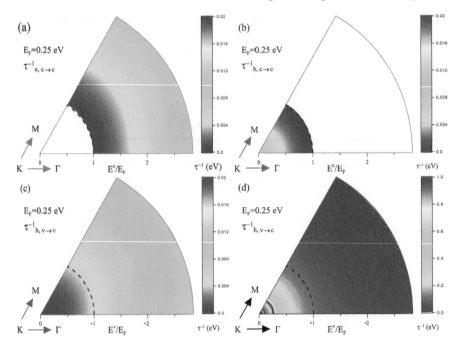

FIGURE 14.9

The similar plot as Figure 14.8, but shown at $E_F = 0.25$ eV.

are revealed at a sufficiently low Fermi level, e.g., excitation spectra at $E_F \leq 0.5$ eV (Chapter 3). These are directly reflected in the Coulomb decay rates (Figures 14.8 and 14.9). For the higher Fermi levels, the available momentum-frequency deexcitation ranges of the strongest acoustic plasmons and the interband single-particle excitations are greatly enhanced, since they could coexist together (Figure 14.10). This leads to a stronger dependence of decay rates on the state energy and direction of **k**, such as a detailed comparison among those in Figure 14.8d at $E_F = 0.5$ eV, Figure 14.9d at $E_F = 0.25$ eV and Figure 14.10d at $E_F = 0.75$ eV. The E_F-induced significant differences are further illustrated by the Coulomb decay rates of the specific states. For example, the largest decay rates arise from the interband hole channels; the efficient decay is mainly contributed by plasmons. Furthermore, the stability of conduction/valence Dirac-point states is held even under heavy dopings, e.g., $E_F = 0.75$ eV in Figure 14.10.

In addition, the Coulomb decay rates of excited conduction and valence states in monolayer germanene have been thoroughly investigated in Ref. [238], as done for graphene systems in Sections 14.1 and 14.2. The screened exchange self-energy in Eqs. (2.14)–(2.20) are suitable for the inelastic Coulomb scatterings in monolayer germanene and silicene with the spin–orbital interactions; that is, similar calculations could be finished under an accurate framework of theoretical mode. It should be noticed that the spin–orbital couplings

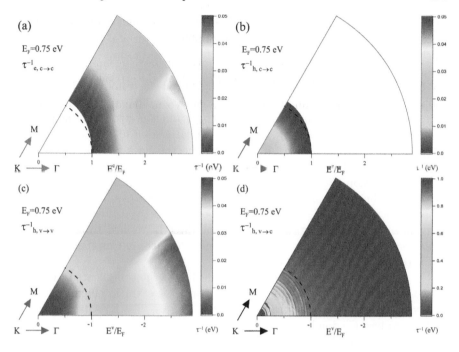

FIGURE 14.10
The similar plot as Figure 14.8, but shown at $E_F = 0.75$ eV.

result in the superposition of the spin-up and the spin-down components. However, it does not need to deal with the spin-up- and spin-down-dependent Coulomb decay rates separately, because they only make the same contribution because of the conservation of spin configurations during Coulomb scatterings. It is sufficient in exploring the wave-vector-, conduction-/valence- and energy-dependent self-energy. The effective deexcitation channels, which are diversified by the composite effects due to the spin–orbital couplings and carrier doping, are predicted to cover the intraband & interband e–h excitations, and the second, third, & fourth kind of plasmon modes. On the theoretical side, the random-phase approximation (RPA) self-energy will be greatly modified at an external electric/magnetic field, mainly owing to the spin–split energy bands and wave functions. This is worthy of a complete investigation on the theoretical models and F- and B_z-dependent Coulomb decay rates, being never studied in previous calculations.

Monolayer germanene, silicene, and graphene, with the hexagonal symmetries, exhibit p_z-orbital-dominated low-lying band structures. The first system possesses the weakest nearest-neighbor hopping integral and the largest spin–orbital couplings, so that the essential properties are relatively easily tuned by the external factors, such as carrier doping, electric field, and magnetic field. Apparently, graphene shows the strongest hopping integral ($\sim$2.6 eV) because of the smallest C–C bond length. This high-symmetry system presents a pair of linearly intersecting valence and conduction band at the gapless Dirac points under the negligible spin–orbital interactions, in which the isotropic Dirac-cone structure is further used to investigate the rich and unique physical properties (Chapters 4 and 14). However, there are important differences between germanene and graphene in electronic excitations and Coulomb decay rates. Germanene is predicted to display strongly anisotropic excitation/deexcitation behaviors, the second and third kinds of plasmons (the available decay channels), the e–h boundaries due to spin–orbital couplings, and F-induced splitting of excitation spectra. Such features are absent in graphene (Chapter 3). The theoretical calculations have been done for the excited conduction and valence electrons in graphene, indicating the isotropic deexcitation behaviors at low doping and a vanishing Coulomb decay rate at the Dirac point [240]. The predicted Coulomb decay rates in monolayer germanene, silicene, and graphene could be directly verified from the high-resolution ARPES measurements on the energy widths of quasiparticle state at low temperatures [41–43, 246, 247].

Up to now, no theoretical models can deal with the complicated dynamic charge screenings using the fully exact manners. The extra comments on RPA are useful in understanding the theoretical progress. The RPA is frequently used to explore Coulomb excitations and deexcitations of condensed-matter systems, especially for high carrier densities in 3D, 2D, and 1D materials [23, 25, 240, 261, 462, 478]. This method might cause poor results at low free carrier densities in certain many-particle properties, mainly owing to insufficient correlation effects. A plenty of approximate models have been proposed to address/modify the e–e Coulomb interactions, e.g., the Hubbard [490] and Singwi–Sjolander models [491, 492] for electronic excitation spectra, and the Ting-Lee-Quinn model for Coulomb decay rates [493]. Concerning the time-dependent first-principles numerical calculations, accompanied with the Bethe–Salpeter equation, they are further developed to investigate the electronic excitations and Coulomb decay rates in detail [483, 484]. Such calculations could account for the experimental measurements on excitation spectra and energy widths only under large energy & momentum scales. However, it would be very difficult to provide much information about the critical mechanisms and physical pictures in determining the significant bare polarization functions, energy loss spectra, and Coulomb decay rates. Whether the calculated results within RPA are suitable and reliable at low energy is worthy of a systematic study.

15

Concluding Remarks and Perspectives

This book clearly presents a fully modified theory on Coulomb excitations/decays in graphene-related systems, in which the theoretical framework combines layer-dependent random-phase approximation (RPA) and the generalized tight-binding model. It can deal with a plenty of critical factors related to different lattice symmetries, layer numbers, dimensions, stacking configurations, orbital hybridizations, intralayer & interlayer hopping integrals, spin–orbital couplings, temperatures, electron/hole dopings, electric field, and magnetic field. Apparently, there exist rich and unique electronic excitation phenomena due to the distinct energy bands and wave functions in various condensed-matter systems, as obviously revealed in the diverse (momentum, frequency)-phase diagrams. The calculated results, with the concise physical pictures, clearly illustrate the important roles of electron–electron (e–e) Coulomb interactions. Of course, they could explain the up-to-date experimental measurements. This model could be generalized to other emergent 2D materials under detailed calculations/investigations, such as the layered silicene [34, 132, 152, 387], germanene [154, 238], tinene [494], phosphorene [29], antimonene [29], bismuthene [29, 389], and MoS_2 [29]. Further studies would provide significant differences among these systems and be useful in thoroughly understanding the close/complicated relations of essential physical properties. On the other hand, the theoretical models should be derived again to solve Coulomb excitations in 1D and 0D systems without good spatial translation symmetry [124, 126, 126, 348, 428–430]. For example, 1D graphene nanoribbons and 0D graphene quantum dots have open boundary conditions, so that they, respectively, possess many energy subbands and discrete energy levels. Maybe, the dielectric function tensor, being characterized by the subband/level index, is an effective way to see the excitation properties [350].

The theoretical framework of Coulomb excitations and decay rates have been fully developed in Chapter 2, in which the experimental progresses, respectively, on electron energy loss spectroscopy (EELS), inelastic X-ray scattering (IXS), and angle resolved photoemission spectroscopy (ARPES) are investigated in detail. The dielectric functions and energy loss functions are, respectively, responsible for the single- and many-particle excitation spectra. They strongly depend on the geometric structure/translation symmetry, directly affecting the longitudinal transferred momenta. As a result, 3D graphite, 2D graphene, and 1D carbon nanotubes, respectively, exhibit dimension-related bare Coulomb interactions, so that such condensed-mattered systems are expected to create a diverse Coulomb excitation phenomena. Specifically, few-layer graphenes possess tensor forms in the dielectric functions, but not scalar quantities. The main reason is that the interlayer hopping integrals and interlayer Coulomb interactions need to be taken into account simultaneously. The straightforward combinations between the modified RPA and generalized tight-binding model have been made; furthermore, the dimensionless energy loss function, directly corresponding to the measured excitation spectrum, is well-defined and rather reliable. The new theoretical model is suitable for the composite effects due to the stacking configurations, the number of layers, the various lattice symmetries, the spin–orbital couplings, the electric field, and the magnetic field, especially for unusual magnetoelectronic excitations arising from magnetic quantization. Also, for the excited quasiparticles in layered graphenes, it is rather reliable in exploring the Coulomb inelastic scattering rates of excited quasiparticles in layered graphenes by linking the screened exchange energy of Matsubara's Green functions. This method could provide clear and concise physical pictures about the effective deexcitation channels. On the other side, the experimental techniques, the up-to-date resolutions, and the whole measured results are other focuses of diverse excitation phenomena. The detailed comparisons with theoretical predictions could be found in the following chapter.

Monolayer graphene exhibits rich and unique Coulomb excitations in the presence/absence of temperature, doping, and interlayer Coulomb coupling. A pristine system is a zero-gap semiconductor, so that the Landau dampings at $T = 0$ only come from inter-π-band transitions of the linear Dirac-cone structure and are too strong to observe the 2D acoustic plasmon modes. The thermal excitations could create a T^2-dependent free electron/hole density. When temperature is higher than the critical one, the intra-π-band excitations and 2D-like plasmons (~ 0.1 eV at room temperature) exist. Compared with thermal effects, carrier doping induces new/extra single-particle and collective excitations more efficiently. Apparently, the higher free carrier density is responsible for the prominent asymmetric peaks in the polarization functions and the energy loss ones. The rich (momentum, frequency)-phase diagram covers the intraband electron–hole (e–h) excitations, the interband ones, the vacuum regions without any excitations, and the undamped/damped acoustic plasmon mode, being sensitive to the variation of the Fermi level/the doping density. However, it hardly depends on the electron or hole doping as a result of an almost symmetric Dirac-cone energy spectrum about $E_F = 0$. In addition to the tight-binding model, the effective-mass approximation is utilized to obtain the analytic formula for polarization/dielectric function, where the conservation of particle numbers needs to be carefully solved during the detailed derivation. It should be noticed that the significant differences between a doped monolayer graphene and 2D electron gas, directly reflecting the conduction & valence Dirac cones and the parabolic conduction band, include the distinct boundaries of the intraband excitations, the existence/absence of interband excitations, the different Landau dampings of the 2D plasmon modes, the stronger momentum dependence of plasmon frequencies and more rich (momentum, frequency)-phase diagram in the former. Up to now, 2D plasmon modes are identified in the EELS measurements of alkali-doped graphene systems [170, 185, 186, 188, 189, 192], while temperature-induced ones require further experimental examinations. The Dirac-cone band structure is approximately reliable in a double-layer system with a sufficiently long interlayer distance. The in-phase and out-of-phase collective oscillations, which, respectively, arise from the symmetric and antisymmetric superpositions of free carriers on the first and second layers, appear in the energy loss spectra. They belong to the optical and acoustic plasmon modes according to the momentum dependence. The superlattice mode, with a Dirac-cone band structure and significant interlayer Coulomb interactions, is very useful in understanding the excitation and deexcitation spectra in graphite intercalation compounds [234, 261].

According to the tight-binding model and first-principles method [31, 53], the trilayer ABA stacking presents composite energy bands, being the super-position of the monolayer- and bilayer-like ones. The layer-dependent polarization functions are 3×3 Hermitian matrix, in which there exist four independent components. A pristine system only exhibits obvious interband e–h excitations and cannot create any plasmon modes, mainly owing to very low free carrier density and nonprominent density of states (DOSs) from the first pair of energy bands. The strong effects, which are due to electron doping, can dramatically alter the boundaries of single-particle excitations, add the new excitation channels, and induce three kinds of plasmon modes. The first acoustic mode, the second, and the third optical ones, respectively, originate from all the intraband $\pi_i^c \to \pi_i^c$ excitations, the $\pi_2^c \to \pi_3^c$ and $\pi_1^c \to \pi_3^c$ interband transitions. The last kind could be observed only for the higher Fermi level crossing the highest conduction band. There exist the diverse $(\mathbf{q}, \omega)$-phase diagrams, being sensitive to the doping carrier density, stacking configuration, and layer number. From the viewpoint of electronic and optical properties [31], the ABA-stacked trilayer graphene could be regarded as the superposition of monolayer and bilayer systems under single-particle schemes. Apparently, it is not suitable for many-particle e–e Coulomb interactions. That is to say, it is impossible to understand the diverse (momentum, frequency)-phase diagrams of a trilayer system from those of the composite ones, since different energy bands will have significant relations during interband Coulomb scatterings. Of course, the single-particle and collective excitations are getting more complicated in the increment of layer number, in which the cross-over behavior between the 2D layered graphene and 3D Bernal graphite might be worthy of a systematic investigation.

Trilayer ABC-stacked graphene has three pairs of unusual energy bands near the Fermi level (details in Section 6.1); therefore, it exhibits rich and unique Coulomb excitations. The layer-indexed bare response functions have nine components, but only four independent ones. There are a lot of single-particle channels and five kinds of plasmon modes during the variation of free carrier densities. The latter cover (I) the intraband plasmon related to the $\pi_1^c \to \pi_1^c$ transitions, (II) the interband mode associated with the $\pi_1^v \to \pi_2^c$ excitations, (III) the interband one arising from the $\pi_1^v \to \pi_1^c$ channels, and (IV) & (V) the multimode collective excitations under various intraband and interband channels. The complicated relations between single-particle and collective excitations create diverse (momentum, frequency)-excitation phase diagrams. The plasmon peaks in the energy loss spectra might decline and even disappear under various Landau dampings. The linear acoustic plasmon is related to the surface states (the partially flat bands at E_F) in pristine system, while it becomes a square-root acoustic mode at any doping. Specially, all the layer-dependent atomic interactions and Coulomb interactions have been included in polarization and dielectric functions. The predicted results could be examined from the high-resolution electron energy-loss spectroscopy [35, 37–40, 155–192, 192–195, 195, 196, 196–208, 208–213] and IXS [165, 214–226]. The magnetoelectronic Coulomb excitations in ABC-stacked few-layer graphene systems are expected to the new magnetoplasmon modes because of the frequent-anticrossing Landau levels (LLs) [339]. Moreover, the theoretical framework of layer-based RPA could be further generalized to study the e–e interactions in emergent 2D materials, e.g., ABC-stacked trilayer silicene and germanene [34, 154].

Apparently, the trilayer AAB stacking exhibits the unique electronic properties and thus the diverse Coulomb excitations. The lower stacking symmetry leads to three pairs of unusual energy dispersions: the oscillatory, sombrero-shaped, and parabolic ones. The former two possess large and special van Hove singularities, especially for the first pair nearest to the Fermi level. As a result, for the pristine system, there exist nine categories of valence→conduction interband transitions. The special structures in the bare response functions cover the square-root asymmetric peaks and the shoulder structures (the pairs of antisymmetric prominent peaks and logarithmically symmetric peaks) in the imaginary (real) part. The threshold channel, $\pi_1^c \to \pi_1^c$, can create significant single-particle excitations and strong collective excitations. The low-frequency acoustic plasmon, being characterized by the pronounced peak in the energy loss spectrum, is purely due to the large DOS in the oscillatory valence and conduction bands and the narrow energy gap; furthermore, its intensity and frequency are somewhat reduced by finite temperatures. A similar plasmon mode is revealed in a narrow-gap carbon nanotube [348]. The critical mechanism about the creation of this plasmon is thoroughly transformed into all intraband conduction-band excitations, in which the effective channels and critical transferred momenta strongly depend on the Fermi level. After the electron/hole doping, the interband e–h excitation regions are drastically modified and the extra intraband ones are generated during the variation of E_F. Moreover, one or two higher-frequency optical plasmon modes survive under various Fermi levels. They are closely related to the specific excitation channels or the strongly overlapped multichannels, being sensitive to the Fermi level and transferred momenta. There are certain important differences among the trilayer AAB, ABC, ABA, and AAA stackings, such as the boundaries of various intraband and interband e–h excitations, and the mechanism, number, strength, frequency, and mode of collective excitations. To fully explore the geometry-enriched Coulomb excitations, the aforementioned theoretical predictions require experimental verifications.

The sliding bilayer graphene systems obviously exhibit the stacking-configuration- and doping-enriched Coulomb excitation phenomena, mainly owing to the unique essential properties. The shift-dependent characteristics of electronic properties cover two distinct Dirac cones with the nontitled/titled axis, the normal parabolic bands, the highly hybridized/distorted energy dispersions, the stateless arc-shaped regions, the Dirac points, the local minima/maxima, the saddle points, the Fermi-momentum states, and the symmetric & antisymmetric/the abnormal superposition of the subenvelope functions on the equivalent/nonequivalent sublattices. The available excitation channels are classified into intrapair intraband, interpair, and intrapair interband transitions, which strongly depend on the relative shift of the two graphene layers. Under small transferred momenta, they, respectively, appear at low frequency, 0.2 eV$< \omega 0.4$ eV and 0.7 eV$< \omega 0.9$ eV. The former two channels are very sensitive to free carrier doping; that is, they might come exist or be partially/fully suppressed under the rigid shift of the Fermi level. The last one hardly depends on E_F. The single-particle excitations, which are associated with the band-edge and Fermi-momentum states, create special structures in the polarization functions. The abnormal imaginary/real parts present in the square-root asymmetric peaks, the shoulders/the logarithmically symmetric peaks, and the inverse structures of the second types. They, respectively, arise from the Fermi momenta with linear energy dispersions, the extremal states of the parabolic bands, and the saddle points. Such prominent single-particle responses are responsible for the serious Landau dampings. Also, both Fermi momenta and band-edge states determine the various e–h boundaries, where the latter correspond to the higher-frequency regions. As to the collective excitations, the acoustic and optical plasmons might exist simultaneously, only one mode survives, or both of them are absent, strongly relying on the relative shift and the free carrier density. The single- and many-particle excitations keep the similar behaviors under a small relative shift of two graphene layers. The $[\delta = 0 \; \delta = 1/8]/[\delta = 6/8, \; \delta = 1]/[\delta = 11/8, \; \delta = 12/8]$ bilayer stackings are similar to each other, while these three sets sharply contrast with one another. The plasmon modes are most easy to be observed in the first set, and the opposite is true for the third set.

A uniform perpendicular electric field has created significant effects on the electronic properties and thus diversifies the Coulomb excitation phenomena. The essential properties, which over the Dirac points, the Fermi velocity, the Fermi momenta, the valence and conduction band overlap/the free electron and hole densities, the bandgap, the band-edge states with large DOSs, the energy dispersions, the symmetric & antisymmetric linear superposition of the tight-binding functions, and the equivalence of the layer-dependent A and

B sublattices, clearly display the drastic changes or the dramatic transformations under layer-related Coulomb potential on-site energies. This field is further responsible for the unusual single-particle and collective excitations in layered graphene systems. Apparently, the AA-, AB-, and ABC-stacked few-layer graphenes exhibit diverse excitation behaviors. For the N-layer AA stackings, there is two/one acoustic plasmon modes in the presence/absence of an electric field. The higher- and lower-frequency modes, respectively, come from all the free carriers (the intrapair intraband transitions) and the F-induced band disparity, so that the former behaves like that in a 2D electron gas, and the latter exist only under the significant splitting of the intrapair intraband and interband transitions at the large enough F's. More optical plasmon modes are revealed in the energy loss spectrum except for the bilayer stacking, when compared with the $N - 1$ modes in the pristine system. Such optical plasmons mainly arise from the interpair interband channels; furthermore, the extra modes are associated with the nonuniform Fermi velocities and energy spacings between the different Dirac points. Specifically, the AA bilayer stacking can clearly illustrate the concise physical mechanisms by analytically evaluating the electric-field-dependent band structure, wave functions, bare polarization functions (intrapair and interpair parts), and dielectric function matrix. For example, the vanishing determinant of the last one is useful in understanding the plasmon-frequency dispersion relations with the transferred momentum. Apparently, the F-induced electronic excitations are sensitive to the stacking configurations. The significant differences among the trilayer AAA, ABA, and ABC stackings lie in the main features of electron properties, the available transition channels, the singular structures in bare polarization functions; the e–h excitation boundaries and the plasmon modes in the (q, ω)- & (F, ω)-phase diagrams. During the variation of F, three pairs of vertical Dirac-cone structures remain similar in the AAA stacking, while the others present strongly oscillatory energy dispersions, especially for the first pair of valence and conduction bands. Most important, the semimetal–semiconductor transition is only revealed under the specific ABC stacking. The observable single-particle transition channels, which correspond to the AAA-, ABA-, and ABC-stacked systems, are, respectively, divided into three interband and two intraband excitation categories, five interband excitation categories, and four ones. Furthermore, such bare polarization functions, being associated with the band-edge states, display the square-root divergent peaks and the shoulder & asymmetric peaks (composite structures of shoulders and logarithmically divergent peaks) in the AAA and [ABA, ABC] stackings. The most prominent plasmon, respectively, behaves like the acoustic and optical modes for the [AAA, ABA] and ABC stackings. The mechanism is different between AAA and [ABA, ABC] stackings, since this mode mainly originates from the first & third pairs of Dirac cones and the first pair of valence and conduction bands. Within a certain range of (q, F), there coexist the splitting modes, in which the number is, respectively, two, two, and three. As to the higher-frequency optical plasmons, it is relatively easy to observe them in the trilayer AAA stacking because of three splitting modes. However, only one mode appears in the ABA and ABC systems.

The entire LL energy spectrum is included in the calculations. This ensures the correctness of the dielectric function within the RPA and consequently the energy loss function, and the intensity and frequency of magnetoplasmon modes. The exact diagonalization method in efficiently solving the LL wave functions and the Coulomb matrix elements could also be applicable to multilayer graphene or bulk graphite over a wide range of magnetic and electric fields. As to monolayer graphene, there are a lot of observable and weak magnetoplasmon peaks, being sensitive to the strength of the e–h dampings. Their frequencies present the unusual q-dependence, in which the critical transferred momentum directly reflects the rather strong competition between the longitudinal Coulomb forces and the transverse cyclotron ones. The positive and negative group velocities, which, respectively, occur at $q < q_B$ and $q > q_B$, correspond to the dominance of the former and the latter. q_B and the magnetoplasmon frequencies increases with the magnetic field, especially for $\omega_p \propto \sqrt{B_z}$. Temperature effectively induces certain low-lying intraband excitations due to the conduction/valence LLs. The intraband magnetoplasmons would come to exist only under the sufficiently high temperature, in which the critical temperature grows with the increasing magnetic-field strength. The magneto-Coulomb excitations are greatly diversified by the stacking configurations. The low-energy excitations in the bilayer AB/AA stacking are only created by the interband/intrabands of the intragroup inter-LL channels. This directly reflects the spatial symmetric or antisymmetric distributions on four sublattices of the magnetic LL wave functions, as observed in zero-field cases. The former and the latter, respectively, exhibit a lot of prominent peaks in the layer-dependent polarization functions and the energy loss spectra. The discrete magnetoplasmon frequencies of the AB stacking are similar to those in monolayer graphene, while they present more complicated momentum dependence. The critical momenta and the serious Landau dampings are responsible for the discontinuous but monotonous B_z-dependence. Specially, the 2D-like plasmon, accompanied with some discrete modes, appears in the AA stacking, clearly illustrating free holes/electrons in the unoccupied valence LLs of the first group/the occupied conduction ones of the second group. The dramatic variation of the highest occupied LL leads to the discontinuous and oscillatory B_z-dependence. Moreover, the significant differences between the Coulomb and electromagnetic (EM) wave perturbations clearly indicate the diverse magentoelectronic excitation phenomena.

Apparently, the three kinds of 3D graphites show the diverse Coulomb excitation phenomena in the low- and middle-frequency ranges, mainly owing to the dimensionality and stacking configurations. That is, the single-particle and collective excitation fully reflect the main features of band structures, the strong wave-vector dependence, the highly anisotropic behavior, the special symmetry, and the 3D dimensional band overlap. Among the well-stacked graphites, the simple hexagonal (rhombohedral) graphite possesses the higher (lower) geometric symmetry, the largest (smallest) energy width along $\hat{k}_z$, the strongest (weakest) band overlap, and the heaviest (lightest) free electron and hole densities. The unusual geometric and electronic properties are responsible for the unique electronic excitations. As to the AA-stacked graphite, both low-frequency e–h excitations and plasmon modes could survive for any directions of transferred momenta, in which the optical modes are similar to those in a 3D electron gas. The significant differences between the parallel and perpendicular transferred momenta cover the higher plasmon frequency, the lower peak intensity, and the smaller critical momentum under the former case. It is relatively easy to observe the low-frequency optical plasmons in the perpendicular case because of the almost vanishing inter-π-band e–h excitations. On the other hand, both AB- and ABC-stacked graphite display the low-frequency optical plasmons only for the perpendicular transferred momenta, in which the dispersion relations with q_z are weak. Their low-frequency excitations are strongly affected by temperature because of low free carrier densities, especially for the existence of the optical plasmons in the latter. The predicted T-dependent plasmon frequencies in Bernal graphite are roughly consistent with those measured by the high-resolution EELS [37, 38, 167, 168, 215, 231–233] and optical reflectance spectroscopy [231–233, 406]. Doping in the AA-stacked graphite leads to drastic changes in the π-electronic excitations, such as the great enhancement of plasmon frequency and strength. The theoretical predictions could account for the EELS and optical measurements on the optical plasmons in LiC$_6$ [171, 172]. The simple hexagonal graphite sharply contrasts with a monolayer graphene in terms of low- and middle-frequency excitation channels. For example, concerning graphene system, the low-frequency plasmon modes are absent for the pristine case and $T = 0$. They are created by the doping or finite temperatures, but their behaviors at long wavelength limit belong to the acoustic modes. Moreover, the π-plasmon frequency of graphene is much lower than those of the 3D systems at small transferred momenta, owing to the lower and narrower saddle-point DOS, and the weaker Coulomb interactions at small momenta. Specifically, the diversified π plasmons are revealed in three kinds of graphites, in which the main features, the existence, intensity and frequency, depend on the direction and magnitude of the 3D transferred momenta, and the stacking configurations. The π plasmons, the collective oscillation mode of the whole valence π electrons, are present under any parallel momenta, while those for the perpendicular momenta are present only at sufficiently large parallel components. The dependences on $q_\parallel$ and q_z, respectively, increase rapidly and decline slowly with the increment of momentum. The azimuthal anisotropy exists when $q_\parallel$ is high enough. The predicted π-plasmon frequencies for the $q_\parallel$-component approximately agree with the EELS measured results on Bernal graphite [37, 38, 167, 168, 215, 231–233]. In addition, a simple superlattice model is also utilized to explain the significant features of π-electronic collective excitations. Apparently, such π plasmons are revealed in the sp^2-bonding carbon-based materials, but regardless of sp and sp^3 bondings. There exist certain important differences among graphites [25, 37, 38], layered graphenes [23, 27, 40], carbon nanotubes [19, 48, 124, 126, 173, 174, 176, 177], and C_{60}-related fullerenes [211], such as momentum- and angular-momentum-decoupled collective oscillation modes.

The 1D carbon nanotubes have rich electronic properties and many-particle excitation phenomena, being sensitive to the changes in the transferred momenta and angular moments (q, L), nanotube geometry (radii & chiral angles), temperatures, doping levels, and magnitudes & directions of magnetic fields. There are three types of (m, n) carbon nanotubes under the curvature effects according to the concise rule. (1) type-I metals for $m = n$ (armchair), (2) type-II narrow-gap systems for $m \neq n$ & $2m + n = 3I$, and (3) type-III moderate-gap ones for $2m + n \neq 3I$. Only the first type exhibits a low-frequency acoustic plasmon with the specific $q|ln(qr)|^{1/2}$ dependence at small momenta. Such plasmon comes from interband excitations of a pair of linearly intersecting valence and conduction bands near the Fermi level. This plasmon mode hardly depends on finite temperatures, while the sufficiently high Ts can create intraband low-frequency plasmon in a narrow-gap large carbon nanotube and a zero-gap semiconducting monolayer graphene. The free carriers can induce a low-frequency intraband plasmon of $L = 0$ in all carbon nanotubes. Furthermore, the high-level doping even leads to the creation of the $L = 1$ optical plasmon mode. As to the magnetic field, its strong effects are to drastically/dramatically alter the energy dispersions & bandgaps of electronic structures and wave functions, mainly arising from the obvious variation of $J \rightarrow J + \phi/\phi_0$ and the coupling of distinct J's. The former mechanism induces the periodical Aharnov–Bohm effect under negligible Zeeman splittings, and the latter one results in irregular standing waves on a cylindrical surface. Each armchair nanotube only shows one magnetoplasmon mode due to interband excitations under the nonperpendicular magnetic field. However, the low-frequency magnetoplasmon becomes the composite intraband and interband mode at $\alpha = 90°$ or high T's. The nonarmchair narrow-gap. nanotubes exhibit two magnetoplasmon, while the nonarmchair metallic nanotubes present one interband magnetoplasmon and one interband & intraband magnetoplasmon. The Zeeman splitting plays an important role on the significant differences among these magnetoplasmon modes. In addition to the low-frequency π-electronic excitations, there exist inter-π band and π plasmons in various carbon nanotubes, in which they, respectively, correspond to the specific valence bands and whole π-valence electrons. Their dependence on angular momentum, radius, and chiral angle is significant, while the momentum dispersion relation might be strong or weak. Among three types of nanotube systems, armchair/zigzag ones have the smallest/largest mode number of inter-π-band plasmons. The π-plasmon frequencies of $\omega_p > 5$ eV, which increases with an increase in (q, L), are suitable in identifying the intrinsic L-decoupled modes of cylindrical systems. Up to now, the higher-frequency inter-π-band and π plamsons have been confirmed by accurate EELS measurements [173–179]. However, the low-frequency ones, being due to the geometry,

temperature, magnetic field, and doping, require further experimental verifications. Apparently, the Coulomb excitation phenomena are greatly diversified
by the 1D cylindrical surface and 2D plane (the different geometric symmetries). The distinct bare Coulomb interactions and electronic properties can
create sharp differences for carbon nanotubes and layered graphenes [495,496],
covering the conserved quantities in e–e scatterings, the existence/absence of
low-frequency plasmons, and the composite effects coming from the critical
factors (temperature, strength and direction of magnetic field, and doping
density).

Obviously, group-IV monolayer systems exhibit rich and unique Coulomb
excitation spectra, in which there are certain important differences among silicene, germanene, and graphene. For pristine systems, the near-neighboring
hopping integral, spin–orbital coupling, and temperature play critical roles
in creating the available single-particle transitions and observable plasmon
modes. The first and second factors are, respectively, responsible for gap
and Fermi velocity; therefore, they codetermine the critical temperature and
threshold excitation frequency. When the temperature is sufficiently high, the
T-induced intraband transitions will strongly suppress and even replace the
original interband ones. The intraband and interband e–h excitations, respectively, display the square-root divergent peak and shoulder structure in the
imaginary part of dielectric function. There are three types of collective excitations, as clearly illustrated from (q, ω) phase diagrams under specific temperatures. For example, silicene, with a very narrow gap, shows (I) the undamped
plasmon between intraband and interband boundaries, (II) the similar one and
another weaker plasmon, and (III) the undamped and damped plasmon, in
which the first and third kinds of plasmon modes (the second kind) experience
the interband (intraband) Landau dampings. The diverse behaviors strongly
depend on the range of temperature: (I) $T <75$ K, (II) 75 K$\leq T \leq$ 120 K, and
(III) $T > 120$ K. The electric field in a buckled system can destroy the mirror
symmetry about the $z = 0$ plane and create the splitting of spin-up and spin-
down energy bands, resulting in diversified Coulomb excitation phenomena
in the (q, ω)-, (T, ω)- and (F, ω)-phase diagrams. They clearly present different critical momenta, temperature, and electric-field strengths under various
cases, indicating the dramatic changes of the intensities of energy loss functions and thus the emergence/disappearance of plasmon mode and significant
Landau dampings. Furthermore, four types of single-particle and collective
excitations, which depend on the specific T and F, are characterized as follows. (I) The excitation gap exists between the intraband and interband transitions of the spin-degenerate valence and conduction bands (without F), and
an intraband plasmon is undamped at small q's and then disappears at large
q's because of serious Landau dampings near the intraband boundary. (II)
Three e–h boundaries, the lower, middle, and upper ones, are induced by
the intraband transitions from the spin-down-dominated energy bands, the
interband transitions of the same pair, and the interband transitions due to
the spin-up-dominated pair. The most prominent intraband plasmon survives

between the former two boundaries. It experiences the interband e–h damping near the middle boundary and accompanies with another weaker plasmon. This mode is absent at large q's under intraband dampings (close to the lower boundary). (III) The merged intraband and interband boundaries related to the linear energy bands intersecting at E_F and the higher-frequency interband boundary are associated with the spin-up-dominated energy bands. The intraband plasmon has a strong e–h damping even at long wavelength limit. (IV) The e–h boundaries are the same with those in (II), while this plasmon is undamped only near $q \to 0$, and then it enters into the distinct interband Landau dampings related to the first and second pairs of spin-split energy bands. Concerning the complex composite effects arising from spin–orbital interactions, magnetic field, electric field, and electron doping, they further diversify the single-particle and collective excitations. The magnetic field creates interband magnetoplasmons with discrete frequency dispersions restricted to the quantized LL states. An intraband magnetoplasmon, with a higher intensity and continuous dispersion relation, exists in the presence of free conduction carriers. This mode is dramatically transformed into an interband plasma excitation when the magnetic field is increased, leading to abrupt changes in plasma frequency and intensity. Specifically, an electric field could separate the spin and valley polarizations and induce additional magnetoplasmon modes, a unique feature arising from the buckled structure and the existence of significant spin–orbit couplings. The intraband and interband magnetoplasmons, respectively, belong to the continuous and localized modes. In short, it is relatively easy to observe the split valleys and spin configurations in monolayer germanene, since it has the largest spin–orbital couplings among three group-IV systems (C, Si, Ge).

Apparently, monolayer graphene presents a rich and unique Coulomb deexcitation phenomena, being very different from 2D electron gas, 1D carbon nanotubes, silicene, and germanene. The inelastic Coulomb scatterings could be investigated from the screened exchange self-energy method and the Fermi golden rule. Concerning the Coulomb decay rates in pristine graphene, they directly reflect the characteristics of band structure, the zero-gap semiconductor, and the strong dependence on wave vector/state energy. At $T = 0$, only the interband e–h excitations exist, and their contributions to the Coulomb decay rates are negligible. The temperature-induced free carriers create intraband e–h excitations and an acoustic plasmon. The strength of plasmon mode increases (declines) as temperature (momentum) increases. The Coulomb inelastic scatterings of the Fermi-momentum state (the Dirac point) only utilize intraband e–h excitations, with any transferred momenta, leading to the nonspecific T-dependence. The calculated decay rate is close to the measured results of layered graphite [240, 479]. The intraband e–h excitations also make important contributions to other states. Both interband e–h excitations and plasmon belong to the critical deexcitation channels for the small-k states. They, respectively, result in the special shoulder and peak structures in the wave-vector-dependent decay rates of the conduction- and valence-band states. The e–e interactions are more efficient in carrier deexcitations, when compared with the electron–phonon interactions. The important differences between graphene and 2D electron gas lie in the temperature and wave-vector dependences, since the band structure of the latter can only generate the intraband e–h excitations and its acoustic plasmon survives even at $T = 0$. The plasmon is not the effective deexcitation channel and the small-(q, ω_{de}) intraband e–h excitations predominate over the relaxation process, so that the temperature- and energy-dependent Coulomb decay rates could be expressed in the analytic formulas [240]. The absence of specific formulas in monolayer graphene clearly indicates that the large-(q, ω_{de}) intraband e–h excitations, the plasmon, and the interband e–h excitations are not negligible in inelastic Coulomb scatterings. These three kinds of electronic excitations are expected to play an important role on the low-energy quasiparticle properties of few-layer graphene systems. There also exist significant differences between 2D graphene and 1D gapless carbon nanotubes, covering the magnitude, temperature dependence, and wave-vector dependence of Coulomb decay rates. The Dirac point in armchair nanotube shows the linear T-dependence, as observed in a 1D electron gas [237]. Apparently, the Coulomb decay rates of monolayer graphene are greatly diversified by the electron (or hole) doping. The deexcitation processes cover the intraband single-particle excitations, the interband e–h excitations, and the damped or undamped collective excitations, depending on the quasiparticle states and the Fermi energies. The low-lying valence holes can decay through undamped collective excitations; therefore, they present the fast Coulomb deexcitations, nonmonotonous energy dependence, and anisotropic behavior. However, the low-energy conduction electrons and holes are similar to those in a two-dimensional electron gas. In addition, the Fermi-momentum states display the $T^2 ln[T]$ dependence. The higher-energy conduction states and the deeper-energy valence ones behave similarly in the available deexcitation channels and have a strong dependence on the wave vector $\mathbf{k}$, since the contributions due to intraband (interband) e–h excitations quite differ from one another along the distinct directions. Moreover, there exist significant differences among the extrinsic graphene, silicene, and germanene in terms of available deexcitation channels and certain Coulomb decay phenomena. Specially, a doped germanene shows composite effects due to spin–orbital couplings and electron doping. The second and third kinds of plasmon modes might be the effective Coulomb scattering channels. Moreover, the deexcitation mechanisms sharply contrast for the separated valence and conduction Dirac points, in which the former could decay by the acoustic plasmon and thus has a faster Coulomb decay rate.

The generalized tight-binding model and the modified RPA are developed to deal with Coulomb excitations and decay rates under various stacking configurations, layer numbers, dimensionalities, electric fields, and magnetic fields. The bare and screened response functions are thoroughly clarified using the dynamically e–e inelastic scatterings. Specifically, the latter is useful in the further understanding of time-dependent carrier oscillation/propagation in real graphene-related systems. These energy loss spectra are sensitive to the magnitude and direction of the transferred momenta under various frequencies. From the theoretical point of view, one must use the 3D Fourier transform for them to explore the time- and position-dependent collective oscillation phenomena accompanied with Landau dampings. The calculated results in Chapters 3–14 clearly show that there are three kinds of plasmon modes due to the carbon $2p_z$ orbitals in graphene-related sp^2-bonding systems. The π and inter-π-band plasmons possess oscillational frequencies more than 1 eV, corresponding to a rapid oscillation/propagation with a period below 4 fs. Apparently, the time resolution is too high for the up-to-date fomtosecond optical spectroscopies; that is, it is impossible to observe and identify the oscillation behaviors from these two kinds of plasmon modes using the current high-resolution equipment. On the other hand, the low-frequency plasmon modes, with different momentum dispersions, are expected to reveal all the graphene-related materials. For example, the $\sim$ 10-meV plasmon modes, with oscillation periods $\sim$ 400 fs, are suitable for the experimental verifications on the diverse plasmon waves accompanied with various Landau dampings. Such collective excitations are deduced to originate from temperatures, lattice symmetry, dimensionalities, stacking configurations, numbers of layers, spin–orbital interactions, electric fields, and magnetic fields. That is to say, a lot of critical mechanisms could also be illustrated by the dynamic behavior of excited carriers. The time-dependent charge oscillations/propagations in the layered structures will become an interesting and emergent research topic; furthermore, the dynamics of plasmon modes might have high potentials in electronic and optical devices [43, 435]. Such work is under current investigation.

The aforementioned two models are useful in fully exploring the static charge screening of graphene-related systems, especially for the spatial variations (distributions) of effective Coulomb potentials and the induced carrier density on the distinct planes. The momentum- and frequency-dependent Coulomb potentials, which have been thoroughly investigated in the current work for the 1D, 2D, and 3D sp^2-bonding systems, could be recovered to the static effective Coulomb interactions in the real space by doing the Fourier transform at zero frequency. A similar method is applied to the induced charge density (the product of the effective potential and the bare response function in the momentum space). The free carrier density and the Fermi surface (the Fermi momenta), or the energy gap, are expected to play critical roles on the unusual screening behaviors. Whether the long-range effective Coulomb interactions decay quicker than the inverse of distance is the standard criteria for the charge screening ability. The semiconducting carbon nanotubes have been predicted to exhibit a long-distance Coulomb potential similar to the bare one [125]. However, metallic armchair nanotubes [126], layered graphenes [497], and graphite intercalation compounds [498] obviously display an exponential decay at short distance and the well-known Friedel oscillations at long distance. The characteristic decay length and the rapid charge oscillations are very sensitive to the concentration of free carriers and spatial dimensions. From the theoretical point of view, the strong Friedel oscillations mainly arise from the divergent derivative of the static bare response functions versus the transferred momentum at the double of Fermi momentum, as identified from 3D and 2D electron gases [237]. On the other hand, scanning tunneling spectroscopy (STM) has been demonstrated to be a powerful tool for visualizing Friedel oscillations in real space of 2D graphene-related systems in response to the atomic-scale impurities [499]. In general, both pristine few-layer graphene systems and 3D graphites, which belong to semimetals, are expected to show marginal (partial) screening ability in between the full and almost vanishing ones. Also, the doped layered graphene systems, with the distinct stacking configurations and layer numbers, are suitable for exploring the diversified charge screening phenomena, e.g., the theoretical [23, 125, 236, 268, 287, 288, 291, 364, 500, 501] and experimental studies on 2D Friedel charge oscillations [64, 153, 243]. In addition, the effective Coulomb potentials due to the charged impurities could investigate the residual resistivities in layered materials, e.g., the zero-temperature elastic scatterings for residual resistivities in alkali-doped graphite compounds [498].

The Friedel oscillations can be studied with the use of static dielectric screening in the generalized tight-binding model. In a monolayer graphene, the static dielectric function $\epsilon(q, 0)$ is calculated by specifying the general results of the full charge screening due to the massless-Dirac quasiparticle, which is incorporated by the development of the modified RPA, summarized in Chapter 2. As shown in Figure 15.1, the induced charge distribution and the screened potential are presented for the static case of a charge impurity e at $r = 0$ in the doped graphene under various Fermi levels. The most outstanding feature of Friedel oscillations is the strong oscillation behavior in real space, a phenomenon being deduced from a discontinuity of the second derivative of the static dielectric function. At higher dopings, the screening charges become rather extended in r-space, with a $1/r$ long-wavelength decay of oscillations at a distance from the impurity [497]. It should be noticed that there is no interband contribution from the long-wavelength behavior of polarization, since the corresponding polarization approaches zero at $q \sim 0$. This indicates that the nonoscillating part at small-r distances comes from the intraband polarization with a characteristic decaying length that can be evaluated within the Thomas-Fermi approximation. The previous work has indicated that as q increases the interband transitions greatly enhance the large-q screening when compared with the intraband transitions, and consequently, make the small-r screening effects rather effective from an impurity [498]. However, and very importantly, at a long distance from the impurity, the decaying oscillation behavior with a unique period given by $\pi/2k_F$ comes from the discontinuity at $\hbar v_F q = 2E_F$ as a result of singularities in the second derivative of the static dielectric function. The aforementioned features are expected to have important influences on, for example, the ordering of impurities and the resistivity in graphene-related systems [498].

The dimensionless energy loss function, which characterizes the intrinsic charge screening of a layered condensed-matter system, is well developed in this work. Furthermore, the modified RPA (Section 2.1) is suitable and reliable in evaluating the effective Coulomb potentials for intralayer and interlayer many-particle interactions. As a result, such potentials could be directly linked with the self-energy method of the Matibara Green functions (Section 2.6). The decay rate, being associated with the e–e Coulomb inelastic scatterings on all the graphene layers, could be derived and even expressed in the analytic form. The phenomenological formulas will be investigated fully and delicately. The effects arising from the number of layers and stacking configurations are taken into consideration simultaneously. The rich and unique deexcitation phenomena are expected to reveal few-layer graphene systems, and they need to be thoroughly explored within this new theoretical framework. For example, the AAA-/ABA-/ABC-stacked N-layer graphenes are present in an electromagnetic field, and the excited conduction electrons and valence holes in N pairs of energy bands will relax through effective Coulomb excitation channels. The numerical calculations become very difficult, since the anisotropic multidimensional integrations appear in the band structure, polarization functions, and transferred momenta and energies. Some efficient methods need to be introduced to overcome the serious issues. On the experimental side, the femtosecond pump-probe optical spectroscopies of absorption and fluorescence are available to identify the diverse deexcitation phenomena in few-layer graphene systems. Apparently, such studies could be generalized to other emergent 2D materials. In short, the near-future researches on femtosecond dynamics of charge carriers are predicted to open a new studying category.

In addition to the Coulomb scatterings, the electron–phonon scatterings at finite temperatures might play important roles in the e–e effective interactions, and thus have a strong effect electronic excitation spectra and quasiparticle lifetimes. These two kinds of inelastic scatterings could be taken into account simultaneously under the RPA of the same order, as clearly illustrated in Ref. [237]. This method has been clearly verified to be suitable and reliable for the doped GaAs semiconductor by measuring the plasmon branches using Raman scattering [502]. Apparently, the combination of the longitudinal plasmon modes and optical phonons creates well-known anticrossing phenomenon in the spectrum of collective oscillation frequencies. In the near-future studies on few-layer graphene systems, the couplings of acoustic plasmons and longitudinal optical phonons are expected to diversify the (momentum, frequency)-phase diagrams and greatly modify the inelastic scattering rates of excited quasiparticle states [503–506]. On the other side, certain physical barriers in theoretical models need to be overcome, such as the delicate phonon spectra of layered graphenes calculated by the oscillator model [507, 508], the effective amplitude of the electron–phonon scatterings using the phenomenological method [509–511], the modifications of the layer-dependent RPA [512], and the modified screened exchange energies [513]. They will become challenges and chances for a full understanding of the overall quasiparticle phenomena in 2D emergent materials. For example, there are some theoretical predictions on phonon spectra, especially for the first-principles calculations [514–516], in which how to transform the evaluated results into the useful information in establishing the electron–phonon scattering amplitudes is one of the high barriers.

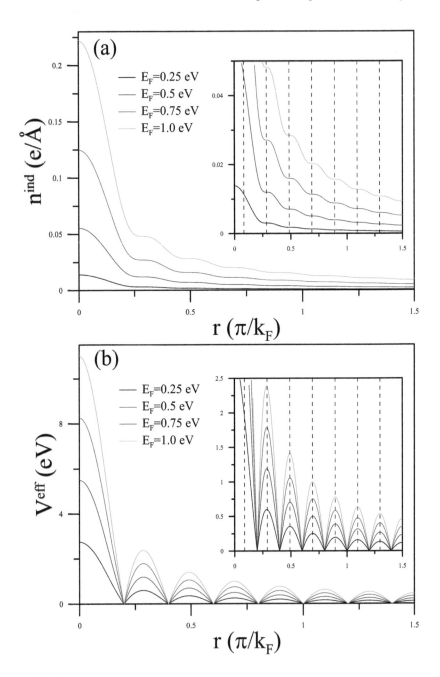

FIGURE 15.1
(a) The static induced charge distributions and (b) effective potentials on a doped monolayer graphene.

16

Problems

The following problems could be solved by the generalized tight-binding model, the effective-mass method, the random-phase approximation (RPA), the Hubbard approximation, and the screened exchange self-energy.

1. Consider the momentum-dependent bare Coulomb interactions during the electron–electron scatterings for the following condensed-matter systems: (a) 3D, (b) 2D & (c) 1D electron gases, (d) two electron-gas planes with the interlayer distance I_z, (e) two coaxial carbon nanotubes of radii r_1 and r_2, and (f) electron gases on two concentric spherical shells.

2. Calculate the dielectric functions within the RPA in the analytic forms for (a) 3D, (b) 2D, and (c) 1D electron gases, especially for the thorough discussions on the responses of the Fermi-surface states. (d) Plot the electron–hole (e–h) excitation boundaries. (e) The long wavelength limit is very useful in understanding the collective charge oscillations; that is, one could obtain the approximate dispersion relations of the plasmon frequencies with the transferred momenta from the vanishing real part of dielectric function. (f) The (momentum, frequency)-phase diagrams provide the full information on Landau dampings. (g) The static cases could be calculated simultaneously, and the significant differences among the dimension-different systems are investigated fully. The Fermi momenta are closely related to the well-known Friedel oscillations in the long-distance charge screenings.

3. Evaluate the static bare and screened response functions for the pristine and doped monolayer graphene systems in the numerical methods using (a) the tight-binding model, (b) the effective-mass approximation, and (c) the Hubbard model. (d) A detailed comparison with those of 2D electron gas is made in terms of the static dielectric function, the effective Colombo potential in the real space, and the spatial screening charge density. (e) Similar discussions could be done for the dynamic properties, e.g., the main features of energy loss spectra and (q, ω)-phase diagrams.

4. Using the superlattice model in the absence of interlayer hopping integrals, investigate the (a) Coulomb excitations and (b) deexcitations for the layered graphene and electron gas with infinite layers

under the numerical evaluations, such as the definition of dimensionless energy loss functions, the unusual plasmon bands, the rich Landau dampings, and the unique Coulomb decay rates. This model is roughly suitable for (c) the stage-1 and (d) stage-2 graphite intercalation compounds (details for geometric structures in Adv. Phys. 30, 139(1981)), in which the unusual plasmon spectra will display an unusual phenomena.

5. If the intensities of energy loss spectra is fully dominated by the plasmon mode, derive the analytic formula for the Coulomb decay rates in the dimension-dependent electron gases: (a) 3D, (b) 2D, and (c) 1D. (d) The thorough comparisons with the numerical results from the RPA self-energy are also made.

6. The band structures of few-layer graphene systems are calculated by the tight-binding model with vertical and tilted atomic interactions. Furthermore, the intralayer and interlayer Coulomb interactions are covered in the modified RPA. (a) Use the Dyson equation to get the effective Coulomb potential between any two layers for trilayer AAA, ABA, ABC, and AAB, and then (b) utilize the screened exchange self-energy to obtain the Coulomb decay rates for the excited states under the layer-decomposed scheme. (c) One could further explore the temperature, doping, and energy/wave-vector dependences.

7. The similar methods in (6) are available for exploring (a) the static screenings if one charge impurity is situated at the middle layer, such as the effective Coulomb potentials and the induced charge densities at any layers. And then, (b) evaluate the residual resistivities from Boltzmann transport formulas. (c) The impurity screening problems could also be applied to monolayer silicene and germanene.

8. The dynamic charge oscillations are studied using the dimensionless screened response function, $\text{Im}[-1\epsilon]$ in Eq. (2.8). Calculate the time-dependent plasmon waves in 2D real space for (a) the temperature-induced plasmon mode in monolayer graphene, (b) the low-doping extrinsic systems ($E_F \sim 25$ meV at zero temperature, and (c) the middle-doping case ($E_F \sim 0.5$ eV. Furthermore, (d) both pristine silicene and germanene will exhibit different propagation behaviors and (e) an armchair (a metallic) carbon nanotube does so. (f) Further studies could be done for the undoped and doped few-layer graphene systems: AAA, ABA, ABC, and AAB stackings.

9. The magnetoelectronic excitations, single-particle, and collective excitations in a uniform perpendicular magnetic field are worthy of a symmetric investigation for trilayer graphene systems under the generalized tight-binding model and the modified RPA: (a) ABC, (b) AAB, (c) ABA, and (d) AAA, especially for the former two.

Such typical stackings have rich and unique electronic properties, and their Coulomb interactions due to novel Landau levels are thus expected to exhibit a diverse phenomena, such as the abnormal excitation behaviors due to frequent anticrossing and crossing behaviors and the nonmonotonous magnetic-field dependence.

10. For a coaxial carbon nanotube, the intralayer and interlayer hopping integrals and Coulomb interactions are taken into consideration simultaneously (details in PRB 76, 115422 (2006)). Under an axial uniform magnetic field, calculate and discuss (a) the magnetoelectronic properties, e.g., the Aharnov-Bohm effect and the dramatic changes of energy dispersions and bandgap, and (b) the magnetoelectron–hole excitations and magnetoplasmons for angular-momentum-decoupled modes.

11. Using a single-walled armchair carbon nanotube in the presence of a transverse electric/magnetic field, evaluate (a) band structure and wave function by changing the on-site energies/adding the Peierls phases in the hopping integrals. (b) Using the decomposition of the decoupled angular momenta (Ls), explore the main feature of standing waves. And then, (c) calculate the L-dependent polarization function and energy loss spectra, especially for the magnetoplasmon modes for pristine and doping cases.

12. Consider a concentric spherical shell with the inner and outer radii of r_1 and r_2, respectively, and (a) calculate the bare Coulomb interactions for electron gases during e–e scatterings. And then (b) evaluate the dielectric function and (c) discuss the decoupled plasmon modes.

13. (a) Calculate the various Coulomb matrix elements for an electron gas in an 1D nanoribbon, (b) derive the subband-based dielectric function, and (c) utilize its zero points to analyze various plasmon modes. (d) Do the similar calculations for 1D graphene nanoribbons and (e) compare the significant differences between zigzag and armchair systems.

14. Compare between AA-stacked graphene systems and 3D hexagonal graphite in (a) the transferred momentum, (b) van Hove singularities of the low-lying energy bands, (c) polarization functions (the special structures), (d) energy loss spectra, and (e) momentum-dependent plasmon modes.

15. A detailed comparison could also be done for (a) AB-stacked bilayer graphene and Bernal graphite, and (b) ABC-stacked graphene and rhombohedral graphite.

16. Describe and explain the significant differences among trilayer AAA, ABA, ABC, and AAB stackings in (a) hopping integrals,

(b) electronic properties, and (c) e–h excitations and plasmon modes.

17. Explore the magnetic quantization in trilayer ABC stacking (details in [30]): (a) the spatial probability distributions and (b) the B_z-dependent energy spectra. And then, calculate (c) the layer-based polarization functions, (d) the energy loss functions, and (e) the (momentum, frequency)-phase diagram under undoped and doped cases.

References

[1] K. S. Novoselov, A. K. Geim, S. V. Morozov, D. Jiang, Y. Zhang, S. V. Dubonos, I. V, Grigorieva, and A. A. Firsov, Electric field effect in atomically thin carbon film, *Science* 306, 666, 2004.

[2] A. H. Castro Neto, F. Guinea, N. M. R. Peres, K. S. Novoselov, and A. K. Geim, The electronic properties of graphene, *Rev. Mod. Phys.* 81, 109, 2009.

[3] A. C. Ferrari, J. C. Meyer, V. Scardaci, C. Casiraghi, M. Lazzeri, F. Mauri, S. Piscanec, D. Jiang, K. S. Novoselov, S. Roth, and A. K. Geim, Raman spectrum of graphene and graphene layers, *Phys. Rev. Lett.* 97, 187401, 2006.

[4] Y. Shao, J. Wang, H. Wu, J. Liu, I. A. Aksay, and Y. Lin, Graphene based electrochemical sensors and biosensors: A review, *Electroanalysis* 22, 1027, 2010.

[5] Y. Zhang, L. Zhang, and C. Zhou, Review of chemical vapor deposition of graphene and related applications, *Acc. Chem. Res.* 46, 2329, 2013.

[6] Y. Wang, Y. Shao, D. W. Matson, J. Li, and Y. Lin, Nitrogen-doped graphene and its application in electrochemical biosensing, *ACS Nano* 4, 1790, 2010.

[7] Y. Wang, Y. Li, L. Tang, J. Lu, and J. Li, Application of graphene-modified electrode for selective detection of dopamine, *Electrochem. Commun.* 11, 889, 2000.

[8] Z. Fan, J. Yan, L. Zhi, Q. Zhang, T. Wei, J. Feng, M. Zhang, W. Qian, and F. Wei, A three-dimensional carbon nanotube/graphene sandwich and its application as electrode in supercapacitors, *Adv. Mater.* 22, 3723, 2010.

[9] A. L. M. Reddy, A. Srivastava, S. R. Gowda, H. Gullapalli, M. Dubey, and P. M. Ajayan, Synthesis of nitrogen-doped graphene films for lithium battery application, *ACS Nano*, 4, 6337, 2010.

[10] M. J. Allen, V. C. Tung, and R. B. Kaner, Honeycomb carbon: A review of graphene, *Chem. Rev.* 110, 132, 2010.

[11] H. Wang, T. Maiyalagan, and X. Wang, Review on recent progress in nitrogen-doped graphene: Synthesis, characterization, and its potential applications, *ACS Catal.* 2, 781, 2012.

[12] E. McCann, and M. Koshino, The electronic properties of bilayer graphene, *Rep. Prog. Phys.* 76, 056503, 2013.

[13] H. G. Yan, X. S. Li, B. Chandra, G. Tulevski, Y. Q. Wu, M. Freitag, W. J. Zhu, P. Avouris, and F. N. Xia, Tunable infrared plasmonic devices using graphene/insulator stacks, *Nat. Nanotechnol.* 7, 330, 2012.

[14] H. G. Yan, T. Low, W. J. Zhu, Y. Q. Wu, M. Freitag, X. S. Li, F. Guinea, P. Avouris, and F. N. Xia, Damping pathways of mid-infrared plasmons in graphene nanostructures, *Nat. Photonics* 7, 394, 2013.

[15] T. Low, and P. Avouris, Graphene plasmonics for terahertz to mid-infrared applications, *ACS Nano* 8, 1086, 2014.

[16] Y. C. Fan, N. H. Shen, T. Koschny, and C. M. Soukoulis, Tunable ter-ahertz meta-surface with graphene cut-wires, *ACS Photonics* 2, 151, 2015.

[17] Y. C. Fan, N. H. Shen, F. L. Zhang, Q. Zhao, Z. Y. Wei, P. Zhang, J. J. Dong, Q. H. Fu, II. Q. Li, and C. M. Soukoulis, Electrically tunable slow light using graphene metamaterials, *ACS Photonics* 5, 1612, 2018.

[18] K. W. -K. Shung, Lifetime effects in low-stage intercalated graphite systems, *Phys. Rev. B* 34, 2, 1986.

[19] M. F. Lin, and D. S. Chuu, Elementary excitations in a carbon nanotube, *J. Phys. Soc. Jpn.* 66, 757, 1997.

[20] Y. C. Chuang, J. Y. Wu, M. F. Lin, Analytical calculations on low-frequency excitations in AA-stacked bilayer graphene, *J. Phys. Soc. Jpn.* 81, 124713, 2012.

[21] M. F. Lin, Y. C. Chuang, and J. Y. Wu, Electrically tunable plasma excitations in AA-stacked multilayer graphene, *Phys. Rev. B* 86, 125434, 2012.

[22] Y. C. Chuang, J. Y. Wu, M. F. Lin, Electric-field-induced plasmon in AA-stacked bilayer graphene, *Ann. Phys.* 339, 298, 2013.

[23] J. H. Ho, C. L. Lu, C. C. Hwang, C. P. Chang, and M. F. Lin, Coulomb excitations in AA- and AB-stacked bilayer graphites, *Phys. Rev. B* 74, 085406, 2006.

[24] Y. C. Chuang, J. Y. Wu, and M. F. Lin, Electric field dependence of excitation spectra in AB-stacked bilayer graphene, *Sci. Rep.* 3, 1368, 2013.

[25] J. H. Ho, C. P. Chang, and M. F. Lin, Electronic excitations of the multilayered graphite, *Phys. Lett. A* 352, 446, 2006.

[26] Y. C. Chuang, J. Y. Wu, and M. F. Lin, Electric field dependence of excitation spectra in AB-stacked bilayer graphene, *Sci. Rep.* 3, 1368, 2013.

[27] C. Y. Lin, M. H. Lee, and M. F. Lin, Coulomb excitations in ABC-stacked trilayer graphene, *Phys. Rev. B* 98, 041408, 2018.

[28] C. Y. Lin, T. N. Do, Y. K. Huang, and M. F. Lin, *Electronic and Optical Properties of Graphene in Magnetic and Electric Fields, IOP Concise Physics*. San Raefel, CA: Morgan & Claypool Publishers, 2017.

[29] S. C. Chen, J. Y. Wu, C. Y. Lin, and M. F. Lin, *Theory of Magnetoelectric Properties of 2D Systems, IOP Concise Physics*. San Raefel, CA: Morgan & Claypool Publishers, 2017.

[30] C. Y. Lin, J. Y. Wu, Y. J. Ou, Y. H. Chiu, and M. F. Lin, Magneto-electronic properties of multilayer graphenes, *Phys. Chem. Chem. Phys.* 17, 26008, 2015.

[31] C. Y. Lin, R. B. Chen, Y. H. Ho, and M. F. Lin, *Electronic and Optical Properties of Graphite-Related Systems*. CRC Press, Boca Raton, FL, 2017.

[32] J. Y. Wu, S. C. Chen, O. Roslyak, G. Gumbs, and M. F. Lin, Plasma excitations in graphene: Their spectral intensity and temperature dependence in magnetic field, *ACS Nano* 5, 1026, 2011.

[33] J. Y. Wu, G. Gumbs, and M. F. Lin, Combined effect of stacking and magnetic field on plasmon excitations in bilayer graphene, *Phys. Rev. B* 89, 165407, 2014.

[34] J. Y. Wu, S. C. Chen, G. Gumbs, and M. F. Lin, Feature-rich electronic excitations of silicene in external fields, *Phys. Rev. B* 94, 205427, 2016.

[35] R. F. Egerton, *Electron Energy-Loss Spectroscopy in the Electron Microscope*, 370. New York and London: Plenum, 1996.

[36] W. Schülke, *Electron Dynamics by Inelastic X-Ray Scattering* Oxford: Oxford University Press, 2007.

[37] K. Zeppenfeld, Wavelength dependence and spatial dispersion of the dielectric constant in graphite by electron spectroscopy, *Opt. Com.* 1, 119, 1969.

[38] K. Zeppenfeld, Nonvertical interband transitions in graphite by intrinsic electron scattering, *Z. Phys.* 243, 229, 1971.

[39] C. Kramberger, R. Hambach, C. Giorgetti, M. H. Rummeli, M. Knupfer, J. Fink, B. Buchner, Lucia Reining, E. Einarsson, S. Maruyama, F. Sottile, K. Hannewald, V. Olevano, A. G. Marinopoulos, and T. Pichler, Linear plasmon dispersion in single-wall carbon nanotubes and the collective excitation spectrum of graphene, *Phys. Rev. Lett.* 100, 196803, 2008.

[40] P. Wachsmuth, R. Hambach, M. K. Kinyanjui, M. Guzzo, G. Benner, and U. Kaiser, High-energy collective electronic excitations in free-standing single-layer graphene, *Phys. Rev. B* 88, 075433, 2013.

[41] T. Valla, J. Camacho, Z.-H. Pan, A. V. Fedorov, A. C. Walters, C. A. Howard, and M. Ellerby, Anisotropic electron-phonon coupling and dynamical nesting on the graphene sheets in superconducting CaC_6 using angle-resolved photoemission spectroscopy, *Phys. Rev. Lett.* 102, 107007, 2009.

[42] A. V. Fedorov, N. I. Verbitskiy, D. Haberer, C. Struzzi, L. Petaccia, D. Usachov, O. Y. Vilkov, D. V. Vyalikh, J. Fink, M. Knupfer, B. Buchner, and A. Gruneis, Observation of a universal donor-dependent vibrational mode in graphene, *Nat. Commun.* 5, 3257, 2014.

[43] A. Bostwick, T. Ohta, T. Seyller, K. Horn, E. Rotenberg, Quasiparticle dynamics in graphene, *Nat. Phys.* 3, 36, 2007.

[44] M. S. Dresselhaus, and G. Dresselhaus, Intercalation compounds of graphite, *Adv. Phys.* 51, 1, 2002.

[45] S. Tongay, J. Hwang, D. B. Tanner, H. K. Pal, D. Maslov, and A. F. Hebard, Supermetallic conductivity in bromine-intercalated graphite, *Phys. Rev. B* 81, 115428, 2010.

[46] X. Meng, S. Tongay, J. Kang, Z. Chen, F. Wu, S. S. Li, J. B. Xia, J. Li, and J. Wu, Stable p- and n-type doping of few-layer graphene/graphite, *Carbon* 57, 507, 2013.

[47] S. Tongay, K. Berke, M. Lemaitre, Z. Nasrollahi, D. B. Tanner, A. F. Hebard, and B. R. Appleton, Stable hole doping of graphene for low electrical resistance and high optical transparency, *Nanotechnology* 22, 425701, 2011.

[48] M. F. Lin, and K. W. -K. Shung, Plasmons and optical properties of carbon nanotubes, *Phys. Rev. B Rapid Commun.* 50, 17744, 1994.

[49] G. Borghi, M. Polini, R. Asgari, and A. H. MacDonald, Dynamical response functions and collective modes of bilayer graphene, *Phys. Rev. B* 80, 241402, 2009.

[50] S. D. Sarma, and E. H. Hwang, Plasmons in coupled bilayer structures, *Phys. Rev. Lett.* 81, 4216, 1998.

[51] J. C. Charlier, X. Gonze, and J. P. Michenaud, First principles study of the stacking effect on the electronic properties of graphite, *Carbon* 32, 289, 1994.

[52] Z. Liu, K. Suenaga, P. J. F. Harris, and S. Iijima, Open and closed edges of graphene layers, *Phys. Rev. Lett.* 102, 015501, 2009.

[53] N. T. T. Tran, S. Y. Lin, C. Y. Lin, and M. F. Lin, *Geometric and Electronic Properties of Graphene-Related Systems: Chemical bondings.* CRC Press, Boca Raton, FL, 2017.

[54] K. S. Kim, A. L. Walter, L. Moreschini, T. Seyller, K. Horn, E. Rotenberg, and A. Bostwick, Coexisting massive and massless Dirac fermions in symmetry-broken bilayer graphene, *Nat. Mater.* 12, 887, 2013.

[55] C. Bao, W. Yao, E. Wang, C. Chen, J. Avila, M. C. Asensio, and S. Zhou, Stacking-dependent electronic structure of trilayer graphene resolved by nanospot angle-resolved photoemission spectroscopy, *Nano Lett.* 17, 1564, 2017.

[56] S. Hattendorf, A. Georgi, M. Liebmann, and M. Morgenstern, Networks of ABA and ABC stacked graphene on mica observed by scanning tunneling microscopy, *Surf. Sci.* 610, 53, 2013.

[57] H. L. Guo, X. F. Wang, Q. Y. Qian, F. B. Wang, and X. H. Xia, A green approach to the synthesis of graphene nanosheets, *ACS Nano* 3, 2653, 2009.

[58] L. Tang, Y. Wang, Y. Li, H. Feng, J. Lu, and J. Li, Preparation, structure, and electrochemical properties of reduced graphene sheet films, *Adv. Funct. Mater.* 19, 2782, 2009.

[59] P. Xu, Y. R. Yang, S. D. Barber, J. K. Schoelz, D. Qi, M. L. Ackerman, L. Bellaiche, and P. M. Thibado, New scanning tunneling microscopy technique enables systematic study of the unique electronic transition from graphite to graphene, *Carbon* 50, 4633, 2012.

[60] P. Xu, Y. R. Yang, D. Qi, S. D. Barber, J. K. Schoelz, M. L. Ackerman, L. Bellaiche, and P. M. Thibado, Electronic transition from graphite to graphene via controlled movement of the top layer with scanning tunneling microscopy, *Phys. Rev. B* 86, 085428, 2012.

[61] J. D. Bernal, The structure of graphite, *Proc. R. Soc. London, Ser. A* 106, 749, 1924.

[62] H. Lipson, and A. R. Stokes, The structure of graphite, *Proc. R. Sot. London, Ser. A* 181, 101, 1942.

[63] E. Moreau, S. Godey, X. Wallart, I. Razado-Colambo, J. Avila, M. C. Asensio, and D. Vignaud, Highresolution angle-resolved photoemission spectroscopy study of monolayer and bilayer graphene on the C-face of SiC, *Phys. Rev. B* 88, 075406, 2013.

[64] T. Ohta, A. Bostwick, and J. L. McChesney, Interlayer interaction and electronic screening in multilayer graphene investigated with angle-resolved photoemission spectroscopy, *Phys. Rev. Lett.* 98, 206802, 2007.

[65] C. Coletti, S. Forti, A. Principi, K. V. Emtsev, A. A. Zakharov, K. M. Daniels, B. K. Daas, M. V. S. Chandrashekhar, T. Ouisse, D. Chaussende, A. H. MacDonald, M. Polini, and U. Starke, Revealing the electronic band structure of trilayer graphene on SiC: An angle-resolved photoemission study, *Phys. Rev. B* 88, 155439, 2013.

[66] K. Sugawara, N. Yamamura, K. Matsuda, W. Norimatsu, M. Kusunoki, T. Sato, and T. Takahashi, Selective fabrication of free-standing ABA and ABC trilayer graphene with/without Dirac-cone energy bands, *NPG Asia Mater.* 10, 466, 2018.

[67] C. Y. Lin, M. C. Lin, J. Y. Wu, and M. F. Lin, Unusual electronic excitations in ABA trilayer graphene, arXiv:1803.10715.

[68] S. Das Sarma and Qiuzi Li, Intrinsic plasmons in two-dimensional Dirac materials, *Phys. Rev. B* 87, 235418, 2013.

[69] R. Xiao, F. Tasnadi, K. Koepernik, J. W. F. Venderbos, M. Richter, and M. Taut, Density functional investigation of rhombohedral stacks of graphene: Topological surface states, nonlinear dielectric response, and bulk limit, *Phys. Rev. B* 84, 165404, 2011.

[70] W. T. Pong, J. Bendall, and C. Durkan, Observation and investigation of graphite superlattice boundaries by scanning tunneling microscopy, *Surf. Sci.* 601, 498, 2007.

[71] L. B. Biedermann, A. C. Michael, Z. Dmitry, and G. R. Ronald, Insights into few-layer epitaxial graphene growth on 4H-SiC(000$\bar{1}$) substrates from STM studies, *Phys. Rev. B* 79, 125411, 2009.

[72] Y. Que, W. Xiao, H. Chen, D. Wang, S. Du, and H. J. Gao, Stacking-dependent electronic property of trilayer graphene epitaxially grown on Ru(0001), *Appl. Phys. Lett.* 107, 263101, 2015.

[73] S. K. Asieh, S. Crampin, and A. Ilie, Stacking-dependent superstructures at stepped armchair interfaces of bilayer/trilayer graphene, *Appl. Phys. Lett.* 102, 163111, 2013.

[74] X. Feng, S. Kwon, J. Y. Park, and M. Salmeron, Superlubric sliding of graphene nanoflakes on graphene, *ACS Nano* 7, 1718, 2013.

[75] Z. Liu, J. Yang, F. Grey, J. Z. Liu, Y. Liu, Y. Wang, Y. Yang, Y. Cheng, and Q. Zheng, Observation of microscale superlubricity in graphite, *Phys. Rev. Lett.* 108, 205503, 2012.

[76] L. Jiang, S. Wang, Z. Shi, C. Jin, M. I. B. Utama, S. Zhao, Y. R. Shen, H. J. Gao, G. Zhang, and F. Wang, Manipulation of domain-wall solitons in bi- and trilayer graphene, *Nat. Nanotech.* 13, 204, 2018.

[77] Z. Y. Rong, and P. Kuiper, Electronic effects in scanning tunneling microscopy: Moire pattern on a graphite surface, *Phys. Rev. B* 48, 17427, 1993.

[78] Z. Y. Rong. Extended modifications of electronic structures caused by defects: Scanning tunneling microscopy of graphite, *Phys. Rev. B* 50, 1839, 1994.

[79] P. Xu, M. L. Ackerman, S. D. Barber, J. K. Schoelz, D. Qi, P. M. Thibado, V. D. Wheeler, L. O. Nyakiti, R. L. Myers-Ward, C. R. Eddy, Jr., and D. K. Gaskill, Graphene manipulation on 4H-SiC (0001) using scanning tunneling microscopy, *Jpn. J. Appl. Phys.* 52, 035104, 2013.

[80] D. Pierucci, T. Brumme, J.-C. Girard, M. Calandra, M. G. Silly, F. Sirotti, A. Barbier, F. Mauri, and A. Ouerghi, Atomic and electronic structure of trilayer graphene/SiC(0001): Evidence of strong dependence on stacking sequence and charge transfer, *Sci. Rep.* 6, 33487, 2016.

[81] T. N. Do, P. H. Shih, C. P. Chang, C. Y. Lin, and M. F. Lin, Rich magneto-absorption spectra in AAB-stacked trilayer graphene, *Phys. Chem. Chem. Phys.* 18, 17597, 2016.

[82] T. N. Do, C. Y. Lin, Y. P. Lin, P. H. Shih, and M. F. Lin, Configuration-enriched magnetoelectronic spectra of AAB-stacked trilayer graphene, *Carbon* 94, 619, 2015.

[83] J. S. Alden, A. W. Tsen, P. Y. Huang, R. Hovden, L. Brown, J. Park, D. A. Muller, and P. L. McEuen, Strain solitons and topological defects in bilayer graphene, *Proc. Natl. Acad. Sci.* 110, 11256, 2013.

[84] K. W. Lee, and C. E. Lee, Extreme sensitivity of the electric-field-induced band gap to the electronic topological transition in sliding bilayer graphene, *Sci. Rep.* 5, 17490, 2015.

[85] N. T. T. Thuy, S. Y. Lin, O. Glukhova, and M. F. Lin, Configuration-induced rich electronic properties of bilayer graphene, *J. Phys. Chem. C.* 119, 10623, 2015.

[86] A. Daboussi, L. Mandhour, J. N. Fuchs, and S. Jaziri, Tunable zero-energy transmission resonances in shifted graphene bilayer, *Phys. Rev. B* 89, 085426, 2014.

[87] A. M. Popov, I. V. Lebedeva, A. A. Knizhnik, Y. E. Lozovik, and B. V. Potapkin, Commensurate-incommensurate phase transition in bilayer graphene, *Phys. Rev. B* 84, 045404, 2011.

[88] Y. W. Son, S. M. Choi, Y. P. Hong, S. Woo, and S. H. Jhi, Electronic topological transition in sliding bilayer graphene, *Phys. Rev. B* 84, 155410, 2011.

[89] B. L. Huang, C. P. Chuu, and M. F. Lin, Asymmetry-enriched electronic and optical properties of bilayer graphene, arXiv:1805.10775v2.

[90] I. Martin, Y. M. Blanter, and A. F. Morpurgo, Topological confinement in bilayer graphene, *Phys. Rev. Lett.* 100, 036804, 2008.

[91] Y. K. Huang, S. C. Chen, Y. H. Ho, C. Y. Lin, and M. F. Lin, Feature-rich magnetic quantization in sliding bilayer graphenes, *Sci. Rep.* 4, 7509, 2014.

[92] C. W. Chiu, and R. B. Chen, Infuence of electric fields on absorption spectra of AAB-stacked trilayer graphene, *Appl. Phys. Express* 9, 065103, 2016.

[93] S. C. Chen, C. W. Chiu, C. L. Wu, and M. F. Lin, Shift-enriched optical properties in bilayer graphene, *RSC Adv.* 4, 63779, 2014.

[94] M. Koshino, Electronic transmission through AB-BA domain boundary in bilayer graphene, *Phys. Rev. B* 88, 115409, 2013.

[95] J. Zheng, P. Guo, Z. Ren, Z. Jiang, J. Bai, and Z. Zhang, Mono-bi-monolayer graphene junction introduced quantum transport channels, *Appl. Phys. Lett.* 103, 173519, 2013.

[96] S. Bhattacharyya, and A. K. Singh, Lifshitz transition and modulation of electronic and transport properties of bilayer graphene by sliding and applied normal compressive strain, *Carbon* 99, 432, 2016.

[97] S. M. Choi, S. H. Jhi, and Y. W. Son, Anomalous optical phonon splittings in sliding bilayer graphene, *ACS Nano* 7, 7151, 2013.

[98] V. Perebeinos, J. Tersoff, and P. Avouris, Phonon-mediated interlayer conductance in twisted graphene bilayers, *Phys. Rev. Lett.* 109, 236604, 2012.

[99] A. I. Cocemasov, D. L. Nika, and A. A. Balandin, Phonons in twisted bilayer graphene, *Phys. Rev. B* 88, 035428, 2013.

[100] J. M. B. Lopes dos Santos, N. M. R. Peres, and A. H. Castro Neto, Graphene bilayer with a twist: Electronic structure, *Phys. Rev. Lett.* 99, 256802, 2007.

[101] X. J. Wang, H. Y. Meng, S. Liu, S. Y. Deng, T. Jiao, Z. C. Wei, F. Q. Wang, C. H. Tan, and X. G. Huang, Tunable graphene-based mid-infrared plasmonic multispectral and narrow band-stop filter, *Mater. Res. Exp.* 5, 045804, 2018.

[102] G. Y. Gao, B. S. Wan, X. Q. Liu, Q. J. Sun, X. N. Yang, L. F. Wang, C. F. Pan, and Z. L. Wang, Tunable tribotronic dual-gate logic devices based on 2D MoS$_2$ and black phosphorus, *Adv. Mater.* 30, 1705088, 2018.

[103] A. Dey, A. Singh, D. Das, and P. K. Iyer, Photosensitive organic field effect transistors: The influence of ZnPc morphology and bilayer dielectrics for achieving a low operating voltage and low bias stress effect, *Phys. Chem. Chem. Phys.* 18, 32602, 2016.

[104] Y. S. Ang, S. Y. A. Yang, C. Zhang, Z. S. Ma, and L. K. Ang, Valleytronics in merging Dirac cones: All-electric-controlled valley filter, valve, and universal reversible logic gate, *Phys. Rev. B* 96, 245410, 2017.

[105] T. Khodkov, I. Khrapach, M. F. Craciun, and Saverio Russo, Direct observation of a gate tunable band gap in electrical transport in ABC-trilayer graphene, *Nano Lett.* 15, 4429, 2015.

[106] Y. Zhang , T. T. Tang , C. Girit, Z. Hao, M. C. Martin, A. Zettl, M. F. Crommie, Y. R. Shen, and F. Wang, Direct observation of a widely tunable bandgap in bilayer graphene, *Nature* 459, 820, 2009.

[107] M. S. Dresselhaus, and G. Dresselhaus, Intercalation compounds of graphite, *Adv. Phys.* 30, 139, 1980.

[108] J. K. Lee, S. C. Lee, J. P. Ahn, S. C. Kim, J. I. B. Wilson, and P. John, The growth of AA graphite on (111) graphite, *J. Chem. Phys.* 129, 234709, 2008.

[109] A. Gruneis, C. Attaccalite, T. Pichler, V. Zabolotnyy, H. Shiozawa, S. L. Molodtsov, D. Inosov, A. Koitzsch, M. Knupfer, J. Schiessling, R. Follath, R. Weber, P. Rudolf, L. Wirtz, and A. Rubio, Electron-electron correlation in graphite: a combined angle-resolved photoemission and first-principles study, *Phys. Rev. Lett.* 100, 037601, 2008.

[110] S. Y. Zhou, G.-H. Gweon, J. Graf, A. V. Fedorov, C. D. Spataru, R. D. Diehl, Y. Kopelevich, D.-H. Lee, Steven G. Louie, and A. Lanzara, First direct observation of Dirac fermions in graphite, *Nat. Phys.* 2, 595, 2006.

[111] C. M. Cheng, C. J. Hsu, J. L. Peng, C. H. Chen, J. Y. Yuh, and K. D. Tsuei, Tight-binding parameters of graphite determined with angle-resolved photoemission spectra, *Appl. Surf. Sci.* 354, 229, 2015.

[112] C. S. Leem, Chul Kim, S. R. Park, M. K. Kim, H. J. Choi, and C. Kim, High-resolution angle-resolved photoemission studies of quasiparticle dynamics in graphite, *Phys. Rev. B* 79, 125438, 2009.

[113] F. L. Shyu, and M. F. Lin, Plasmons and optical properties of semimetal graphite, *J. Phys. Soc. Jpn. Lett.* 69, 3781, 2000.

[114] F. L. Shyu, and M. F. Lin, Low-frequency π-electronic excitations of simple hexagonal graphite, *J. Phys. Soc. Jpn.* 70, 897, 2001.

[115] C. W. Chiu, F. L. Shyu, M. F. Lin, G. Gumbs, and O. Roslyak, Anisotropy of π-plasmon dispersion relation of AA-stacked graphite, *J. Phys. Soc. Jpn.* 81, 104703, 2012.

[116] S. Iijima, Helical microtubules of graphitic carbon, *Nature* 354, 56, 1991.

[117] S. Iijima, and T. Ichihashi, Single-shell carbon nanotubes of 1-nm diameter, *Nature* 363, 603, 1993.

[118] F. L. Shyu, C. P. Chang, R. B. Chen, C. W. Chiu, and M. F. Lin, Magnetoelectronic and optical properties of carbon nanotubes, *Phys. Rev. B* 67, 045405, 2003.

[119] R. Saito, M. Fujita, G. Dresselhaus, and M. S Dresselhaus, Electronic structure of chiral graphene tubules, *Appl. Phys. Lett.* 60, 2204, 1998.

[120] X. Blase, L. X. Benedict, E. L. Shirley, and S. G. Louie, Hybridization effects and metallicity in small radius carbon nanotubes, *Phys. Rev. Lett.* 72, 1878, 1994.

[121] J. W. G. Wildor, L. C. Venema, A. G. Rinzler, R. E. Smalley, and C. Dekker, Electronic structure of atomically resolved carbon nanotubes, *Nature* 391, 59, 1998.

[122] S. Zaric, G. N. Ostojic, J. Kono, J. Shaver, V. C. Moore, and M. S. Strano, Optical signatures of the Aharonov-Bohm phase in single-walled carbon nanotubes, *Science* 304, 1129, 2004.

[123] A. Bachtold, C. Strunk, J.-P. Salvetat, J.-M. Bonard, L. Forro, T. Nussbaumer, and C. Schonenberger, AharonovVBohm oscillations in carbon nanotubes, *Nature* 397, 673, 1999.

[124] M. F. Lin, and K. W. -K. Shung, Elementary excitations in cylindrical tubules, *Phys. Rev. B* 47, 6617, 1993.

[125] M. F. Lin, and D. S. Chuu, Impurity screening in carbon nanotubes, *Phys. Rev. B* 56, 4996, 1997.

[126] Y. H. Ho, G. W. Ho, S. C. Chen, J. H. Ho, and M. F. Lin, Low-frequency excitation spectra in double-walled armchair carbon nanotubes, *Phys. Rev. B* 76, 115422, 2007.

[127] N. Takagi, C. L. Lin, K. Kawahara, E. Minamitani, N. Tsukahara, M. Kawai, and R. Arafune, Silicene on Ag(1 1 1): Geometric and electronic structures of a new honeycomb material of Si, *Prog. Surf. Sci.* 90, 1, 2015.

[128] M. Derivaz, D. Dentel, R. Stephan, M. Hanf, A. Mehdaoui, P. Sonnet, and C. Pirri, Continuous germanene layer on Al(111), *Nano Lett.* 15, 2510, 2015.

[129] F. F. Zhu, W. J. Chen, Y. Xu, C. L. Gao, D. D. Guan, C. H. Liu, D. Qian, S. C. Zhang, and J. F. Jia, Epitaxial growth of two-dimensional stanene, *Nat. Mater.* 14, 1020, 2015.

[130] M. Ezawa, Valley-polarized metals and quantum anomalous Hall effect in silicene, *Phys. Rev. Lett.* 109, 055502, 2012.

[131] C. C. Liu, H. Jiang, and Y. Yao, Low-energy effective Hamiltonian involving spin-orbit coupling in silicene and two-dimensional germanium and tin, *Phys. Rev. B* 84, 195430, 2011.

[132] J. Y. Wu, S. C. Chen, and M. F. Lin, Temperature-dependent Coulomb excitations in silicene, *New J. Phys.* 16, 125002, 2014.

[133] S. C. Chen, C. L. Wu, J. Y. Wu, and M. F. Lin, Magnetic quantization of sp^3 bonding in monolayer gray tin, *Phys. Rev. B* 94, 045410, 2016.

[134] N. D. Drummond, V. Zolyomi, and V. I. Falko, Electrically tunable band gap in silicene, *Phys. Rev. B* 85, 075423, 2012.

[135] Y. Xu, B. Yan, H. J. Zhang, J. Wang, G. Xu, P. Tang, W. Duan, and S. C. Zhang, Large-Gap quantum spin Hall insulators in tin films, large-gap quantum spin Hall insulators in tin films, *Phys. Rev. Lett.* 111, 136804, 2013.

[136] T. P. Kaloni, M. Modarresi, M. Tahir, M. R. Roknabadi, G. Schreckenbach, and M. S. Freund, Electrically engineered band gap in two-dimensional Ge, Sn, and Pb: A first-principles and tight-binding approach, *J. Phys. Chem. C* 119, 11896, 2015.

[137] C. J. Tabert, and E. J. Nicol, Valley-spin polarization in the magneto-optical response of silicene and other similar 2D crystals, *Phys. Rev. Lett.* 110, 197402, 2013.

[138] C. J. Tabert, J. P. Carbotte, and E. J. Nicol, Magnetic properties of Dirac fermions in a buckled honeycomb lattice, *Phys. Rev. B* 91, 035423, 2015.

[139] L. Tao, E. Cinquanta, D. Chiappe, C. Grazianetti, M. Fanciulli, M. Dubey, A. Molle, and D. Akinwande, Silicene field-effect transistors operating at room temperature, *Nat. Tech.* 10, 227, 2014.

[140] F. Al-Dirini1, F. M. Hossain, M. A. Mohammed, A. Nirmalathas, and E. Skafidas, Highly effective conductance modulation in planar silicene field effect devices due to buckling, *Sci. Rep.* 5, 14815, 2015.

[141] Kh. Shakouri, P. Vasilopoulos, V. Vargiamidis, and F. M. Peeters, Integer and half-integer quantum Hall effect in silicene: Influence of an external electric field and impurities, *Phys. Rev. B* 90, 235423, 2014.

[142] C. J. Tabert, and E. J. Nicol, Magneto-optical conductivity of silicene and other buckled honeycomb lattices, *Phys. Rev. B* 88, 085434, 2013.

[143] A. Molle1, J. Goldberger, M. Houssa, Y. Xu, S. C. Zhang, and D. Akinwande, Buckled two-dimensional Xene sheets, *Nat. Mater.* 16, 163, 2017.

[144] A. Kara, H. Enriquez, A. P. Seitsonen, L. C. L. Y. Voon, S. Vizzini, B. Aufray, and H. Oughaddou, A review on silicone-New candidate for electronics, *Sur. Sci. Rep.* 67, 1, 2012.

[145] P. Vogt, P. De Padova, C. Quaresima, J. Avila, E. Frantzeskakis, M. C. Asensio, A. Resta, B. Ealet, and G. Le Lay, Silicene: Compelling experimental evidence for graphenelike two-dimensional silicon, *Phys. Rev. Lett.* 108, 155501, 2012.

[146] L. Meng, Y. Wang, L. Zhang, S. Du, R. Wu, L. Li, Y. Zhang, G. Li, H. Zhou, W. A. Hofer, and H. J. Gao, Buckled silicene formation on Ir(111), *Nano Lett.* 13, 685, 2013.

[147] A. Fleurence, R. Friedlein, T. Ozaki, H. Kawai, Y. Wang, and Y. Y. Takamura, Experimental evidence for epitaxial silicene on diboride thin films, *Phys. Rev. Lett.* 108, 245501, 2012.

[148] L. Li, S. Z. Lu, J. Pan, Z. Qin, Y. Wang, Y. Q. Wang, G. Cao, S. Du, and H. Gao, Buckled germanene formation on Pt(111), *Adv. Mater.* 26, 4820, 2014.

[149] M. E. Davila, L. Xian, S. Cahangirov, A. Rubio, and G. Le Lay, Germanene: A novel two-dimensional germanium allotrope akin to graphene and silicene, *New J. Phys.* 16, 095002, 2014.

[150] N. B. M. Schrtöer, M. D. Watson, L. B. Duffy, M. Hoesch, Y. Chen, T. Hesjedal, and T. K. Kim, Emergence of Dirac-like bands in the monolayer limit of epitaxial Ge films on Au(111), *2D Mater.* 4, 031005, 2017.

[151] M. Ezawa, A topological insulator and helical zero mode in silicene under an inhomogeneous electric field, *New J. Phys.* 14, 033003, 2012.

[152] J. Y. Wu, C. Y. Lin, G. Gumbs, and M. F. Lin, Temperature-induced Plasmon excitations for intrinsic silicene and effect of perpendicular electric field, *RSC Adv.* 5, 51912, 2015.

[153] C. J. Tabert, and E. J. Nicol, Dynamical polarization function, plasmons, and screening in silicene and other buckled honeycomb lattices, *Phys. Rev. B* 89, 195410, 2014.

[154] P. Shih, Y. Chiu, J. Wu, F. Shyu, and M. Lin, Coulomb excitations of monolayer germanene, *Sci. Rep.* 7, 40600, 2017.

[155] R. F. Egerton, Limits to the spatial, energy and momentum resolution of electron energy-loss spectroscopy, *Ultramicroscopy* 107, 575, 2007.

[156] H. Ibach, and D. L. Mills, *Electron Energy-Loss Spectroscopy and Surface Vibrations*. New York: Academic, 1982.

[157] H. Claus, A. Bussenschutt, and M. Henzler, Low-energy electron diffraction with energy resolution, *Rev. Sci. Instrum.* 63, 2195, 1992.

[158] D. S. Su, H. W. Zandbergen, P. C. Tiemeijer, G. Kothleitner, M. Havecker, C. Hebert, A. Knop-Gericke, B. H. Freitag, F. Hofer, and R. Schlogl, High resolution EELS using monochromator and high performance spectrometer: Comparison of V2O5 ELNES with NEXAFS and band structure calculations, *Micron* 34, 235, 2003.

[159] M. Terauchi, M. Tanaka, K. Tsuno, M. Ishida, Development of a high energy resolution electron energy-loss spectroscopy microscope, *J. Microsc.* 194, 203, 1999.

[160] S. Lazar, G. A. Botton, and H. W. Zandbergen, Enhancement of resolution in core-loss and low-loss spectroscopy in a monochromated microscope, *Ultramicroscopy* 106, 1091, 2006.

[161] F. Roth, A. Konig, J. Fink, B. Buchner, and M. Knupfer, Electron energy-loss spectroscopy: A versatile tool for the investigations of plasmonic excitations, *J. Electron. Spectrosc. Relat. Phenom.* 195, 85–95, 2014.

[162] R. F. Egerton, Electron energy-loss spectroscopy in the TEM, *Rep. Prog. Phys.* 72, 016502, 2009.

[163] F. J. G. de Abajo, Optical excitations in electron microscopy, *Rev. Mod. Phys.* 82, 209–275, 2010.

[164] M. K. Kinyanjui, G. Benner, G. Pavia, F. Boucher, H.-U. Habermeier, B. Keimer, and U. Kaiser, Spatially and momentum resolved energy electron loss spectra from an ultra-thin $PrNiO_3$ layer, *App. Phys. Lett.* 106, 203102, 2015.

[165] P. Cudazzo, K. O. Ruotsalainen, C, J. Sahle, A. Al-Zein, H. Berger, E. Navarro-Moratalla, S. Huotari, M. Gatti, and A. Rubio, High-energy collective electronic excitations in layered transition-metal dichalcogenides *Phys. Rev. B* 90, 125125, 2014.

[166] U. Buchner, Wave-vector dependence of the electron energy losses of boron nitride and graphite, *Phys. Status. Solidi B* 81, 227, 1977.

[167] N. Papageorgiou, M. Portail, and J. M. Layet, Dispersion of the interband π electronic excitation of highly oriented pyrolytic graphite measured by high resolution electron energy loss spectroscopy, *Sur. Sci.* 454, 462, 2000.

[168] A. G. Marinopoulos, L. Reining, V. Olevano, A. Rubio, T. Pichler, X. Liu, M. Knupfer, and J. Fink, Anisotropy and interplane interactions in the dielectric response of graphite, *Phys. Rev. Lett.* 89, 076402, 2002.

[169] J. J. Ritsko, and M. J. Rice, Plasmon spectra of ferric-chloride-intercalated graphite, *Phys. Rev. Lett.* 42, 666, 1979.

[170] L. A. Grunes, and J. J. Ritsko, Valence and core excitation spectra in K, Rb, and Cs alkali-metal stage-1 intercalated graphite, *Phys. Rev. B* 28, 3439, 1983.

[171] A. Hightower, C. C. Ahn, B. Fultz, and P. Rez, Electron energy-loss spectrometry on lithiated graphite, *Appl. Phys. Lett.* 77, 238, 2000.

[172] J. E. Fischer, J. M. Bloch, C. C. Shieh, M. E. Preil, and K. Jelley, Reflectivity spectra and dielectric function of stage-1 donor intercalation compounds of graphite, *Phys. Rev. B* 31, 4773, 1985.

[173] F. S. Hage, T. P. Hardcastle, A. J. Scott, R. Brydson, and Q. M. Ramasse, Momentum- and space-resolved high-resolution electron energy loss spectroscopy of individual single-wall carbon nanotubes, *Phys. Rev. B* 95, 195411, 2017.

[174] M. Knupfera, T. Pichlera, M. S. Goldena, J. Finka, A. Rinzlerb, and R. E. Smalley, Electron energy-loss spectroscopy studies of single wall carbon nanotubes, *Carbon* 37, 733, 1999.

[175] B. W. Reed, and M. Sarikaya, Electronic properties of carbon nanotubes by transmission electron energy-loss spectroscopy, *Phys. Rev. B* 64, 195404, 2001.

[176] T. Stockli, J. M. Bonard, A. Chatelain, Z. L. Wang, and P. Stadelmann, Collective oscillations in a single-wall carbon nanotube excited by fast electrons, *Phys. Rev. B* 64, 115424, 2001.

[177] T. Pichler, M. Knupfer, M. S. Golden, J. Fink, A. Rinzler, and R. E. Smalley, Localized and delocalized electronic states in single-wall carbon nanotubes, *Phys. Rev. Lett.* 80, 4729, 1998.

[178] R. Kuzuo, M. Terauchi, and M. Tanaka, Electron energy-loss spectra of carbon nanotubes, *Jpn. J. Appl. Phys.* 31, L1484, 1992.

[179] P. M. Ajayan, S. Ijima, and T. Ichihashi, Surface plasmon observed for carbon nanotubes, *Phys. Rev. B* 49, 2882, 1994.

[180] M. R. Went, M. Vos, and W. S. M. Werner, Extracting the Ag surface and volume loss functions from reflection electron energy loss spectra, *Sur. Sci.* 602, 2069, 2008.

[181] W. S. M. Werner, Dielectric function of Cu, Ag, and Au obtained from reflection electron energy loss spectra, optical measurements, and density functional theory, *Appl. Phys. Lett.* 89, 213106, 2006.

[182] W. S. M. Werner, M. R. Went, and M. Vos, Surface plasmon excitation at a Au surface by 150V40000 eV electrons, *Surf. Sci.* 601, L109, 2007.

[183] A. Politanoa. I. Radović, D. Borkab, Z. L. Mišković, H. K. Yu, D. Farías, and G. Chiarello, Dispersion and damping of the interband π plasmon in graphene grown on Cu(111) foils, *Carbon* 114, 70, 2017.

[184] S. C. Liou, C.-S. Shie, C. H. Chen, R. Breitwieser, W. W. Pai, G. Y. Guo, and M.-W. Chu, π-plasmon dispersion in free-standing graphene by momentum-resolved electron energy-loss spectroscopy, *Phys. Rev. B* 91, 045418, 2015.

[185] F. S. Hage, T. P. Hardcastle, M. N. Gjerding, D. M. Kepaptsoglou, C. R. Seabourne, K. T. Winther, R. Zan, J. A. Amani, H. C. Hofsaess, U. Bangert, K. S. Thygesen, and Q. M. Ramasse, Local plasmon engineering in doped graphene, *ACS Nano* 12, 1837, 2018.

[186] V. B. Jovanović, I. Radović, D. Borka, and Z. L. Mišković, High-energy plasmon spectroscopy of freestanding multilayer graphene, *Phys. Rev. B* 84, 155416, 2011.

[187] B. Diaconescu, K. Pohl, L. Vattuone, L. Savio, P. Hofmann, V. M. Silkin, J. M. Pitarke, E. V. Chulkov, P. M. Echenique, D. Farias, and M. Rocca, Low-energy acoustic plasmons at metal surfaces, *Nature* 448, 57, 2007.

[188] X. Luo, T. Qiu, W. Lu, and Z. Ni, Plasmons in graphene: Recent progress and applications, *Mater. Sci. Eng. R* 74, 351, 2013.

[189] Z. Fei, E. G. Iwinski, G. X. Ni, L. M. Zhang, W. Bao, A. S. Rodin, Y. Lee, M. Wagner, M. K. Liu, S. Dai, M. D. Goldflam, M. Thiemens, F. Keilmann, C. N. Lau, A. H. Castro-Neto, M. M. Fogler, and D. N. Basov, Tunneling plasmonics in bilayer graphene, *Nano Lett.* 15, 4973, 2015.

[190] Z. Fei, A. S. Rodin, G. O. Andreev, W. Bao, A. S. McLeod, M. Wagner, L. M. Zhang, Z. Zhao, M. Thiemens, G. Dominguez, M. M. Fogler, A. H. Castro Neto, C. N. Lau, F. Keilmann, and D. N. Basov, Gate-tuning of graphene plasmons revealed by infrared nano-imaging, *Nature* 487, 82, 2012.

[191] T. Eberlein, U. Bangert, R. R. Nair, R. Jones, M. Gass, A. L. Bleloch, K. S. Novoselov, A. Geim, and P. R. Briddon, Plasmon spectroscopy of free-standing graphene films, *Phys. Rev. B* 77, 233406, 2008.

[192] S. Y. Shin, N. D. Kim, J. G. Kim, K. S. Kim, D. Y. Noh, Kwang S. Kim, and J. W. Chung, Control of the π plasmon in a single layer graphene by charge doping, *Appl. Phys. Lett.* 99, 082110, 2016.

[193] J. Lu, K. P. Loh, H. Huang, W. Chen, and A. T. S. Wee, Plasmon dispersion on epitaxial graphene studied using high-resolution electron energy-loss spectroscopy, *Phys. Rev. B* 80, 113410, 2009.

[194] F. J. Nelson, Juan-Carlos Idrobo, J. D. Fite, Z. L. Miskovic, S. J. Pennycook, S. T. Pantelides, J. U. Lee, and A. C. Diebold, Electronic excitations in graphene in the 1–50 eV Range: The π and $\pi + \sigma$ peaks are not plasmons, *Nano Lett.* 14, 3827, 2014.

[195] S. Y. Shin, C. G. Hwang, S. J. Sung, N. D. Kim, H. S. Kim, and J. W. Chung, Observation of intrinsic intraband π-plasmon excitation of a single-layer graphene, *Phys. Rev. B* 83, 161403, 2011.

[196] A. Cupolillo, N. Ligato, and L. S. Caputi, Plasmon dispersion in quasi-freestanding graphene on Ni(111), *Appl. Phys. Lett.* 102, 111609, 2013.

[197] A. Politano, A. R. Marino, V. Formoso, D. Farías, R. Miranda, G. Chiarello, Quadratic dispersion and damping processes of π plasmon in monolayer graphene on Pt(111), *Plasmonics* 7, 369, 2012.

[198] A. V. Generalov, and Yu. S. Dedkov, EELS study of the epitaxial graphene/Ni(1 1 1) and graphene/Au/Ni(1 1 1) systems, *Carbon* 50, 183, 2012.

[199] T. Langer, H. Pfnür, Christoph Tegenkamp, Stiven Forti, Konstantin Emtsev and Ulrich Starke, Manipulation of plasmon electronVhole coupling in quasi-free-standing epitaxial graphene layers, *New J. Phys.* 14, 103045, 2002.

[200] A. Politano, V. Formoso, and G. Chiarello, Evidence of composite plasmonVphonon modes in the electronic response of epitaxial graphene, *J. Phys.: Condens. Matter* 25, 345303, 2013.

[201] A. Cupolillo, N. Ligato, and L. Caputi, Low energy two-dimensional plasmon in epitaxial graphene on Ni (111) surface, *Science* 608, 88, 2013.

[202] M. K. Kinyanjui, C. Kramberger, T. Pichler, J. C. Meyer, P. Wachsmuth, G. Benner, and U. Kaiser, Direct probe of linearly dispersing 2D interband plasmons in a free-standing graphene monolayer, *EPL* 97, 57005, 2012.

[203] H. Pfnür, T. Langer, J. Baringhaus, and C. Tegenkamp, Multiple plasmon excitations in adsorbed two-dimensional systems, *J. Phys.: Condens. Matter* 23, 112204, 2011.

[204] Y. Liu, R. F. Willis, K. V. Emtsev and Th. Seyller, Plasmon dispersion and damping in electrically isolated two-dimensional charge sheets, *Phys. Rev. B* 78, 201403, 2008.

[205] A. Politano, A. R. Marino, V. Formoso, D. Farías, R. Miranda, and G. Chiarello, Evidence for acoustic-like plasmons on epitaxial graphene on Pt(111), *Phys. Rev. B* 84, 033401, 2011.

[206] H. Kato, K. Suenaga, M. Mikawa. M. Okumura, N. Miwa, A. Yashiro, H. Fujimura, A. Mizuno, Y. Nishida, K. Kobayashi, and H. Shinohar, Syntheses and EELS characterization of water-soluble multi-hydroxyl Gd@C$_{82}$ fullerenols, *Chem. Phys. Lett.* 324, 255, 2000.

[207] T. Oku, T. Hirano, M. Kuno, T. Kusunose, K. Niihara, and K. Suganuma, Synthesis, atomic structures and properties of carbon and boron nitride fullerene materials, *Mater. Sci. Eng. B* 74, 206, 2000.

[208] T. Stockli, J.-M. Bonard, A. Chatelain, Z. L. Wang, and P. Stadelmann, Plasmon excitations in graphitic carbon spheres measured by EELS, *Phys. Rev. B* 61, 5751, 2000.

[209] S. Tomita, M. Fujii, S. Hayashi, and K. Yamamoto, Electron energy-loss spectroscopy of carbon onions, *Chem. Phys. Lett.* 305, 225, 1999.

[210] S. Tomita, Structure and electronic properties of carbon onions, *J. Chem. Phys.* 114, 7477, 2001.

[211] L. Henrard, F. Malengreau, P. Rudolf, K. Hevesi, R. Caudano, P. Lambin, and T. Cabioc'h, Electron-energy-loss spectroscopy of plasmon excitations in concentric-shell fullerenes, *Phys. Rev. B* 59, 5832, 1999.

[212] P. Redlich, F. Banhart, Y. Lyutovich, and P. M. Ajayan, EELS study of the irradiation-induced compression of carbon onions and their transformation to diamond, *Carbon* 36, 561, 1998.

[213] K. Suenaga, and M. Koshino, Atom-by-atom spectroscopy at graphene edge, *Nature* 468, 1088, 2010.

[214] X. Gao, C. Burns, D. Casa, M. Upton, T. Gog, J. Kim, and C. Li, Development of a graphite polarization analyzer for resonant inelastic X-ray scattering, *Rev. Sci. Instrum.* 82, 113108, 2011.

[215] W. Schulke, U. Bonse, H. Nagasawa, A. Kaprolat, and A. Berthold, Interband transitions and core excitation in highly oriented pyrolytic graphite studied by inelastic synchrotron X-ray scattering: Band-structure information, *Phys. Rev. B* 38, 2112, 1988.

[216] S. Huotari, F. Albergamo, Gy. Vanko, R. Verbeni, and G. Monaco, Resonant inelastic hard X-ray scattering with diced analyzer crystals and positionsensitive detectors, *Rev. Sci. Instrum.* 77, 053102, 2006.

[217] A. Kotani, and S. Shin, Resonant inelastic X-ray scattering spectra for electrons in solids, *Rev. Mod. Phys.* 73, 203, 2001.

[218] T. P. Devereaux, Inelastic light scattering from correlated electrons, *Rev. Mod. Phys.* 79, 175, 2007.

[219] L. J. P. Ament, M. van Veenendaal, T. P. Devereaux, J. P. Hill, and J. van den Brink, Resonant inelastic X-ray scattering studies of elementary excitations, *Rev. Mod. Phys.* 83, 705, 2001.

[220] L. Zhang, N. Schwertfager, T. Cheiwchanchamnangij, X. Lin, P.-A. Glans-Suzuki, L. F. J. Piper, S. Limpijumnong, Y. Luo, J. F. Zhu, W. R. L. Lambrecht, and J. -H. Guo, Electronic band structure of graphene from resonant soft X-ray spectroscopy: The role of core-hole effects, *Phys. Rev. B* 86, 245430, 2012.

[221] M. Mohr, J. Maultzsch, E. Dobardžić, S. Reich, I. Milošević, M. Damn-janovićq3, A. Bosak, M. Krisch, and C. Thomsen, Phonon dispersion of graphite by inelastic X-ray scattering, Phonon dispersion of graphite by inelastic X-ray scattering, *Phys. Rev. B* 76, 035439, 2007.

[222] K. Kimura, K. Matsuda, N. Hiraoka, T. Fukumaru, Y. Kajihara, M. Inui, and M. Yao, Inelastic X-ray scattering study of plasmon dispersions in solid and liquid Rb, *Phys. Rev. B* 89, 014206, 2014.

[223] G. Tirao, G. Stutz, V. M. Silkin, E. V. Chulkov, and C. Cusatis, Plasmon excitation in beryllium: Inelastic X-ray scattering experiments and first-principles calculations, *J. Phys.: Condens. Matter* 19, 046207, 2007.

[224] K. Kimura, K. Matsuda, N. Hiraoka, Y. Kajihara, T. Miyatake, Y. Ishiguro, T. Hagiya, M. Inui, and M. Yao, Inelastic X-ray scattering study on plasmon dispersion in liquid Cs, *J. Phys. Soc. Jpn.* 84, 084701, 2015.

[225] S. Galambosi, J. A. Soininen, A. Mattila, S. Huotari, S. Manninen, Gy. Vanko, N. D. Zhigadlo, J. Karpinski, and K. Hamalainen, Inelastic X-ray scattering study of collective electron excitations in MgB_2, *Phys. Rev. B* 71, 060504, 2005.

[226] R. Hambach, C. Giorgetti, N. Hiraoka, Y. Q. Cai, F. Sottile, A. G. Marinopoulos, F. Bechstedt and Lucia Reining, Anomalous angular dependence of the dynamic structure factor near Bragg reflections: Graphite, *Phys. Rev. Lett.* 101, 266406, 2008.

[227] C. Kittel, *Introduction to Solid State Physics*, 8th Edition. Hoboken, NJ: John Wiley & Sons, 2012.

[228] H. A. Brink, M. M. G. Barfels, R. P. Burgner, and B. N. Edwards, A sub-50 meV spectrometer and energy filter for use in combination with 200 kV monochromated (S)TEMs, *Ultramicroscopy* 96, 367, 2003.

[229] V. N. Strocov, T. Schmitt, U. Flechsig, T. Schmidt, A. Imhof, Q. Chen, J. Raabe, R. Betemps, D. Zimoch, J. Krempasky, X. Wang, M. Grioni, A. Piazzalunga, and L. Patthey, High-resolution soft X-ray beamline ADRESS at the Swiss Light Source for resonant inelastic X-ray scattering and angle-resolved photoelectron spectroscopies, *J. Synchrotron Radiat.* 17, 631, 2010.

[230] R. Qiao, Q. Li, Z. Zhuo, S. Sallis1, O. Fuchs, M. Blum, L. Weinhardt, C. Heske, J. Pepper, M. Jones, A. Brown, A. Spucces, K. Chow, B. Smith, P.-A. Glans, Y. Chen, S. Yan, F. Pan, L. F. J. Piper, J. Denlinger, J. Guo, Z. Hussain, Y.-D. Chuang, and W. Yang, High-efficiency in situ resonant inelastic X-ray scattering (iRIXS) endstation at the Advanced Light Source, *Rev. Sci. Instrum.* 88, 033106, 2017.

[231] E. A. Taft, and H. R. Philipp, Analytic calculation of the optical properties of graphite, *Phys. Rev.* 138, A197, 1965.

[232] L. G. Johnson, and G. Dresselhaus, Optical properies of graphite, *Phys. Rev. B* 7, 2275, 1973.

[233] D. L. Greenaway, G. Harbeke, F. Bassani, and E. Tosatti, Anisotropy of the optical constants and the band structure of graphite, *Phys. Rev.* 178, 1340, 1969.

[234] M. F. Lin, C. S. Huang, and D. S. Chuu, Plasmons in graphite and stage-1 graphite intercalation compounds, *Phys. Rev. B* 55, 13961, 1997.

[235] S. Y. Lin, N. T. T. Tran, S. L. Chang, W. P. Su, and M. F. Lin, *Structure- and Adatom-Enriched Essential Properties of Graphene Nanoribbons.* ISBN 978-0-367-00229-9, 2018.

[236] A. Scholz, T. Stauber, and J. Schliemann, Dielectric function, screening, and plasmons of graphene in the presence of spin-orbit interactions, *Phys. Rev. B* 86, 195424, 2012.

[237] G. D. Mahan, *Many-Particle Physics.* New York: Plenum, 1990.

[238] P. H. Shih, C. W. Chiu, J. Y. Wu, T. N. Do, and M. F. Lin, Coulomb scattering rates of excited states in monolayer electron-doped germanene, *Phys. Rev. B* 97, 195302, 2018.

[239] C. W. Chiu, Y. H. Ho, S. C. Chen, C. H. Lee, C. S. Lue, and M. F. Lin, Electronic decay rates in semiconducting carbon nanotubes, *Physica E* 34, 658–661, 2006.

[240] J. H. Ho, C. P. Chang, R. B. Chen, and M. F. Lin, Electron decay rates in a zero-gap graphite layer, *Phys. Lett. A* 357, 401, 2006.

[241] P. Ruffieux, J. Cai, N. C. Plumb, L. Patthey, D. Prezzi, A. Ferretti, and R. Fasel, Electronic structure of atomically precise graphene nanoribbons, *ACS Nano* 6, 6930, 2012.

[242] K. Sugawara, T. Sato, S. Souma, T. Takahashi, and H. Suematsu, Fermi surface and edge-localized states in graphite studied by high-resolution angle-resolved photoemission spectroscopy, *Phys. Rev. B* 73, 045124, 2006.

[243] D. A. Siegel, W. Regan, A. V. Fedorov, A. Zettl, and A. Lanzara, Charge-carrier screening in single-layer graphene, *Phys. Rev. Lett.* 110, 146802, 2013.

[244] K. S. Kim, A. L. Walter, L. Moreschini, T. Seyller, K. Horn, E. Roten- berg, A. Bostwick, Coexisting massive and massless Dirac fermions in symmetry-broken bilayer graphene. *Nat. Mater.* 12, 887, 2013.

[245] M. Papagno, S. Rusponi, P. M. Sheverdyaeva, S. Vlaic, M. Etzkorn, D. Pacile, P. Moras, C. Carbone, and H. Brune, Large band gap opening between graphene Dirac cones induced by Na adsorption onto an Ir superlattice, *ACS Nano* 6, 199, 2012.

[246] T. Hertel, and G. Moos, Electron-phonon interaction in single-wall carbon nanotubes: A time-domain study, *Phys. Rev. Lett.* 84, 5002, 2000.

[247] M. Ichida, Y. Hamanaka, H. Kataura, Y. Achiba, and A. Nakamura, Ultrafast relaxation dynamics of photoexcited carriers in metallic and semiconducting single-walled carbon nanotubes, *J. Phys. Soc. Jpn.* 73, 3479, 2004.

[248] S. Xu, J. Cao, C. C. Miller, D. A. Mantell, R. J. D. Miller, and Y. Gao, Energy dependence of electron lifetime in graphite observed with femtosecond photoemission spectroscopy, *Phys. Rev. Lett.* 76, 483, 1996.

[249] G. Moos, C. Gahl, R. Fasel, M. Wolf, and T. Hertel, Anisotropy of quasiparticle lifetimes and the role of disorder in graphite from ultrafast time-resolved photoemission spectroscopy, *Phys. Rev. Lett.* 87, 267402, 2001.

[250] J-S. Lauret, C. Voisin, G. Cassabois, C. Delalande, Ph. Roussignol, O. Jost, and L. Capes, Ultrafast carrier dynamics in single-wall carbon nanotubes, *Phys. Rev. Lett.* 90, 057404, 2003.

[251] O. J. Korovyanko, C.-X. Sheng, Z. V. Vardeny, A. B. Dalton, and R. H. Baughman, Ultrafast spectroscopy of excitons in single-walled carbon nanotubes, *Phys. Rev. Lett.* 92, 017403, 2004.

[252] G. N. Ostojic, S. Zaric, J. Kono, M. S. Strano, V. C. Moore, R. H. Hauge, and R. E. Smalley, Interband recombination dynamics in resonantly excited single-walled carbon nanotubes, *Phys. Rev. Lett.* 92, 117402, 2004.

[253] L. Huang, H. N. Pedrosa, and T. D. Krauss, Ultrafast ground-state recovery of single-walled carbon nanotubes, *Phys. Rev. Lett.* 93, 017403, 2004.

[254] Y. Bai, J.-H. Olivier, G. Bullard, C. Liu, and M. J. Therien, Dynamics of charged excitons in electronically and morphologically homogeneous single-walled carbon nanotubes, *Proc. Natl. Acad. Sci. USA* 115, 674, 2018.

[255] K. Maekawa, K. Yanagi, Y. Minami, M. Kitajima, I. Katayama, and J. Takeda, Bias-induced modulation of ultrafast carrier dynamics in metallic single-walled carbon nanotubes, *Phys. Rev. B* 97, 075435, 2018.

[256] X.-P. Tang, A. Kleinhammes, H. Shimoda, L. Fleming, K. Y. Bennoune, S. Sinha, C. Bower, O. Zhou, Y. Wu, Electronic structures of single-walled carbon nanotubes determined by NMR, *Science* 288, 492–494, 2000.

[257] K. Seibert, G. C. Cho, W. Kutt, H. Kurz, D. H. Reitze, J. I. Dadap, H. Ahn, M. C. Downer, and A. M. Malvezzi, Femtosecond carrier dynamics in graphite, *Phys. Rev. B* 42, 2842, 1990.

[258] A. Hagen, M. Steiner, M. B. Raschke, C. Lienau, T. Hertel, H. Qian, A. J. Meixner, and A. Hartschuh, Exponential decay lifetimes of excitons in individual single-walled carbon nanotubes, *Phys. Rev. Lett.* 95, 197401, 2005.

[259] F. Wang, G. Dukovic, L. E. Brus, and T. F. Heinz, Time-resolved fluorescence of carbon nanotubes and its implication for radiative lifetimes, *Phys. Rev. Lett.* 92, 177401, 2004.

[260] J. C. Charlier, J. P. Michenaud, and X. Gonze, First-principles study of the electronic properties of simple hexagonal graphite, *Phys. Rev. B* 46, 4531, 1992.

[261] Kenneth W. -K. Shung, Dielectric function and plasmon structure of stage-1 intercalated graphite, *Phys. Rev. B* 34, 979, 1986.

[262] J. D. Jackson, *Classical Electrodynamics*. New York: John Wiley & Sons, 1975.

[263] G. Gumbs, A. Balassis, and V. M. Silkin, Combined effect of doping and temperature on the anisotropy of low-energy plasmons in monolayer graphene, *Phys. Rev. B* 96, 045423, 2017.

[264] M. F. Lin, and F. L. Shyu, Temperature-induced plasmons in a graphite sheet, *J. Phys. Soc. Jpn.* 69, 607, 2000.

[265] D. K. Patel, S. S. Z. Ashraf, and A. C. Sharma, Finite temperature dynamicalpolarizat ion and plasmons in gapped graphene, *Phys. Status Solidi B* 252, 1817, 2015.

[266] A. Iurov, G. Gumbs, D. Huang, and G. Balakrishnan, Thermal plasmons controlled by different thermal-convolution paths in tunable extrinsic Dirac structures, *Phys. Rev. B* 96, 245403, 2017.

[267] A. F. Page, J. M. Hamm, and O. Hess, Polarization and plasmons in hot photoexcited graphene, *Phys. Rev. B* 97, 045428, 2018.

[268] M. R. Ramezanali, M. M. Vazifeh, R. Asgari, M. Polini, and A. H. MacDonald, Finite-temperature screening and the specific heat of doped graphene sheets, *J. Phys. A: Math. Theor.* 42, 214015, 2009.

[269] G. Grosso, and G. Parravicini, *Solid State Physics*, 2nd Edition, Barakhamba: Reed Elsevier India Private Limited, 2014.

[270] R. A. Jishi, *Feynman Diagram Techniques in Condensed Matter Physics*. Cambridge: Cambridge University Press, 2013.

[271] M. Q. Zhao, Q. Zhang, J. Q. Huang, G. L. Tian, J. Q. Nie, H. J. Peng, and F. Wei, Unstacked double-layer templated graphene for high-rate lithiumVsulphur batteries, *Nat. Comm.* 5, 3410, 2014.

[272] D. Rodrigo, A. Tittl, O. Limaj, F. J. G. de Abajo, V. Pruneri, and H. Altug, Double-layer graphene for enhanced tunable infrared plasmonics, *Light Sci. Appl.* 6, e16277, 2017.

[273] R. E. V. Profumo, R. Asgari, M. Polini, and A. H. MacDonald, Double-layer graphene and topological insulator thin-film plasmons, *Phys. Rev. B* 85, 085443, 2012.

[274] T. Stauber, and G. Gomez-Santos, Plasmons and near-field amplification in double-layer graphene, *Phys. Rev. B* 85, 075410, 2012.

[275] J. Borysiuk, J. Soltys, and J. Piechota, Stacking sequence dependence of graphene layers on SiC (000-1): Experimental and theoretical investigation, *J. Appl. Phys.* 109, 093523, 2011.

[276] M. Sanderson, and Y. S. Ang, Klein tunneling and cone transport in AA-stacked bilayer graphene, *Phys. Rev. B* 88, 245404, 2013.

[277] A. Kasry, M. A. Kuroda, G. J. Martyna, G. S. Tulevski, and A. A. Bol, Chemical doping of large-area stacked graphene films for use as transparent, conducting electrodes, *ACS Nano* 4, 3839, 2010.

[278] S. J. Tsai, Y. H. Chiu, Y. H. Ho and M. F. Lin, Gate-voltage-dependent Landau levels in AA-stacked bilayer graphene, *Chem. Phys. Lett.* 550, 104, 2012.

[279] Y. H. Ho, J. Y. Wu, R. B. Chen, Y. H. Chiu, and M. F. Lin, Optical transitions between Landau levels: AA-stacked bilayer graphene, *Appl. Phys. Lett.* 97, 101905, 2010.

[280] C. P. Chang, Exact solution of the spectrum and magneto-optics of multilayer hexagonal graphene, *J. Appl. Phys.* 110, 013725, 2011.

[281] K. S. Kim, A. L. Walter, L. Moreschini, T. Seyller, K. Horn, E. Rotenberg, et al. Coexisting massive and massless Dirac fermions in symmetry-broken bilayer graphene, *Nat. Mater.* 12, 887, 2013.

[282] C. W. Chiu, S. C. Chen, Y. C. Huang, F. L. Shyu, and M. F. Lin, Critical optical properties of AA-stacked multilayer graphenes, *Appl. Phys. Lett.* 103 041907, 2013.

[283] I. Lobato, and B. Partoens, Multiple Dirac particles in AA-stacked graphite and multilayers of graphene, *Phys. Rev. B* 83, 165429, 2011.

[284] K. F. Mak, J. Shan, and T. F. Heinz, Electronic structure of few-layer graphene: Experimental demonstration of strong dependence on stacking sequence, *Phys. Rev. Lett.* 104, 176404, 2010.

[285] T. Ohta, A. Bostwick, T. Seyller, K. Horn, and E. Rotenberg, Controlling the electronic structure of bilayer graphene, *Science* 313, 951, 2006.

[286] Y. C. Chuang, J. Y. Wu, M. F. Lin, Dynamical conductivity and zero-mode anomaly in honeycomb lattices, *J. Phys. Soc. Jpn.* 71, 1318, 2002.

[287] O. V. Gamayun, Dynamical screening in bilayer graphene, *Phys. Rev. B* 84, 085112, 2011.

[288] R. Sensarma, E. H. Hwang, and S. Das Sarma, Dynamic screening and low-energy collective modes in bilayer graphene, *Phys. Rev. B* 82, 195428, 2010.

[289] M. B. Roman, and S. M. Bose, Low energy intraband plasmons and electron energy loss spectra of single and multilayered graphene, *Plasmonics* 12, 145, 2017.

[290] B. Wunsch, T. Stauber, F. Sols, and F. Guinea, Dynamical polarization of graphene at finite doping, *New J. Phys.* 8, 318, 2006.

[291] E. H. Hwang, and S. Das Sarma, Dielectric function, screening, and plasmons in two-dimensional graphene, *Phys. Rev. B* 75, 205418, 2007.

[292] O. Roslyak, G. Gumbs, and D. Huang, Plasma excitations of dressed Dirac electrons in graphene layers, *J. App. Phys.* 109, 113721, 2011.

[293] Y. C. Chuang, J. Y. Wu, and M. F. Lin, Electric field dependence of excitation spectra in AB-stacked bilayer graphene, *Sci. Rep.* 3, 1368, 2013.

[294] V. N. Kotov, B. Uchoa, V. M. Pereira, F. Guinea, and A. H. Castro Neto, Electron-electron interactions in graphene: Current status and perspectives, *Rev. Mod. Phys.* 84, 1067, 2012.

[295] A. Hill, S. A. Mikhailov, and K. Ziegler, Dielectric function and plasmons in graphene, *EPL* 87, 27005, 2009.

[296] C. W. Chiu, S. H. Lee, S. C. Chen, and M. F. Lin, Electronic excitations in doped monolayer graphenes, *J. App. Phys.* 106, 113711, 2008.

[297] Z. Q. Li, E. A. Henriksen, Z. Jiang, Z. Hao, M. C. Martin, P. Kim, H. L. Stormer, and D. N. Basov, Dirac charge dynamics in graphene by infrared spectroscopy, *Nat. Phys.* 4, 532, 2008.

[298] L. Ju, B. Geng, J. Horng, C. Girit, M. Martin, Z. Hao, H. A. Bechtel, X. Liang, and A. Zettl, Y. R. Shen, and F. Wang, Graphene plasmonics for tunable terahertz metamaterials, *Nat. Nanotechnol.* 6, 630, 2011.

[299] S. Berciaud, M. Potemski, and C. Faugeras, Probing electronic excitations in mono- to pentalayer graphene by micro magneto-Raman spectroscopy, *Nano Lett.* 14, 4548, 2014.

[300] I. G. Gurtubay, J. M. Pitarke, W. Ku, A. G. Eguiluz, B. C. Larson, J. Tischler, P. Zschack, and K. D. Finkelstein. Electron-hole and plasmon excitations in 3d transition metals: Ab initio calculations and inelastic X-ray scattering measurements, *Phys. Rev. B* 72, 125117, 2005.

[301] J.-C. Charlier, X. Gonze, and J.-P. Michenaud, First-principles study of the electronic properties of graphite, *Phys. Rev. B* 43, 4579, 1991.

[302] D. S. L. Abergel, and V. I. Fal'ko, Optical and magneto-optical far-infrared properties of bilayer graphene, *Phys. Rev. B* 75, 155430, 2007.

[303] J. C. Slonczewski, and P. R. Weiss, Band structure of graphite, *Phys. Rev.* 109, 272, 1958.

[304] J. W. McClure, Theory of diamagnetism of graphite, *Phys. Rev.* 119, 606, 1960.

[305] G. Li, and E. Y. Andrei, Observation of Landau levels of Dirac fermions in graphite, *Nat. Phys.* 3, 623, 2007.

[306] T. Matsui, H. Kambara, Y. Niimi, K. Tagami, M. Tsukada, and H. Fukuyama, STS observations of Landau levels at graphite surfaces, *Phys. Rev. Lett.* 94, 226403, 2005.

[307] Y. H. Ho, J. Wang, Y. H. Chiu, M. F. Lin, and W. P. Su, Characterization of Landau subbands in graphite: A tight-binding study, *Phys. Rev. B* 83, 121201, 2011.

[308] K. Nakao, Landau level structure and magnetic breakthrough in graphite, *J. Phys. Soc. Jpn.* 40, 761, 1976.

[309] M. Inoue, Landau levels and cyclotron resonance in graphite, *J. Phys. Soc. Jpn.* 17, 808, 1962.

[310] M. Orlita, C. Faugeras, J. M. Schneider, G. Martinez, D. K. Maude, and M. Potemski, Graphite from the viewpoint of Landau level spectroscopy: An effective graphene bilayer and monolayer, *Phys. Rev. Lett.* 102, 166401, 2009.

[311] N. A. Goncharuk, L. Nádvornik, C. Faugeras, M. Orlita, and L. Smrčka, Infrared magnetospectroscopy of graphite in tilted fields, *Phys. Rev. B* 86, 155409, 2012.

[312] M. Orlita, C. Faugeras, A.-L. Barra, G. Martinez, M. Potemski, D. M. Basko, M. S. Zholudev, F. Teppe, W. Knap, V. I. Gavrilenko, N. N. Mikhailov, S. A. Dvoretskii, P. Neugebauer, C. Berger, and W. A. de Heer, Infrared magneto-spectroscopy of two-dimensional and three-dimensional massless fermions: A comparison, *J. App. Phys.* 117, 112803, 2015.

[313] M. Orlita, C. Faugeras, G. Martinez, D. K. Maude, M. L. Sadowski, and M. Potemski, Dirac fermions at the H Point of graphite: Magnetotransmission studies, *Phys. Rev. Lett.* 100, 136403, 2008.

[314] W. W. Toyt, and M. S. Dresselhaus, Minority carriers in graphite and the H-point magnetoreflection spectra, *Phys. Rev. B* 15, 4077, 1977.

[315] M. L. Sadowski, G. Martinez, M. Potemski, C. Berger, and W. A. de Heer, Landau level spectroscopy of ultrathin graphite layers, *Phys. Rev. Lett.* 97 266405, 2006.

[316] L. J. Yin, S. Y. Li, J. B. Qiao, J. C. Nie, and L. He, Landau quantization in graphene monolayer, Bernal bilayer, and Bernal trilayer on graphite surface, *Phys. Rev. B* 91, 115405, 2015.

[317] Y. H. Ho, Y. H. Chiu, D. H. Lin, C. P. Chang, and M. F. Lin, Magneto-optical selection rules in bilayer Bernal graphene, *ACS Nano* 4, 1465, 2010.

[318] M. Koshino, and E. McCann, Landau level spectra and the quantum Hall effect of multilayer graphene, *Phys. Rev. B* 83, 165443, 2011.

[319] E. V. Castro, K. S. Novoselov, S. V. Morozov, N. M. R. Peres, J. M. B. Lopes dos Santos, Johan Nilsson, F. Guinea, A. K. Geim, and A. H. Castro Neto, Biased bilayer graphene: Semiconductor with a gap tunable by the electric field effect. *Phys. Rev. Lett.* 99, 216802, 2007.

[320] C. L. Lu, C. P. Chang, Y. C. Huang, R. B. Chen, and M. L. Lin, Influence of an electric field on the optical properties of few-layer graphene with AB stacking, *Phys. Rev. B* 73, 144427, 2006.

[321] C. L. Lu, C. P. Chang, Y. C. Huang, J. H. Ho, C. C. Hwang, and M. F. Lin, Electronic properties of AA- and ABC-stacked few-layer graphites. *J. Phys. Soc. Jpn.* 7, 024701, 2007.

[322] E. McCann, Asymmetry gap in the electronic band structure of bilayer graphene, *Phys. Rev. B* 74, 161403, 2006.

[323] B. Datta, H. Agarwal, A. Samanta, A. Ratnakar, K. Watanabe, T. Taniguchi, R. Sensarma, and M. M. Deshmukh, Landau level diagram and the continuous rotational symmetry breaking in trilayer graphene, *Phys. Rev. Lett.* 121, 056801, 2018.

[324] L. M. Zhang, Z. Q. Li, D. N. Basov, and M. M. Fogler, Determination of the electronic structure of bilayer graphene from infrared spectroscopy, *Phys. Rev. B* 78, 235408, 2008.

[325] Z. Q. Li, E. A. Henriksen, Z. Jiang, Z. Hao, M. C. Martin, P. Kim, H. L. Stormer, and D. N. Basov, Band structure asymmetry of bilayer graphene revealed by infrared spectroscopy, *Phys. Rev. Lett.* 102, 037403, 2009.

[326] A. B. Kuzmenko, I. Crassee, D. van der Marel, P. Blake, and K. S. Novoselov, Determination of the gate-tunable band gap and tight-binding parameters in bilayer graphene using infrared spectroscopy, *Phys. Rev. B* 80, 165406, 2009.

[327] M. O. Goerbig, Electronic properties of graphene in a strong magnetic field, *Rev. Mod. Phys.* 83, 1193, 2011.

[328] S. Y. Shin, C. G. Hwang, S. J. Sung, N. D. Kim, H. S. Kim, and J. W. Chung, Observation of intrinsic intraband π-plasmon excitation of a single-layer graphene, *Phys. Rev. B* 83, 161403, 2011.

[329] J. Lu, K. P. Loh, H. Huang, W. Chen, and A. T. S. Wee, Plasmon dispersion on epitaxial graphene studied using high-resolution electron energy-loss spectroscopy, *Phys. Rev. B* 80, 113410, 2009.

[330] S. Hattendorf, A. Georgi, M. Liebmann, and M. Morgenstern, Networks of ABA and ABC stacked graphene on mica observed by scanning tunneling microscopy, *Surf. Sci.* 610, 53, 2013.

[331] J. H. Warner, M. Mukai, and A. I. Kirkland, Atomic structure of ABC rhombohedral stacked trilayer graphene, *ACS Nano* 6, 5680, 2012.

[332] W. Norimatsu, and M. Kusunoki, Selective formation of ABC-stacked graphene layers on SiC(0001), *Phys. Rev. B* 81, 161410, 2010.

[333] H. Zhou, W. J. Yu, L. Liu, R. Cheng, Y. Chen, X. Huang, Y. Liu, Y. Wang, Y. Huang, and X. Duan, Chemical vapour deposition growth of large single crystals of monolayer and bilayer graphene, *Nat. Commun.* 4, 2096, 2013.

[334] P. Lauffer, K. V. Emtsev, R. Graupner, Th. Seyller, L. Ley, S. A. Reshanov, and H. B. Weber, Atomic and electronic structure of few-layer graphene on SiC(0001) studied with scanning tunneling microscopy and spectroscopy, *Phys. Rev. B* 77, 155426, 2008.

[335] W. T. Pong, J. Bendall, and C. Durkan, Observation and investigation of graphite superlattice boundaries by scanning tunneling microscopy, *Sur. Sci.* 601, 498, 2007.

[336] C. W. Chiu, Y. C. Huang, F. L. Shyu, and M. F. Lin, Optical absorption spectra in ABC-stacked graphene superlattice, *Syn. Met.* 162, 800, 2012.

[337] C. H. Lui, Z. Li, K. F. Mak, E. Cappelluti, and T. F. Heinz, Observation of an electrically tunable band gap in trilayer graphene, *Nat. Phys.* 7, 944, 2011.

[338] J.-C. Charlier, J.-P. Michenaud, and Ph. Lambin, Tight-binding density of electronic states of pregraphitic carbon, *Phys. Rev. B* 46, 4540, 1992.

[339] C. Y. Lin, J. Y. Wu, Y. H. Chiu, C. P. Chang, and M. F. Lin, Stacking-dependent magnetoelectronic properties in multilayer graphene, *Phys. Rev. B* 90, 205434, 2014.

[340] Y. P. Lin, C. Y. Lin, Y. H. Ho, T. N. Do, and M. F. Lin, Magneto-optical properties of ABC-stacked trilayer graphene, *Phys. Chem. Chem. Phys.* 17, 15921, 2015.

[341] Y. P. Lin, C. Y. Lin, C. P. Chang, and M. F. Lin, Electric-field-induced rich magneto-absorption spectra of ABC-stacked trilayer graphene, *RSC Adv.* 5, 80410, 2015.

[342] A. Kawabata, and R. Kubo, Electronic properties of fine metallic particles. II. Plasma resonance absorption, *J. Phys. Soc. Jpn.* 21, 1765, 1966.

[343] R. Xu, L. J. Yin, J. B. Qiao, K. K. Bai, J. C. Nie, and L. He, Direct probing of the stacking order and electronic spectrum of rhombohedral trilayer graphene with scanning tunneling microscopy, *Phys. Rev. B* 91, 035410, 2015.

[344] D. Pierucci, H. Sediri, M. Hajlaoui, J. C. Girard, T. Brumme, M. Calandra, E. Velez-Fort, G. Patriarche, M. G. Silly, G. Ferro, V. Souliere, M. Marangolo, F. Sirotti, F. Mauri, and A. Ouerghi, Evidence for flat bands near the Fermi level in epitaxial rhombohedral multilayer graphene, *ACS Nano* 9, 5432, 2015.

[345] P. R. Wallace, The band theory of graphite, *Phys. Rev.* 71, 622, 1947.

[346] P. Nozières, and D. Pines, Electron interaction in solids. The nature of the elementary excitations, *Phys. Rev.* 109, 1062, 1958.

[347] W. Yan, M. Liu, R. F. Dou, L. Meng, L. Feng, Z. D. Chu, Y. Zhang, Z. Liu, J. C. Nie, and L. He, Angle-dependent van Hove singularities in a slightly twisted graphene bilayer, *Phys. Rev. Lett.* 109, 126801, 2012.

[348] F. L. Shyu, and M. F. Lin, π-electronic excitations in multiwalled carbon nanotubes, *J. Phys. Soc. Jpn.* 68, 3806–3809, 1999.

[349] T. Taychatanapat, and P. Jarillo-Herrero, Electronic transport in dual-gated bilayer graphene at large displacement fields, *Phys. Rev. Lett.* 105, 166601, 2010.

[350] H. C. Chung, C. P. Chang, C. Y. Lin, and M. F. Lin, Electronic and optical properties of graphene nanoribbons in external fields, *Phys. Chem. Chem. Phys.* 18, 7573, 2016.

[351] C. P. Chang, J. Wang, C. L. Lu, Y. C. Huang, M. F. Lin, and R. B. Chen, Optical properties of simple hexagonal and rhombohedral few-layer graphenes in an electric field, *J. Appl. Phys.* 103, 103109, 2008.

[352] M. H. Lee, H. C. Chung, J. M. Lu, C. P. Chang, and M. F. Lin, Electronic and optical properties in graphene, *Philos. Mag.* 95, 2717, 2015.

[353] C. W. Chiu, and R. B. Chen, Influence of electric fields on absorption spectra of AAB-stacked trilayer graphene, *Appl. Phys. Express* 9, 065103, 2016.

[354] L. Yang, First-principles study of the optical absorption spectra of electrically gated bilayer graphene, *Phys. Rev. B* 81, 155445, 2010.

[355] R. Stein, D. Hughes, and J. A. Yan, Electric-field effects on the optical vibrations in AB-stacked bilayer graphene, *Phys. Rev. B* 87, 100301, 2013.

[356] Y. Lee, J. Velasco, J. D. Tran, F. Zhang, W. Bao, L. Jing, K. Myhro, D. Smirnov, and C. N. Lau, Broken symmetry quantum Hall states in dual-gated ABA trilayer graphene, *Nano Lett.* 13, 1627, 2013.

[357] G. M. Rutter, S. Jung, N. N. Klimov, D. B. Newell, N. B. Zhitenev, and J. A. Stroscio, Microscopic polarization in bilayer graphene, *Nat. Phys.* 7, 649, 2011.

[358] R. T. Weitz, M. T. Allen, B. E. Feldman, J. Martin, and A. Yacoby, Broken-symmetry states in doubly gated suspended bilayer graphene, *Science* 330, 812, 2010.

[359] J. V. Jr, L. Jing, W. Bao, Y. Lee, P. Kratz, V. Aji, M. Bockrath, C. N. Lau, C. Varma, R. Stillwell, D. Smirnov, Fan Zhang, J. Jung, and A. H. MacDonald, Transport spectroscopy of symmetry-broken insulating states in bilayer graphene, *Nat. Nanotechnol.* 7, 156, 2012.

[360] P. Maher, C. R. Dean, A. F. Young, T. Taniguchi, K. Watanabe, K. L. Shepard, J. Hone, and P. Kim, Evidence for a spin phase transition at charge neutrality in bilayer graphene, *Nat. Phys.* 9, 154, 2013.

[361] K. C. Tang, R. Qin, J. Zhou, H. Qu, J. X. Zheng, R. X. Fei, H. Li, Q. Y. Zheng, Z. X. Gao, and J. Lu, Electric-field-induced energy gap in few-layer graphene, *J. Phys. Chem. C*, 115, 9458, 2011.

[362] K, Myhro, S. Che, Y. Shi, Y. Lee, K. Thilahar, K. Bleich, D. Smirnov, and C. N. Lau, Large tunable intrinsic gap in rhombohedral-stacked tetralayer graphene at half filling, *2D Mater.* 5, 045013, 2018.

[363] J. J. Wang, Z. Y. Wang, R. J. Zhang, Y. X. Zheng, L. Y. Chen, S. Y. Wang, C. C. Tsoo, H. J. Huang, and W. S. Su, A first-principles study of the electrically tunable band gap in few-layer penta-graphene, *Phys. Chem. Chem. Phys.* 20, 18110, 2018.

[364] M. Pisarra, A. Sindona, M. Gravina, V. M. Silkin, and J. M. Pitarke, Dielectric screening and plasmon resonances in bilayer graphene, *Phys. Rev. B* 93, 035440, 2016.

[365] E. H. Hwang, and S. Das Sarma, Plasmon modes of spatially separated double-layer graphene, *Phys. Rev. B* 80, 205405, 2009.

[366] K. S. Novoselov, A. K. Geim, S. V. Morozov, D. Jiang, M. I. Katsnelson, I. V. Grigorieva, S. V. Dubonos, and A. A. Firsov, Two-dimensional gas of massless Dirac fermions in graphene, *Nature* 438, 197, 2005.

[367] J. H. Ho, Y. H. Lai, Y. H. Chiu, and M. F. Lin, Landau levels in graphene, *Phys. E* 40, 1722, 2008.

[368] Y. B. Zhang, Y. W. Tan, H. L. Stormer, and P. Kim, Experimental observation of the quantum Hall effect and Berry's phase in graphene, *Nature* 438, 201, 2005.

[369] E. McCann, and V. I. Fal'ko, Landau-level degeneracy and quantum hall effect in a graphite bilayer, *Phys. Rev. Lett.* 96, 086805, 2006.

[370] K. S. Novoselov, E. McCann, S. V. Morozov, V. I. Fal'ko, M. I. Katsnelson, U. Zeitler, D. Jiang, F. Schedin, and A. K. Geim, Unconventional quantum Hall effect and Berry's phase of 2π in bilayer graphene, *Nat. Phys.* 2, 177, 2006.

[371] T. Taychatanapat, K. Watanabe, T. Taniguchi, and P. J. Herrero, Quantum Hall effect and Landau-level crossing of Dirac fermions in trilayer graphene, *Nat. Phys.* 7, 621, 2008.

[372] L. Zhang, Y. Zhang, J. Camacho, M. Khodas, and I. Zaliznyak, The experimental observation of quantum Hall effect of $l = 3$ chiral quasiparticles in trilayer graphene, *Nat. Phys.* 7, 953, 2011.

[373] K. W. Chiu, and J. J. Quinn, Plasma oscillations of a two-dimensional electron gas in a strong magnetic field, *Phys. Rev. B* 9, 4724, 1974.

[374] R. B. Chen, C. W. Chiu, and M. F. Lin, Magnetoplasmons in simple hexagonal graphite, *RSC Adv.* 5, 53736, 2015.

[375] H. Yan, Z. Li, X. Li, W. Zhu, P. Avouris, and F. Xia, Infrared spectroscopy of tunable Dirac terahertz magneto-plasmons in graphene, *Nano Lett.* 12, 3766, 2012.

[376] I. Crassee, M. Orlita, M. Potemski, A. L. Walter, M. Ostler, Th. Seyller, I. Gaponenko, J. Chen, and A. B. Kuzmenko, Intrinsic terahertz plasmons and magnetoplasmons in large scale monolayer graphene, *Nano Lett.* 12, 2470, 2012.

[377] R. Roldan, J.-N. Fuchs, and M. O. Goerbig, Collective modes of doped graphene and a standard two-dimensional electron gas in a strong magnetic field: Linear magnetoplasmons versus magnetoexcitons, *Phys Rev. B* 80, 085408, 2009.

[378] I. Petkovi, F. I. B. Williams, K. Bennaceur, F. Portier, P. Roche, and D. C. Glattli, Carrier drift velocity and edge magnetoplasmons in graphene, *Phys. Rev. Lett.* 110, 016801, 2013.

[379] J. M. Poumirol, W. Yu, X. Chen, C. Berger, W. A. de Heer, M. L. Smith, T. Ohta, W. Pan, M. O. Goerbig, D. Smirnov, and Z. Jiang, Magnetoplasmons in quasineutral epitaxial graphene nanoribbons, *Phys. Rev. Lett.* 110, 246803, 2013.

[380] A. Iyengar, Jianhui Wang, H. A. Fertig, and L. Brey, Excitations from filled Landau levels in graphene, *Phys. Rev. B* 75, 125430, 2007.

[381] Yu. A. Bychkov, and G. Martinez, Magnetoplasmon excitations in graphene for filling factors $\nu = 6$, *Phys. Rev. B* 77, 125417, 2008.

[382] O. L. Berman, G. Gumbs, and Y. E. Lozovik, Magnetoplasmons in layered graphene structures, *Phys. Rev. B* 78, 085401, 2008.

[383] N. Kumada, P. Roulleau, B. Roche, M. Hashisaka, H. Hibino, I. Petković, and D. C. Glattli, Resonant edge magnetoplasmons and their decay in graphene, *Phys. Rev. Lett.* 113, 266601, 2014.

[384] M. Koshino, and E. McCann, Trigonal warping and Berrys phase $N\pi$ in ABC-stacked multilayer graphene, *Phys. Rev. B* 80, 165409, 2009.

[385] H. Min, and A. H. MacDonald, Chiral decomposition in the electronic structure of graphene multilayers, *Phys. Rev. B* 77, 155416, 2008.

[386] S. H. R. Sena, J. M. Pereira, F. M. Peeters, and G. A. Faria, Landau levels in asymmetric graphene trilayers, *Phys. Rev. B* 84, 205448, 2011.

[387] T. N. Do, P. H. Shih, G. Gumbs, D. Huang, C. W. Chiu, and M. F. Lin. Diverse magnetic quantization in bilayer silicene, *Phys. Rev. B* 97, 125416, 2018.

[388] J. Y. Wu, S. C. Chen, G. Gumbs, and M. F. Lin, Field-created diverse quantizations in monolayer and bilayer black phosphorus, *Phys. Rev. B* 95, 115411, 2017.

[389] S. C. Chen, J. Y. Wu, and M. F. Lin, Feature-rich magneto-electronic properties of bismuthene, *New J. Phys.* 20, 062001, 2018.

[390] Z. Jiang, E. A. Henriksen, L. C. Tung, Y. J. Wang, M. E. Schwartz, M. Y. Han, P. Kim, and H. L. Stormer, Infrared spectroscopy of Landau levels of graphene, *Phys. Rev. Lett.* 98, 197403, 2007.

[391] D. L. Miller, K. D. Kubista, G. M. Rutter, M. Ruan, W. A. de Heer, P. N. First, and J. A. Stroscio, Observing the quantization of zero mass carriers in graphene, *Science* 324, 924–927, 2009.

[392] D. L. Miller, K. D. Kubista, G. M. Rutter, M. Ruan, W. A. de Heer, M. Kindermann, P. N. First, and J. A. Stroscio, Real-space mapping of magnetically quantized graphene states. *Nat. Phys.* 6, 811–817, 2010.

[393] K. Hashimoto, T. Champel, S. Florens, C. Sohrmann, J. Wiebe, Y. Hirayama, R. A. Römer, R. Wiesendanger, and M. Morgenstern, Robust nodal structure of Landau level wave functions revealed by Fourier transform scanning tunneling spectroscopy, *Phys. Rev. Lett.* 109, 116805, 2012.

[394] C. Faugeras, M. Amado, P. Kossacki, M. Orlita, M. Kuhne, A. A. L. Nicolet, Yu. I. Latyshev, and M. Potemski, Magneto-Raman scattering of graphene on graphite: Electronic and phonon excitations, *Phys. Rev. Lett.* 107, 036807, 2011.

[395] P. Plochocka, C. Faugeras, M. Orlita, M. L. Sadowski, G. Martinez, M. Potemski, M. O. Goerbig, J. -N. Fuchs, C. Berger, and W. A. de Heer, High-energy limit of massless Dirac fermions in multilayer graphene using magneto-optical transmission spectroscopy, *Phys. Rev. Lett.* 100, 087401, 2008.

[396] Y. Henni, H. P. O. Collado, K. Nogajewski, M. R. Molas, G. Usaj, C. A. Balseiro, M. Orlita, M. Potemski, and C. Faugeras, Rhombohedral multilayer graphene: A magneto-Raman scattering study, *Nano Lett.* 16, 3710, 2016.

[397] J. C. Charlier, J.VP. Michenaud, X. Gonze, and J.VP. Vigneron, Tight-binding model for the electronic properties of simple hexagonal graphite, *Phys. Rev. B* 44, 13237, 1991.

[398] A. G. Marinopoulos, L. Reining, A. Rubio, and V. Olevano, Ab initio study of the optical absorption and wave-vector-dependent dielectric response of graphite, *Phys. Rev. B* 69, 245419, 2004.

[399] T. G. Pedersen, Analytic calculation of the optical properties of graphite, *Phys. Rev. B* 67, 113106, 2003.

[400] C. L. Lu, H. C. Lin, and C. C. Hwang, Absorption spectra of trilayer rhombohedral graphite, *Appl. Phys. Lett.* 89, 221910, 2006.

[401] O. V. Sedelnikova, L. G. Bulusheva, I. P. Asanov, I. V. Yushina, and A. V. Okotrub, Energy shift of collective electron excitations in highly corrugated graphitic nanostructures: Experimental and theoretical investigation, *Appl. Phys. Lett.* 104, 161905, 2014.

[402] C. H. Ho, C. P. Chang, and M. F. Lin, Optical magnetoplasmons in rhombohedral graphite with a three-dimensional Dirac cone structure, *J. Phys.: Condens. Matter* 27, 125602, 2015.

[403] C. W. Chiu, S. H. Lee, S. C. Chen, F. L. Shyu, and M. F. Lin, Absorption spectra of AA-stacked graphite, *New J. Phys.* 12, 083060, 2010.

[404] E. T. Jensen, R. E. Palmer, and W. Allison, Temperature-dependent plasmon frequency and linewidth in a semimetal, *Phys. Rev. Lett.* 66, 492, 1991.

[405] M. Gleeson, B. Kasemo, and D. Chakarov, Thermal and adsorbate induced plasmon energy shifts in graphite, *Sur. Sci.* 524, L77, 2003.

[406] G. I. Dovbeshko, V. R. Romanyuk, D. V. Pidgirnyi, V. V. Cherepanov, E. O. Andreev, V. M. Levin, P. P. Kuzhir, T. Kaplas, and Y. P. Svirko, Optical properties of pyrolytic carbon films versus graphite and graphene, *Nanoscale Res. Lett.* 10, 234, 2015.

[407] J. Geiger, H. Katterwe, und B. Schroder, Electron energy loss spectra of graphite single crystals and evaporated carbon films in the range 0.02V0.4 eV, *Z. Phys.*, 241, 45, 1971.

[408] C. H. Ho, C. P. Chang, W. P. Su, and M. F. Lin, Processing anisotropic Dirac cone and Landau subbands along anodal spiral, *New J. Phys.* 15, 053032, 2013.

[409] N. Hamada, S. Sawada, and A. Oshiyama, New one-dimensional conductors: Graphitic microtubules, *Phys. Rev. Lett.* 68, 1579, 1992.

[410] T. W. Ebbesen, and P. M. Ajayan, Large-scale synthesis of carbon nanotubes, *Nature* 358, 220–222, 1992.

[411] Y. Hernandez, V. Nicolosi, M. Lotya, F. M. Blighe, Z. Sun, S. De, I. T. McGovern, B. Holland, M. Byrne, Y. K. Gun'Ko, J. J. Boland, P. Niraj, G. Duesberg, S. Krishnamurthy, R. Goodhue, J. Hutchison, V. Scardaci, A. C. Ferrari, and J. N. Coleman, High-yield production of graphene by liquid-phase exfoliation of graphite, *Nat. Nanotechnol.* 3, 563, 2008.

[412] Z. F. Ren, Z. P. Huang, J. W. Xu, J. H. Wang, P. Bush, M. P. Siegal, and P. N. Provencio, Synthesis of large arrays of well-aligned carbon nanotubes on glass, *Science* 282, 1105–1107, 1998.

[413] M. Moniruzzaman, and K. I. Winey, Polymer nanocomposites containing carbon nanotubes, *Macromolecules* 39, 5194, 2006.

[414] C. Journet, W. K. Maser, P. Bernier, A. Loiseau, M. Lamy de la Chapelle, S. Lefrant, P. Deniard, R. Lee, and J. E. Fischer, Large-scale production of single-walled carbon nanotubes by the electric-arc technique, *Nature* 388, 756, 1997.

[415] A. M. Cassell, J. A. Raymakers, J. Kong, and H. Dai, Large scale CVD synthesis of single-walled carbon nanotubes, *J. Phys. Chem. B* 103, 6484, 1999.

[416] H. M. Cheng, Large-scale and low-cost synthesis of single-walled carbon nanotubes by the catalytic pyrolysis of hydrocarbons, *Appl. Phys. Lett.* 72, 3282, 1998.

[417] A. Thess, R. Lee, P. Nikolaev, H. Dai, P. Petit, J. Robert, C. Xu, Y. H. Lee, S. G. Kim, A. G. Rinzler, D. T. Colbert, G. E. Scuseria, D. Tomanek, J. E. Fischer, and R. E. Smalley, Crystalline ropes of metallic carbon nanotubes, *Science* 273, 483, 1996.

[418] S. H. Joo, S. J. Choi, I. Oh, J. Kwak, Z. Liu, O. Terasaki, and R. Ryoo, Ordered nanoporous arrays of carbon supporting high dispersions of platinum nanoparticles, *Nature* 412, 169, 2001.

[419] W. Z. Li, S. S. Xie, L. X. Qian, B. H. Chang, B. S. Zou, W. Y. Zhou, R. A. Zhao, and G. Wang, Large-scale synthesis of aligned carbon nanotubes, *Science* 274, 1701, 1996.

[420] M. Chhowalla, K. B. K. Teo, C. Ducati, N. L. Rupesinghe, G. A. J. Amaratunga, A. C. Ferrari, D. Roy, J. Robertson, and W. I. Milne, Growth process conditions of vertically aligned carbon nanotubes using plasma enhanced chemical vapor deposition, *J. App. Phys.* 90, 5308, 2001.

[421] M. Terrones, N. Grobert, J. Olivares, J. P. Zhang, H. Terrones, K. Kordatos, W. K. Hsu, J. P. Hare, P. D. Townsend, K. Prassides, A. K. Cheetham, H. W. Kroto, and D. R. M. Walton, Controlled production of aligned-nanotube bundles, *Nature* 388, 52, 1997.

[422] H. A. Mizes, Sang-il Park, and W. A. Harrison, Multiple-tip interpretation of anomalous scanning-tunneling-microscopy images of layered materials, *Phys. Rev. B* 36, 4491(R), 1987.

[423] L. C. Venema, V. Meunier, Ph. Lambin, and C. Dekker. Atomic structure of carbon nanotubes from scanning tunneling microscopy. *Phys. Rev. B* 61, 2991, 2000.

[424] E. Y. Andrei, G. Li, and X. Du, Electronic properties of graphene: A perspective from scanning tunneling microscopy and magnetotransport, *Rep. Prog. Phys.* 75, 056501, 2012.

[425] L. C. Venema, J. W. G. Wildoer, J. W. Janssen, S. J. Tans, H. L. J. T. Tuinstra, L. P. Kouwenhoven, C. Dekker, Imaging electron wave functions of quantized energy levels in carbon nanotubes, *Science* 283, 52, 1999.

[426] C. L. Kane, and E. J. Mele, Size, shape, and low energy electronic structure of carbon nanotubes, *Phys. Rev. Lett.* 78, 1932, 1997.

[427] S. Uryu, Numerical study of cross-polarized plasmons in doped carbon nanotubes, *Phys. Rev. B* 97, 125420, 2018.

[428] M. F. Lin, D. S. Chuu, C. S. Huang, Y. K. Lin, and K. W. -K. Shung, Collective excitations in a single-layer carbon nanotube, *Phys. Rev. B* 53, 15493, 1996.

[429] M. F. Lin, D. S. Chuu, and K. W. -K. Shung, Low-frequency plasmon in a metallic carbon nanotube, *Phys. Rev. B* 56, 1430, 1997.

[430] M. F. Lin, and D. S. Chuu, The π plasmons in carbon nanotube bundles, *Phys. Rev. B* 57, 10183, 1998.

[431] M. F. Lin, and F. L. Shyu, Electronic excitations in coupled armchair carbon nanotubes, *Phys. Lett. A.* 259, 158, 1999.

[432] F. L. Shyu, and M. F. Lin, π plasmons in two-dimensional arrays of aligned carbon nanotubes, *Phys. Rev. B* 60, 14434, 1999.

[433] G. Gumbs, and G. R. Aĭzin, Collective excitations in a linear periodic array of cylindrical nanotubes, *Phys. Rev. B* 65, 195407, 2002.

[434] A. A. Lucas, L. Henrad, and Ph. Lambin, Computation of the ultraviolet absorption and electron inelastic scattering cross section of multishell fullerenes, *Phys. Rev. B* 49, 2888, 1994.

[435] Q. Zhang, E. H. Haroz, Z. Jin, L. Ren, X. Wang, R. S. Arvidson, A. Luttge, and J. Kono, Plasmonic nature of the terahertz conductivity peak in single-wall carbon nanotubes, *Nano Lett.* 13, 5991, 2013.

[436] G. Ya. Slepyan, M. V. Shuba, S. A. Maksimenko, C. Thomsen, and A. Lakhtakia, Terahertz conductivity peak in composite materials containing carbon nanotubes: Theory and interpretation of experiment, *Phys. Rev. B* 81, 205423, 2010.

[437] A. Pekker, and K. Kamaras, Wide-range optical studies on various single-walled carbon nanotubes: Origin of the low-energy gap, *Phys. Rev. B* 84, 075475, 2011.

[438] M. F. Lin, and K. W. -K. Shung, Optical and magneto-optical properties of carbon nanotube bundles, *J. Phys. Soc. Jpn.* 66, 3294, 1997.

[439] M. F. Lin, F. L. Shyu, and R. B. Chen, Optical properties of multiwalled carbon nanotubes, *Phys. Rev. B* 61, 14114, 2000.

[440] M. F. Lin, Optical spectra of single-walled carbon nanotube bundles, *Phys. Rev. B* 62, 13153, 2000.

[441] F. L. Shyu, and M. F. Lin, Electronic and optical properties of narrow-gap carbon nanotubes, *J. Phys. Soc. Jpn.* 71, 1820–1823, 2002.

[442] Y. H. Ho, C. P. Chang, F. L. Shyu, S. C. Chen, and M. F. Lin, Electronic and optical properties of double-walled armchair carbon nanotubes, *Carbon* 42, 3159, 2004.

[443] X. J. Wang, and S. Yokojima, Electronic structures and optical properties of open and capped carbon nanotubes, *J. Am. Chem. Soc.* 122, 11129, 2000.

[444] I. Miloševi, T. Vukovi, S. Dmitrovi, and M. Damnjanovi, Polarized optical absorption in carbon nanotubes: A symmetry-based approach, *Phys. Rev. B* 67, 165418, 2003.

[445] Y. Murakami, E. Einarsson, T. Edamura, and S. Maruyama, Polarization dependence of the optical absorption of single-walled carbon nanotubes, *Phys. Rev. Lett.* 94, 087402, 2005.

[446] H. Kataura, Y. Kumazawa, Y. Maniwa, I. Umezu, S. Suzuki, Y. Ohtsuka, Y. Achib, Optical properties of single-wall carbon nanotubes, *Syn. Metal.* 103, 2555, 1999.

[447] N. Akima, Y. Iwasa, S. Brown, A. M. Barbour, J. Cao, J. L. Musfeldt, H. Matsui, N. Toyota, M. Shiraishi, H. Shimoda, and O. Zhou, Strong anisotropy in the far infrared absorption spectra of stretch aligned single walled carbon nanotubes, *Adv. Mater.* 18, 1166, 2006.

[448] C. W. Chiu, F. L. Shyu, C. P. Chang, R. B. Chen, and M. F. Lin, Novel magnetoplasmons in armchair carbon nanotubes, *Phys. Lett.* A 311, 53, 2003.

[449] C. W. Chiu, C. P. Chang, F. L. Shyu, R. B. Chen, and M. F. Lin, Magneto electronic excitations in single-walled carbon nanotubes. *Phys. Rev. B* 67, 165421, 2003.

[450] A. Abdikian, and M. Bagheri, Electrostatic waves in carbon nanotubes with an axial magnetic field, *Phys. Plasmas* 20, 102103, 2013.

[451] G. Gumbs, Low-energy magnetoplasmon excitations in semimetallic carbon nanotubes, *Phys. Rev. B* 66, 205413, 2002.

[452] C. H. Lee, C. W. Chiu, F. L. Shyu, and M. F. Lin, Magnetoplasmons in a pair of armchair carbon nanotubes, *J. Vac. Sci. Technol. B* 23, 2266–2272, 2005.

[453] C. W. Chiu, Y. H. Chiu, F. L. Shyu, C. P. Chang, D. S. Chuu, and M. F. Lin, Temperature-dependent carrier dynamics in metallic carbon nanotubes, *Phys. Lett. A* 346, 347, 2005.

[454] T. Morimoto, M. Ichida, Y. Ikemoto, and T. Okazaki, Temperature dependence of plasmon resonance in single-walled carbon nanotubes, *Phys. Rev. B* 93, 195409, 2016.

[455] A. Ugawa, A. G. Rinzler, and D. B. Tanner, Far-infrared gaps in single-wall carbon nanotubes, *Phys. Rev. B* 60, R11305(R), 1999.

[456] F. Borondics, K. Kamaras, M. Nikolou, D. B. Tanner, Z. H. Chen, and A. G. Rinzler, Charge dynamics in transparent single-walled carbon nanotube films from optical transmission measurements, *Phys. Rev. B* 74, 045431, 2006.

[457] O. Jost, A. A. Gorbunov, and W. Pompe, Diameter grouping in bulk samples of single-walled carbon nanotubes from optical absorption spectroscopy, *Appl Phys. Lett.* 75, 2217, 1999.

[458] J-C Charlier, and Ph. Lambin, Electronic structure of carbon nanotubes with chiral symmetry, *Phys. Rev. B* 57, 15037, 1998.

[459] R. S. Lee, H. J. Kim, J. E. Fischer, A. Thess, and R. E. Smalley, Conductivity enhancement in single-walled carbon nanotube bundles doped with K and Br, *Nature* 388, 255, 1997.

[460] A. M. Rao, P. C. Eklund, Shunji Bandow, A. Thess, and R. E. Smalley, Evidence for charge transfer in doped carbon nanotube bundles from Raman scattering, *Nature* 388, 257, 1997.

[461] T. W. Odom, J. L. Huang, P. Kim, and C. M. Lieber, Atomic structure and electronic properties of single-walled carbon nanotubes, *Nature* 391, 62, 1998.

[462] F. L. Shyu, and M. F. Lin, Loss spectra of graphite-related systems: A multiwall carbon nanotube, a single-wall carbon nanotube bundle, and graphite layers, *Phys. Rev. B* 62, 8508, 2000.

[463] M. F. Lin, and K. W. -K. Shung, Magnetoconductance of carbon nanotubes, *Phys. Rev. B* 51, 7592, 1995.

[464] T. S. Li, and M. F. Lin, Conductance of carbon nanotubes in a transverse electric field and an arbitrary magnetic field, *Nanotechnology* 17, 5632, 2006.

[465] L. Chen, C. C. Liu, B. Feng, X. He, P. Cheng, Z. Ding, S. Meng, Y. Yao, and K. Wu, Evidence for Dirac fermions in a honeycomb lattice based on silicon, *Phys. Rev. Lett.* 109, 056804, 2012.

[466] B. Feng, Z. Ding, S. Meng, Y. Yao, X. He, P. Cheng, L. Chen, and K. Wu, Evidence of silicene in honeycomb structures of silicon on Ag(111), *Nano Lett.* 12, 3507, 2012.

[467] R. Yaokawa, T. Ohsuna, T. Morishita, Y. Hayasaka, M. J. S. Spencer, and H. Nakano, Monolayer-to-bilayer transformation of silicenes and their structural analysis, *Nat. Commun.* 7, 10657, 2016.

[468] Z. Ni, Q. Liu, K. Tang, J. Zheng, J. Zhou, R. Qin, Z. Gao, D. Yu, and J. Lu, Tunable bandgap in silicene and germanene, *Nano Lett.* 12, 113, 2012.

[469] S. Sadeddine, H. Enriquez, A. Bendounan, P. K. Das, I. Vobornik, A. Kara, A. J. Mayne, F. Sirotti, G. Dujardin, and H. Oughaddou, Compelling experimental evidence of a Dirac cone in the electronic structure of a 2D Silicon layer, *Sci. Rep.* 7, 44400, 2017.

[470] C. L. Lin, R. Arafune, K. Kawahara, M. Kanno, N. Tsukahara, E. Minamitani, Y. Kim, M. Kawai, and N. Takagi, Substrate-induced symmetry breaking in silicene, *Phys. Rev. Lett.* 110, 076801, 2013.

[471] W. Wang, and R. I. G. Uhrberg, Investigation of the atomic and electronic structures of highly ordered two-dimensional germanium on Au(111), *Phys. Rev. Mater.* 1, 074002, 2017.

[472] C. J. Walhout, A. Acun, L. Zhang, M. Ezawa, and H. J. W. Zandvliet, Scanning tunneling spectroscopy study of the Dirac spectrum of germanene, *J. Phys.: Condens. Mat.* 28, 284006, 2016.

[473] D. Richards, Inelastic light scattering from inter-Landau level excitations in a two-dimensional electron gas, *Phys. Rev. B* 61, 7517, 2000.

[474] M. A. Eriksson, A. Pinczuk, B. S. Dennis, S. H. Simon, L. N. Pfeiffer, and K. W. West, Collective excitations in the dilute 2D electron system, *Phys. Rev. Lett.* 82, 2163, 1999.

[475] M. Tahir, and P. Vasilopoulos, Electrically tunable magnetoplasmons in a monolayer of silicene or germanene, *J. Phys.: Condens. Mat.* 27, 075303, 2015.

[476] H. Zhao, and S. Mazumdar, Electron-electron interaction effects on the optical excitations of semiconducting single-walled carbon nanotubes, *Phys. Rev. Lett.* 93, 157402, 2004.

[477] C. W. Chiu, S. H. Lee, and M. F. Lin, Inelastic Coulomb scatterings of doped armchair carbon nanotubes, *J. Nanosci. Nanotechnol.* 10, 2401–2408, 2010.

[478] C. W. Chiu, F. L. Shyu, C. P. Chang, D. S. Chuu, and M. F. Lin, Coulomb scattering rates of excited carriers in moderate-gap carbon nanotubes, *Phys. Rev. B* 73, 235407, 2006.

[479] Q. Li, and S. D. Sarma, Finite temperature inelastic mean free path and quasiparticle lifetime in graphene, *Phys. Rev. B* 87, 085406, 2013.

[480] J. Gonzalez, F. Guinea, and M. A. H. Vozmediano, Unconventional quasiparticle lifetime in graphite, *Phys. Rev. Lett.* 77, 3589, 1996.

[481] C. D. Spataru, M. A. Cazalilla, A. Rubio, L. X. Benedict, P. M. Echenique, and S. G. Louie, Anomalous quasiparticle lifetime in graphite: Band structure effects, *Phys. Rev. Lett.* 87, 246405, 2001.

[482] C.-H. Park, F. Giustino, C. D. Spataru, M. L. Cohen, and S. G. Louie, Inelastic carrier lifetime in bilayer graphene, *Appl. Phys. Lett.* 100, 032106, 2012.

[483] C.-H. Park, F. Giustino, C. D. Spataru, M. L. Cohen, and S. G. Louie, First-principles study of electron linewidths in graphene, *Phys. Rev. Lett.* 102, 076803, 2009.

[484] L. Luer, C. Gadermaier, J. Crochet, T. Hertel, D. Brida, and G. Lanzani, Coherent phonon dynamics in semiconducting carbon nanotubes: A quantitative study of electron-phonon coupling, *Phys. Rev. Lett.* 102, 127401, 2009.

[485] J. Y. Park, S. Rosenblatt, Y. Yaish, V. Sazonova, H. Ustunel, S. Braig, T. A. Arias, P. W. Brouwer, and Pa. L. McEuen, Electron-phonon scattering in metallic single-walled carbon nanotubes, *Nano Lett.* 4, 517, 2004.

[486] M. F. Lin, and K. W. -K. Shung, The self-energy of electrons in graphite intercalation compounds, *Phys. Rev. B* 53, 1109–1118, 1996.

[487] P. Hawrylak, Effective mass and lifetime of electrons in a layered electron gas, *Phys. Rev. Lett.* 59, 485, 1987.

[488] G. F. Giuliani, and J. J. Quinn, Lifetime of a quasiparticle in a two-dimensional electron gas, *Phys. Rev. B* 26, 4421, 1982.

[489] C. P. Weber, N. Gedik, J. E. Moore, J. Orenstein, J. Stephens, and D. D. Awschalom, Observation of spin Coulomb drag in a two-dimensional electron gas, *Nature* 437, 1330–1333, 2005.

[490] J. Hubbard, The description of collective motions in terms of many-body perturbation theory. II. The correlation energy of a free-electron gas, *Proc. R. Soc. Lond.* 243, 336–352, 1963.

[491] K. S. Singwi, M. P. Tosi, R. H. Land, and A. Sjolander, Electron correlations at metallic densities, *Phys. Rev.* 176, 589, 1968.

[492] P. Vashishta, and K. S. Singwi, Electron correlations at metallic densities, *Phys. Rev. B* 6, 875, 1972.

[493] C. S. Ting, T. K. Lee, and J. J. Quinn, Effective mass and g factor of interacting electrons in the surface inversion layer of silicon, *Phys. Rev. Lett.* 34, 870, 1975.

[494] R. B. Chen, S. C. Chen, C. W. Chiu, M. F. Lin, Optical properties of monolayer tinene in electric fields, *Sci. Rep.* 7, 1849, 2017.

[495] C. F. Chen, C. H. Park, B. W. Boudouris, J. Horng, B. Geng, C. Girit, A. Zettl, M. F. Crommie, R. A. Segalman, S. G. Louie, and F. Wang, Controlling inelastic light scattering quantum pathways in graphene, *Nature* 471, 617, 2011.

[496] C. Kramberger, E. Einarsson, S. Huotari, T. Thurakitseree, S. Maruyama, M. Knupfer, and T. Pichler, Interband and plasma excitations in single-walled carbon nanotubes and graphite in inelastic X-ray and electron scattering, *Phys. Rev. B* 81, 205410, 2010.

[497] Á. Bácsi, and A. Virosztek, Local density of states and Friedel oscillations in graphene, *Phys. Rev. B* 82, 193405, 2010.

[498] M. F. Lin, and K. W. -K. Shung, Screening of charged impurities in graphite intercalation compounds, *Phys. Rev. B* 46, 12656, 1992.

[499] R. Wiesendanger, Spin mapping at the nanoscale and atomic scale, *Rev. Mod. Phys.* 81, 1495, 2009.

[500] X. Dou, A. Jaefari, Y. Barlas, and B. Uchoa, Quasiparticle renormalization in ABC graphene trilayers, *Phy. Rev. B* 90, 161411(R), 2014.

[501] R. Roldan, and L. Brey, Dielectric screening and plasmons in AA-stacked bilayer graphene, *Phy. Rev. B* 88, 115420, 2013.

[502] A. Mooradian, and G. B. Wright, Observation of the Interaction of Plasmons with Longitudinal Optical Phonons in GaAs, *Phys. Rev. Lett.* 16, 999, 1966.

[503] Y. Liu, and R. F. Willis, Plasmon-phonon strongly coupled mode in epitaxial graphene, *Phys. Rev. B* 81, 081406(R), 2010.

[504] A. Gonzalez, I. Amenabar, J Chen, T. H. Bointon, S. Dai, M. M. Fogler, D. N. Basov, R. Hillenbrand, M. F. Craciun, F. J. G. de Abajo, S. Russo, and F. H. L. Koppens, Intrinsic plasmonVphonon interactions in highly doped graphene: A near-field imaging study, *Nano Lett.* 17, 5908, 2017.

[505] M. Settnes, J. R. M. Saavedra, K. S. Thygesen, A.-P. Jauho, F. J. G. de Abajo, and N. A. Mortensen, Strong plasmon-phonon splitting and hybridization in 2D materials revealed through a self-energy approach, *ACS Photonics* 4, 2908, 2017.

[506] A. Bostwick, F. Speck, T. Seyller, K. Horn, M. Polini, R. Asgari, A. H. MacDonald, and E. Rotenberg, Observation of plasmarons in quasi-freestanding doped graphene, *Science* 328, 999, 2010.

[507] Z. Y. Ong, and M. V. Fischetti, Theory of interfacial plasmon-phonon scattering in supported graphene, *Phys. Rev. B* 86, 165422, 2012.

[508] Ge. G. Samsonidze, E. B. Barros, R. Saito, J. Jiang, G. Dresselhaus, and M. S. Dresselhaus, Electron-phonon coupling mechanism in two-dimensional graphite and single-wall carbon nanotubes, *Phys. Rev. B* 75, 155420, 2007.

[509] T. Ando, Anomaly of optical phonons in bilayer graphene, *J. Phys. Soc. Jpn.* 76, 104711, 2007.

[510] T. Ando, Exotic electronic and transport properties of graphene, *Physica E* 40, 213, 2007.

[511] H. J. Zeiger, J. Vidal, T. K. Cheng, E. P. Ippen, G. Dresselhaus, and M. S. Dresselhaus, Theory for displacive excitation of coherent phonons, *Phys. Rev. B* 45, 768, 1992.

[512] E. H. Hwang, Rajdeep Sensarma, and S. Das Sarma, Plasmon-phonon coupling in graphene, *Phys. Rev. B* 82, 195406, 2010.

[513] M. Polini, R. Asgari, G. Borghi, Y. Barlas, T. Pereg-Barnea, and A. H. MacDonald, Plasmons and the spectral function of graphene, *Phys. Rev. B* 77, 081411, 2008.

[514] C. H. Park, F. Giustino, M. L. Cohen, and S. G. Louie, Electron-phonon interactions in graphene, bilayer graphene, and graphite, *Nano Lett.* 8, 4229, 2008.

[515] K. M. Borysenko, J. T. Mullen, E. A. Barry, S. Paul, Y. G. Semenov, J. M. Zavada, M. Buongiorno Nardelli, and K. W. Kim, First-principles analysis of electron-phonon interactions in graphene, *Phys. Rev. B* 81, 121412, 2010.

[516] J. A. Yan, W. Y. Ruan, and M. Y. Chou, Phonon dispersions and vibrational properties of monolayer, bilayer, and trilayer graphene: Density-functional perturbation theory, *Phys. Rev. B* 77, 125401, 2008.

Index

Printed and bound by CPI Group (UK) Ltd, Croydon, CR0 4YY

25/10/2024

01779260-0001